BEGINNING PARTIAL DIFFERENTIAL EQUATIONS

BEGINNING PARTIAL DIFFERENTIAL EQUATIONS

PETER V. O'NEIL
University of Alabama at Birmingham

A Wiley-Interscience Publication
JOHN WILEY & SONS, INC.
New York • Chichester • Weinheim • Brisbane • Singapore • Toronto

Library of Congress Cataloging-in-Publication Data:

O'Neil, Peter V.
    Beginning partial differential equations / Peter V. O'Neil.
        p.   cm.
    "A Wiley-Interscience publication."
    Includes index.
    ISBN 0-471-23887-2 (cloth : alk. paper)
    1. Differential equations. Partial.   I. Title.
QA377.054   1999   98-44621
515'.353—dc21

Printed in the United States of America

10   9   8   7   6   5   4

CONTENTS

PREFACE

This book is intended as a first course in partial differential equations. Topics include characteristics, canonical forms, well-posed problems, and properties of solutions, as well as techniques for writing expressions for solutions.

Our objective is to provide an introduction to this important field of mathematics, as well as an entry point, for those who wish it, to the modern, more abstract elements of partial differential equations.

Because this is an introductory treatment, we attempt a balance between theory and technique. Computational facility and theoretical understanding reinforce each other, and both are important for later work in related areas, such as mathematical physics, differential geometry, or analytic number theory. Although the emphasis is on the mathematics, we also point out physical interpretations, for those circumstances in which an initial-boundary value problem models a physical setting.

The exercises have two purposes. Some are computational, ranging from routine to challenging. Answers for many of these are included at the end of the book. Other exercises provide additional information about partial differential equations, or extensions of the material of the text. Many of these come with hints for the development of a proof or a derivation.

Partial differential equations invite graphical representation and experimentation. Sometimes we can visualize a solution as a surface. Further, some partial differential equations model physical phenomena, and it is interesting and instructive to couple mathematics with physical intuition by observing the evolution with time of a solution that represents heat conduction or wave motion. We can also experiment with the influence of different parameters on the physical phenomenon represented by the partial differential equation, such as the effect of the specific heat of a metal bar on the way it conducts heat energy, or the way the tension and density of a guitar string influence the way it vibrates. Parts of some exercises pursue such issues, and require the use of a computational package, such as MAPLE or Mathematica. Readers who do not have access to suitable hardware and software can skip over these exercises.

Preliminary versions of this book have been tested at the University of Alabama at Birmingham over the past five years.

PREREQUISITES

The reader should be familiar with standard properties of real-valued functions of n real variables, vector calculus (theorems of Green and Gauss), topics from the standard post-calculus course in elementary ordinary differential equations, and convergence of series and improper integrals. Occasional mention is made of uniform convergence. Topics from Fourier analysis are not assumed as background and are included. Several discussions (the use of conformal mappings to solve the Dirichlet problem, contour integration to evaluate certain integrals, and Euler's formula) use complex numbers and complex function theory.

Access to a computational program, such as MAPLE, is useful both in performing calculations and in studying properties of solutions, convergence of Fourier series, and other issues of interest in partial differential equations. Such access is not a prerequisite to reading this book, but some exercises inviting computer study and experimentation have been included for those who have it.

ACKNOWLEDGMENTS

I would like to thank my colleagues at UAB for helpful conversations and suggestions, several classes of students for tolerating the use of preliminary versions of this book in the form of course notes, and the editorial staff of John Wiley & Sons for their professionalism in bringing the project to completion.

BEGINNING PARTIAL DIFFERENTIAL EQUATIONS

1

FIRST ORDER PARTIAL DIFFERENTIAL EQUATIONS

1.1 PRELIMINARY NOTATION AND CONCEPTS

A *partial differential equation* is an equation that contains at least one partial derivative. For example,

$$\frac{\partial u}{\partial x} - x\,\frac{\partial u}{\partial y} = xuy^2,$$

$$\frac{\partial^4 w}{\partial x^4} - 3\,\frac{\partial^2 w}{\partial z \partial u} + \frac{\partial^3 w}{\partial u^3} + \cos(xu)\,\frac{\partial^3 w}{\partial x \partial y \partial z} - xyzu\,\frac{\partial w}{\partial x} = 0$$

and

$$\frac{\partial^2 h}{\partial x^2} + \frac{\partial^2 h}{\partial y^2} + \frac{\partial^2 h}{\partial z^2} = f(x,\,y,\,z)$$

are partial differential equations. Such equations are of interest in mathematics (for example, in the study of curvature and surfaces) and in modeling phenomena in the sciences, engineering, economics, ecology, and other areas. For the moment we will be concerned with developing the vocabulary and notation with which to engage in a discussion of partial differential equations.

There is some economy in using subscripts to denote partial derivatives. In this notion, $u_x = \partial u/\partial x$, $u_{xx} = \partial^2 u/\partial x^2$, $u_{xy} = \partial^2 u/\partial y \partial x$, and so on. The partial differential equations listed above can be written, respectively,

$$u_x - xu_y = xuy^2,$$

$$w_{xxxx} - 3w_{uz} + w_{uuuu} + \cos(xu)w_{zyx} - xyzuw_x = 0$$

and

$$h_{xx} + h_{yy} + h_{zz} = f(x, y, z).$$

A *solution* of a partial differential equation is any function that satisfies the equation. Often we seek a solution satisfying certain conditions, and for the independent variables confined to a specified set of values.

As an example of a solution, the equation

$$4u_x + 3u_y + u = 0 \tag{1.1}$$

has solution

$$u(x, y) = e^{-x/4}f(3x - 4y)$$

in which f can be any differentiable function of a single variable. We will verify this by substituting $u(x, y)$ into the partial differential equation. Chain rule differentiations yield

$$u_x = -\frac{1}{4} e^{-x/4}f(3x - 4y) + e^{-x/4} \frac{d}{d(3x - 4y)} [f(3x - 4y)] \frac{d(3x - 4y)}{dx}$$

$$= -\frac{1}{4} e^{-x/4} f(3x - 4y) + 3e^{-x/4}f'(3x - 4y)$$

and

$$u_y = -4e^{-x/4}f'(3x - 4y).$$

Upon substitution into equation 1.1 we obtain

$$4u_x + 3u_y + u = -e^{-x/4}f(3x - 4y) + 12e^{-x/4}f'(3x - 4y)$$

$$- 12e^{-x/4}f'(3x - 4y) + e^{-x/4}f(3x - 4y) = 0$$

for all (x, y).

As another example, consider the equation

$$u_{xx} - 9u_{yy} = 0. \tag{1.2}$$

It is routine to check that

$$u(x, y) = f(3x + y) + g(3x - y)$$

is a solution for any twice differentiable functions f and g of a single variable. For example, we could choose $f(t) = \sin(t)$ and $g(t) = e^{-t} + t^2 - \cos(t)$ to obtain the solution

$$u(x, y) = \sin(3x + y) + e^{-3x+y} + (3x - y)^2 - \cos(3x - y).$$

In view of the latitude in choosing f and g, equation 1.2 has infinitely many solutions (as does equation 1.1).

The *order* of a partial differential equation is the order of the highest partial derivative occurring in the equation. Equation 1.1 is of order one and equation 1.2 is of order two.

A partial differential equation is *linear* if it is linear in the unknown function and its partial derivatives. An equation that is not linear is *nonlinear*. For example,

$$x^2 u_{xx} - y u_{xy} = u$$

is a linear partial differential equation, while

$$x^2 u_{xx} - y u_{xy} = u^2$$

is nonlinear because of the u^2 term, and

$$(u_{xx})^{1/2} - 4u_{yy} = xu$$

is nonlinear because of the $(u_{xx})^{1/2}$ term.

A partial differential equation is *quasi-linear* if it is linear in its highest order derivative terms. For example, the second order equation

$$u_{xx} + 4y u_{yy} - (u_x)^3 + u_x u_y = \cos(u)$$

is quasi-linear, being linear in its second derivative terms u_{xx} and u_{yy}. This equation is nonlinear because of the $\cos(u)$, $u_x u_y$ and $(u_x)^3$ terms. Of course, any linear equation is also quasi-linear.

We are now ready to begin studying first order partial differential equations.

EXERCISE 1   Show that

$$u(x, y, z) = \frac{1}{\sqrt{x^2 + y^2 + z^2}}$$

is a solution of $u_{xx} + u_{yy} + u_{zz} = 0$ for $(x, y, z) \neq (0, 0, 0)$.

EXERCISE 2   Show that $u(x, t) = f(x + ct) + g(x - ct)$ is a solution of

$$u_{tt} = c^2 u_{xx}$$

for any twice differentiable functions f and g of one variable; c is a positive constant.

EXERCISE 3   Show that

$$u(x, t) = \frac{1}{2} (\varphi(x + ct) + \varphi(x - ct)) + \frac{1}{2c} \int_{x-ct}^{x+ct} \psi(s) \, ds$$

is a solution of $u_{tt} = c^2 u_{xx}$ for any φ that is twice differentiable and ψ that is differentiable for all real x. c is a positive constant. Show that this solution satisfies the conditions

$$u(x, 0) = \varphi(x), \, u_t(x, 0) = \psi(x)$$

for all real x.

EXERCISE 4   Show that, if p is a continuously differentiable function of one variable, then the first order partial differential equation

$$u_t = p(u)u_x$$

has solution implicitly defined by

$$u(x, t) = \varphi(x + p(u)t),$$

in which φ can be any continuously differentiable function of one variable.

     Use this idea to determine (perhaps implicitly) a solution of each of the following equations:

1. $u_t = ku_x$, with k a nonzero constant
2. $u_t = uu_x$
3. $u_t = \cos(u)u_x$
4. $u_t = e^u u_x$
5. $u_t = u \sin(u)u_x$

EXERCISE 5   Show that

$$u(x, y) = \ln((x - x_0)^2 + (y - y_0)^2)$$

satisfies Laplace's equation $u_{xx} + u_{yy} = 0$ for all pairs (x, y) of real numbers except (x_0, y_0).

EXERCISE 6 Let v and w be solutions of

$$a(x, y)u_{xx} + b(x, y)u_{xy} + c(x, y)u_{yy} + d(x, y)u_x + e(x, y)u_y + g(x, y)u = 0.$$

Show that $\alpha w + \beta v$ is also a solution for any numbers α and β.

EXERCISE 7 In each of the following, classify the equation as linear, non-linear but quasi-linear, or not quasi-linear.

1. $u^2 u_{xx} + u_y = \cos(u)$
2. $x^2 u_x + y^2 u_y + u_{xy} = 2xy$
3. $(x - y)u_x^2 + u_{xy} = 1$
4. $(x - y)u_x^2 + 2u_y = 4y$
5. $x^2 u_{yy} - y u_{xx} = \tan(u)$
6. $u_x + u_y^2 - u_{xx} = 4$
7. $u_x - u_x u_y - u_y = 0$
8. $u u_x + u_{xy} = u^2$
9. $u_{xy} - u_x^2 + u_y^2 - \sin(u_x) = 0$
10. $u_y / u_x = x^2$

EXERCISE 8 Let k be a positive constant. Let

$$u(x, t) = \frac{1}{2\sqrt{\pi k t}} \int_{-\infty}^{\infty} e^{-(x-\xi)^2/4kt} f(\xi)\, d\xi,$$

in which f is continuous on the real line. Show that $u_t = k u_{xx}$ for $-\infty < x < \infty$, $t > 0$.

Determine $u(x, t)$ when $f(x) \equiv 1$. *Hint*: Use a change of variables and the standard result that

$$\int_{-\infty}^{\infty} e^{-w^2}\, dw = \sqrt{\pi}.$$

1.2 THE LINEAR EQUATION

Consider the linear first order partial differential equation in two independent variables:

$$\text{Linear} \Rightarrow \boxed{a(x, y)u_x + b(x, y)u_y + c(x, y)u = f(x, y).} \tag{1.3}$$

We assume that a, b, c, and f are continuous in some region of the plane, and that $a(x, y)$ and $b(x, y)$ are not both zero for the same (x, y).

We will show how to solve equation 1.3. The key is to determine a change of variables

$$\xi = \varphi(x, y), \quad \eta = \psi(x, y)$$

which transforms equation 1.3 to the simpler linear equation

$$w_\xi + h(\xi, \eta)w = F(\xi, \eta) \qquad (1.4)$$

Final answer form!!

where $w(\xi, \eta) = u(x(\xi, \eta), y(\xi, \eta))$. We will define this transformation in such a way that it is one-to-one, at least for all (x, y) in some set $\mathcal{D}$ of points in the x, y-plane. On $\mathcal{D}$, then, we can, at least in theory, solve for x and y as functions of ξ and η. To insure this we will require that the Jacobian of the transformation does not vanish in $\mathcal{D}$:

$$J = \begin{vmatrix} \varphi_x & \varphi_y \\ \psi_x & \psi_y \end{vmatrix} = \varphi_x\psi_y - \varphi_y\psi_x \neq 0$$

for (x, y) in $\mathcal{D}$.

Begin the search for a suitable transformation by computing chain rule derivatives:

$$u_x = w_\xi\xi_x + w_\eta\eta_x$$

and

$$u_y = w_\xi\xi_y + w_\eta\eta_y.$$

Substitute these into equation 1.3 to obtain:

$$a(w_\xi\xi_x + w_\eta\eta_x) + b(w_\xi\xi_y + w_\eta\eta_y) + cw = f.$$

Write this equation as

$$(a\xi_x + b\xi_y)w_\xi + (a\eta_x + b\eta_y)w_\eta + cw = f. \qquad (1.5)$$

This is nearly in the form of equation 1.4 if we choose $\eta = \psi(x, y)$ so that

$$a\eta_x + b\eta_y = 0$$

for (x, y) in $\mathcal{D}$. If $\eta_y \neq 0$ this requires that

$$\frac{\eta_x}{\eta_y} = -\frac{b}{a}.$$

Suppose for the moment that there is such an η. Putting $\eta(x, y) = c$, with c an arbitrary constant, then

$$d\eta = \eta_x\, dx + \eta_y\, dy = 0$$

implies that

$$\frac{dy}{dx} = -\frac{\eta_x}{\eta_y} = \frac{b}{a}.$$

This means that $\eta = \psi(x, y)$ is an integral of the ordinary differential equation

$$\boxed{\frac{dy}{dx} = \frac{b(x, y)}{a(x, y)}.} \quad (char) \tag{1.6}$$

Equation 1.6 is called the *characteristic equation* of the linear equation 1.3. The equation $\eta(x, y) = c$ defines a family of curves in the plane called *characteristic curves*, or *characteristics*, of equation 1.3. We will say more about these in the next section.

Thus far we have found that we can make the coefficient of w_η in the transformed equation 1.5 vanish if we choose $\eta = \psi(x, y)$, with $\psi(x, y) = c$ an equation defining the general solution of the characteristic equation 1.6. With this step alone equation 1.5 comes very close to the transformed equation 1.4 we want to achieve. We can now choose ξ to suit our convenience and the condition that $J \neq 0$. One simple choice is

$$\xi = \varphi(x, y) = x.$$

With this choice,

$$J = \begin{vmatrix} 1 & 0 \\ \eta_x & \eta_y \end{vmatrix} = \eta_y$$

and this is nonzero in $\mathcal{D}$ by previous assumption.

Now from equation 1.5, the change of variables

$$\boxed{\xi = x,}\ \eta = \psi(x, y)$$

transforms equation 1.3 to

$$a(x, y)w_\xi + c(x, y)w = f(x, y).$$

To complete the transformation to the form of equation 1.4, first write $a(x, y)$, $c(x, y)$, and $f(x, y)$ in terms of ξ and η to obtain

$$A(\xi, \eta)w_\xi + C(\xi, \eta)w = p(\xi, \eta).$$

Finally, restricting the variables to a set in which $A(\xi, \eta) \neq 0$, we have

$$w_\xi + \frac{C}{A} w = \frac{p}{A}$$

and this is in the form of equation 1.4 with

$$h(\xi, \eta) = \frac{C(\xi, \eta)}{A(\xi, \eta)} \text{ and } F(\xi, \eta) = \frac{p(\xi, \eta)}{A(\xi, \eta)}.$$

Example 1   Consider the linear equation

$$x^2 u_x + y u_y + xyu = 1. \qquad (1^{st} \text{ order (linear)})$$

This is equation 1.3 with $a(x, y) = x^2$, $b(x, y) = y$, $c(x, y) = xy$ and $f(x, y) = 1$. We will transform this equation to the simpler equation 1.4.

The characteristic equation is

$$\frac{dy}{dx} = \frac{b}{a} = \frac{y}{x^2}.$$

Write

$$\frac{1}{y} dy = \frac{1}{x^2} dx,$$

integrate and rearrange terms to obtain

$$\ln(y) + \frac{1}{x} = c$$

for $y > 0$ and $x \neq 0$. This is an integral of the characteristic equation and we choose

$$\eta = \psi(x, y) = \ln(y) + \frac{1}{x}. \qquad (\text{char})$$

Graphs of $\ln(y) + 1/x = c$ are the characteristics of this partial differential equation.

Upon choosing $\xi = x$ we have the Jacobian

$$J = \eta_y = \frac{1}{y} \neq 0$$

as required.

Since $\xi = x$,

$$\eta = \ln(y) + \frac{1}{\xi},$$

so

$$\ln(y) = \eta - \frac{1}{\xi}$$

and

$$y = e^{\eta - 1/\xi}.$$

Now apply the transformation

$$\xi = x, \quad \eta = \ln(y) + \frac{1}{x},$$

with

$$w(\xi, \eta) = u(x, y).$$

$\frac{\partial u}{\partial x} = \frac{\partial w}{\partial \xi} \frac{\partial \xi}{\partial x}$

Compute

$$u_x = w_\xi \xi_x + w_\eta \eta_x = w_\xi + w_\eta \left(-\frac{1}{x^2} \right) = w_\xi - \frac{1}{\xi^2} w_\eta$$

This is all unnecessary for some reason.

and

$$u_y = w_\xi \xi_y + w_\eta \eta_y = w_\eta \frac{1}{y} = w_\eta \frac{1}{e^{\eta - 1/\xi}}$$

The partial differential equation transforms to

$$\xi^2 \left(w_\xi - \frac{1}{\xi^2} w_\eta \right) + e^{\eta - 1/\xi} w_\eta \frac{1}{e^{\eta - 1/\xi}} + \xi e^{\eta - 1/\xi} w = 1,$$

$x u_x \qquad y u_y \qquad x y u$

or

$$\xi^2 w_\xi + \xi e^{\eta - 1/\xi} w = 1.$$

Then

$$w_\xi + \frac{1}{\xi} e^{\eta - 1/\xi} w = \frac{1}{\xi^2},$$

and this has the form of equation 1.4, in any region of the ξ, η-plane with $\xi \neq 0$. ∎

The point to transforming equation 1.3 to the form of equation 1.4 is that we can solve this transformed equation. Think of

$$w_\xi + h(\xi, \eta)w = F(\xi, \eta)$$

as a linear first order ordinary differential equation in ξ, with η carried along as a parameter. Following the method for ordinary differential equations, multiply the differential equation by

$$e^{\int h(\xi, \eta) d\xi}$$

to obtain

$$e^{\int h(\xi, \eta) d\xi} w_\xi + h(\xi, \eta) e^{\int h(\xi, \eta) d\xi} w = F(\xi, \eta) e^{\int h(\xi, \eta) d\xi}.$$

Recognize this as

$$\frac{\partial}{\partial \xi} (e^{\int h(\xi, \eta) d\xi} w) = F(\xi, \eta) e^{\int h(\xi, \eta) d\xi}.$$

Integrate with respect to ξ. Since η is being carried through this process as a parameter, the constant of integration may depend on η. We obtain

$$e^{\int h(\xi, \eta) d\xi} w = \int F(\xi, \eta) e^{\int h(\xi, \eta) d\xi} d\xi + g(\eta),$$

in which g is any differentiable function of one variable. Then

$$w(\xi, \eta) = e^{-\int h(\xi, \eta) d\xi} \int F(\xi, \eta) e^{\int h(\xi, \eta) d\xi} d\xi + g(\eta) e^{-\int h(\xi, \eta) d\xi}. \tag{1.7}$$

This is the *general solution* of the transformed equation (by general solution,

we mean one that contains an arbitrary function). Now obtain the general solution of the original equation 1.3 by substituting $\xi = \xi(x, y)$, $\eta = \eta(x, y)$. This
general solution will have the form

$$u(x, y) = e^{\alpha(x,y)}[M(x, y) + g(\psi(x, y))]. \tag{1.8}$$

in which g is any differentiable function of one variable.

Example 2   Consider the constant coefficient equation

$$au_x + bu_y + cu = 0$$

in which a, b, and c are numbers. Assume that $a \neq 0$. The characteristic equation
is

$$\frac{dy}{dx} = \frac{b}{a}$$

with general solution defined by the equation

$$bx - ay = c.$$

Put

$$\xi = x, \quad \eta = bx - ay.$$

The characteristics of this differential equation are the straight line graphs of
$bx - ay = c$.

   With this transformation, we find by a routine calculation that the partial
differential equation transforms to

$$aw_\xi + cw = 0$$

or

$$w_\xi + \frac{c}{a} w = 0.$$

Multiply this equation by $e^{\int (c/a)d\xi}$, or $e^{c\xi/a}$, to get

$$e^{c\xi/a}w_\xi + \frac{c}{a} we^{c\xi/a} = 0,$$

hence

$$\frac{\partial}{\partial \xi} (e^{c\xi/a}w) = 0.$$

Integrate with respect to ξ to get

$$e^{c\xi/a}w = g(\eta)$$

in which g can be any differentiable function of one variable. Then

$$w(\xi, \eta) = e^{-c\xi/a}g(\eta).$$

Finally, transform this solution back in terms of x and y:

$$u(x, y) = e^{-cx/a}g(bx - ay).$$

This solution is readily verified by substitution into the partial differential equation.   ∎

Observe that the solution in Example 2 has the form specified by equation 1.8.

Example 3   Consider

$$u_x + \cos(x)u_y + u = xy.$$

The characteristic equation is

$$\frac{dy}{dx} = \cos(x)$$

with general solution defined by

$$y - \sin(x) = c.$$

Put

$$\xi = x, \eta = y - \sin(x).$$

Graphs of $y - \sin(x) = c$ are the characteristics of this partial differential equation.
   Now we have

$$y = \eta + \sin(x) = \eta + \sin(\xi)$$

and the partial differential equation transforms to

$$w_\xi + w = \xi[\eta + \sin(\xi)].$$

Multiply this equation by $e^{\int d\xi}$, which is e^{ξ}, to obtain

$$e^{\xi}w_{\xi} + we^{\xi} = \eta\xi e^{\xi} + \xi e^{\xi}\sin(\xi).$$

Write this equation as

$$\frac{\partial}{\partial\xi}(we^{\xi}) = \eta\xi e^{\xi} + \xi e^{\xi}\sin(\xi).$$

Integrate to obtain

$$we^{\xi} = \int \eta\xi e^{\xi}\,d\xi + \int \xi e^{\xi}\sin(\xi)\,d\xi$$

$$= \eta e^{\xi}(\xi - 1) + \frac{1}{2}\xi e^{\xi}(\sin(\xi) - \cos(\xi)) + \frac{1}{2}e^{\xi}\cos(\xi) + g(\eta).$$

Then

$$w(\xi, \eta) = \eta(\xi - 1) + \frac{1}{2}\xi(\sin(\xi) - \cos(\xi)) + \frac{1}{2}\cos(\xi) + e^{-\xi}g(\eta).$$

Finally

$$u(x, y) = (y - \sin(x))(x - 1) + \frac{1}{2}x(\sin(x) - \cos(x))$$

$$+ \frac{1}{2}\cos(x) + e^{-x}g(y - \sin(x)),$$

in which g is any differentiable function of a single variable. ■

Contrast the idea of the general solution for the linear first order ordinary differential equation with that for the linear first order partial differential equation. In the former case, the general solution of

$$y' + d(x)y = p(x)$$

contains an arbitrary constant. Graphs of the solutions obtained by making choices of the constant are curves in the (x, y) plane. If we require that $y(x_0) = y_0$, then we pick out the unique solution corresponding to the curve passing through (x_0, y_0).

By contrast, if u is the general solution of the linear first order partial differential equation 1.3, then $z = u(x, y)$ defines a family of surfaces in 3-space, each surface corresponding to a choice of the arbitrary function g in equation

1.8. In the next section we will investigate the kind of information that should be given in order to pick out one of these surfaces and determine a unique solution.

EXERCISE 9   For each of the following partial differential equations, (a) solve the characteristic equation and sketch graphs of some of the characteristics, (b) define a transformation of the partial differential equation to the form of equation 1.4 and obtain the transformed equation, (c) find the general solution of the transformed equation, (d) find the general solution of the given equation, and (e) verify the solution by substituting it into the partial differential equation.

1. $3u_x + 5u_y - xyu = 0$
2. $u_x - u_y + yu = 0$
3. $u_x + 4u_y - xu = x$
4. $-2u_x + u_y - yu = 0$
5. $xu_x - yu_y + u = x$
6. $x^2u_x - 2u_y - xu = x^2$
7. $u_x - xu_y = 4$
8. $x^2u_x + xyu_y + xu = x - y$
9. $u_x + u_y - u = y$
10. $u_x - y^2u_y - yu = 0$
11. $u_x + yu_y + xu = 0$
12. $xu_x + yu_y + 2 = 0$

EXERCISE 10   Find the general solution of

$$u_x + \alpha(y - 1)u_y = \frac{1}{2}\beta f(x)(y - 1)u$$

in which α and β are real numbers and f is continuous on the real line. Use the general solution to find the solution satisfying

$$u(0, y) = y^n,$$

in which n is a nonnegative integer.

1.3   THE SIGNIFICANCE OF CHARACTERISTICS

In the preceding section we mentioned characteristics but did not actually do anything with them. Now we will investigate their significance, beginning with an example that will be instructive for the point we want to make.

Consider

$$2u_x + 3u_y + 8u = 0.$$

The characteristic equation is

$$\frac{dy}{dx} = \frac{3}{2}$$

and the characteristics are the straight line graphs of $3x - 2y = c$.

Using the method of Section 1.2, we find that this partial differential equation has general solution

$$u(x, y) = e^{-4x}g(3x - 2y)$$

in which g can be any differentiable function defined over the real line.

Notice that simply specifying that the solution is to have a given value at a particular point does not uniquely determine g, and hence does not determine a unique solution, as occurs with ordinary differential equations.

Now suppose we specify values of $u(x, y)$ along a curve Γ in the plane. To be specific for this example, suppose we choose Γ as the x-axis and give values of $u(x, y)$ at points on Γ, say

$$u(x, 0) = \sin(x).$$

We need

$$u(x, 0) = e^{-4x}g(3x) = \sin(x)$$

so

$$g(3x) = e^{4x}\sin(x).$$

Putting $t = 3x$,

$$g(t) = e^{4t/3}\sin(t/3).$$

This determines g and the solution satisfying the condition $u(x, 0) = \sin(x)$ on Γ is

$$u(x, y) = e^{-4x}g(3x - 2y) = e^{-4x}e^{4(3x-2y)/3}\sin\left(\frac{1}{3}(3x - 2y)\right)$$

$$= e^{-8y/3}\sin\left(x - \frac{2}{3}y\right).$$

In this example, specifying values of u along the x-axis uniquely determined the arbitrary function in the general solution, and hence determined the unique solution of the partial differential equation having these given values.

Next seek a solution having given values along the line $y = x$, say

$$u(x, x) = x^4.$$

From the general solution, this requires that

$$u(x, x) = e^{-4x}g(x) = x^4$$

and we can choose

$$g(x) = x^4 e^{4x}$$

to obtain the unique solution

$$u(x, y) = e^{-4x}g(3x - 2y) = e^{8(x-y)}(3x - 2y)^4$$

satisfying $u(x, x) = x^4$.

Despite these two successes, not every curve in the plane can be used to determine g. Suppose we choose Γ to be the line $3x - 2y = 1$, and prescribe values $u(x, y)$ is to have along Γ, say

$$u\left(x, \ \frac{1}{2}(3x - 1)\right) = x^2.$$

Now we must choose g so that

$$e^{-4x}g\left(3x - 2\frac{1}{2}(3x - 1)\right) = x^2$$

and this requires that

$$g(1) = e^{4x}x^2.$$

This is impossible, hence there is no solution taking the value x^2 at points (x, y) on this line.

Why did some choices of Γ give a solution, and another choice no solution? The difference was that the x-axis and the line $y = x$ are not characteristics of the partial differential equation, while the line $3x - 2y = 1$ is a characteristic.

To understand the significance of characteristics in the context of existence and uniqueness of solutions, go back to the general solution 1.8 of the linear first order partial differential equation 1.3. This general solution is

$$u(x, y) = e^{\alpha(x,y)}[M(x, y) + g(\psi(x, y))].$$

Suppose we prescribe $u(x, y) = q(x)$ along a characteristic. Now a characteristic is specified by $\psi(x, y) = k$. If $y = y(x)$ along this characteristic, then

$$q(x) = e^{\alpha(x,y(x))}[M(x, y(x)) + g(k)]$$

or

$$q(x) = e^{\alpha(x,y(x))}[M(x, y(x)) + C], \tag{1.9}$$

in which C is constant. The functions $M(x, y)$ and $\alpha(x, y)$ are determined by the partial differential equation, and are not under our control, so equation 1.9 places a constraint on the given data function $q(x)$. If $q(x)$ is not of this form for any constant C, then there is no solution taking on these prescribed values on Γ. On the other hand, if $q(x)$ is of this form for some C, then there are infinitely many such solutions, because we can choose for g any differentiable function such that $g(k) = C$.

Example 4 Consider

$$xu_x + 2x^2u_y - u = x^2e^x. \tag{1.10}$$

First we will find the general solution. The characteristic equation is

$$\frac{dy}{dx} = 2x$$

and this has general solution defined by $y - x^2 = k$. The characteristics are parabolas. Let

$$\xi = x, \eta = y - x^2$$

to obtain

$$\xi w_\xi - w = \xi^2 e^\xi,$$

which we write as

$$w_\xi - \frac{1}{\xi} w = \xi e^\xi.$$

Multiply this equation by $e^{\int(-1/\xi)d\xi}$, which is $1/\xi$, to obtain

$$\frac{1}{\xi}\, w_\xi - \frac{1}{\xi^2}\, w = e^\xi,$$

or

$$\frac{\partial}{\partial \xi}\left(\frac{1}{\xi}\, w\right) = e^\xi.$$

Integrate with respect to ξ to get

$$\frac{1}{\xi}\, w = e^\xi + g(\eta)$$

so

$$w = \xi e^\xi + \xi g(\eta).$$

The general solution of equation 1.10 is

$$u(x, y) = xe^x + xg(y - x^2).$$

We will now attempt to find solutions satisfying given conditions along various curves.

Suppose first we seek a solution such that $u(x, y) = \sin(x)$ on the curve $y = x^2 + 4$. Notice that information is being specified along a characteristic. We will need

$$u(x, x^2 + 4) = xe^x + xg(4) = \sin(x).$$

We must be able to find a constant C such that

$$xe^x + Cx = \sin(x)$$

for all x, and this is impossible. There is no solution satisfying the requested condition.

Next suppose we want a solution such that $u(x, y) = xe^x - x$ on the parabola $y = x^2 + 4$. Now we need

$$u(x, x^2 + 4) = xe^x + xg(4) = xe^x - x.$$

This equation requires that $g(4) = -1$. This problem has infinitely many solutions because we can choose g to be any differentiable function of one variable such that $g(4) = -1$. Even though data is specified on a characteristic, the form of the data allows infinitely many solutions.

Finally, suppose we want a solution such that $u(x, y) = \cos(x)$ along the (noncharacteristic) parabola $y = x^2 + 4x$. Now we need

$$u(x, x^2 + 4x) = xe^x + xg(4x) = \cos(x).$$

This requires that

$$g(4x) = \frac{\cos(x) - xe^x}{x}.$$

Choose

$$g(t) = 4 \frac{\cos(t/4) - \dfrac{t}{4} e^{t/4}}{t}$$

for, say, $t > 0$. The solution of the problem (for $x > 0$) is

$$u(x, y) = xe^x + xg(y - x^2)$$

$$= xe^x + 4x \left(\frac{\cos\left(\dfrac{y - x^2}{.4}\right) - \dfrac{1}{4}(y - x^2)e^{(y - x^2)/4}}{y - x^2} \right). \qquad \blacksquare$$

The problem of finding a solution of equation 1.3 taking on prescribed values on a given curve is called a *Cauchy problem* (for the linear equation), and the given information on the curve is called *Cauchy data*. Our examples suggest that we can expect a unique solution of a Cauchy problem if the curve is not characteristic, and no solution or infinitely many solutions if the curve is characteristic.

EXERCISE 11 For each of the following partial differential equations, solve the characteristic equation and sketch graphs of some of the characteristics, find the general solution of the partial differential equation, and attempt to find solutions satisfying the Cauchy data on the given curves.

1. $3yu_x - 2xu_y = 0$
   (a) Find a solution satisfying $u(x, y) = x^2$ on the line $y = x$.
   (b) Find a solution satisfying $u(x, y) = 1 - x^2$ on the line $y = -x$.
   (c) Find a solution satisfying $u(x, y) = 2x$ on the ellipse $3y^2 + 2x^2 = 4$.

2. $u_x - 6u_y = y$

    (a) Find a solution satisfying $u(x, y) = e^x$ on the line $y = -6x + 2$.

    (b) Find a solution satisfying $u(x, y) = 1$ on the parabola $y = -x^2$.

    (c) Find a solution satisfying $u(x, y) = -4x$ on the line $y = -6x$.

3. $4u_x + 8u_y - u = 1$

    (a) Find a solution satisfying $u(x, y) = \cos(x)$ on the line $y = 3x$.

    (b) Find a solution satisfying $u(x, y) = x$ on the line $y = 2x$.

    (c) Find a solution satisfying $u(x, y) = 1 - x$ on the curve $y = x^2$.

4. $-4yu_x + u_y - yu = 0$

    (a) Find a solution satisfying $u(x, y) = x^3$ on the line $x + 2y = 3$.

    (b) Find a solution satisfying $u(x, y) = -y$ on $y^2 = x$.

    (c) Find a solution satisfying $u(x, y) = 2$ on $x + 2y^2 = 1$.

5. $yu_x + x^2u_y = xy$

    (a) Find a solution satisfying $u(x, y) = 4x$ on the curve $y = (1/3)x^{3/2}$.

    (b) Find a solution satisfying $u(x, y) = x^3$ on the curve $3y^2 = 2x^3$.

    (c) Find a solution satisfying $u(x, y) = \sin(x)$ on the line $y = 0$.

6. $y^2u_x + x^2u_y = y^2$

    (a) Find a solution satisfying $u(x, y) = x$ on $y = 4x$.

    (b) Find a solution satisfying $u(x, y) = -2y$ on $y^3 = x^3 - 2$.

    (c) Find a solution satisfying $u(x, y) = y^2$ on $y = -x$.

1.4   THE QUASI-LINEAR EQUATION AND THE METHOD OF CHARACTERISTICS

Consider the first order quasi-linear partial differential equation

$$f(x, y, u)u_x + g(x, y, u)u_y = h(x, y, u) \tag{1.11}$$

in which we seek a solution u as a function of the independent variables x and y. This is linear in the highest partial derivatives (which in the first order case are of order one), but may be nonlinear in u.

In the linear case we defined characteristics to be certain curves in the x, y-plane. Given a solution $u(x, y)$ of a linear equation and a characteristic Γ, we can define a curve on the surface $u = u(x, y)$ whose projection in the x, y-plane is Γ (Figure 1.1). In this context the curve in x, y, u-space is also called a characteristic, and its projection into the x, y-plane is called a *characteristic trace*.

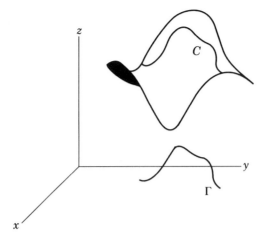

FIGURE 1.1 A characteristic C on a solution surface, projecting onto a characteristic trace.

For the quasi-linear equation 1.11, characteristics are curves in x, y, u-space defined by

$$\frac{dx}{dt} = f(x, y, u), \quad \frac{dy}{dt} = g(x, y, u), \quad \frac{du}{dt} = h(x, y, u). \tag{1.12}$$

We will show that a solution $u(x, y)$ of a quasi-linear equation may be interpreted as a surface made up of characteristics. This observation can be used to obtain solutions containing given (noncharacteristic) curves, and hence provides a way of solving the Cauchy problem when the partial differential equation is quasi-linear.

We need two observations.

Observation One   Suppose $u = \varphi(x, y)$ is a solution of equation 1.11 defining a surface Σ, and that $P_0 : (x_0, y_0, u_0)$ is a point on Σ, so $u_0 = \varphi(x_0, y_0)$. Then the characteristic passing through P_0 lies entirely on Σ.

To see why this is true, suppose the characteristic has parametric equations

$$x = x(t), \quad y = y(t), \quad u = u(t).$$

Then for some t_0,

$$x(t_0) = x_0, \quad y(t_0) = y_0, \quad u(t_0) = u_0.$$

Because this curve is characteristic, we can use equations 1.12 to compute

$$\frac{dt}{dt}\,\varphi(x(t),\,y(t)) = \varphi_x\,\frac{dx}{dt} + \varphi_y\,\frac{dy}{dt}$$

$$= \varphi_x f(x,\,y,\,u) + \varphi_y g(x,\,y,\,u) = h(x,\,y,\,u) = \frac{du}{dt}.$$

Therefore

$$u(t) = \varphi(x(t),\,y(t)) + k$$

for some constant k. But P_0 is on the surface, so

$$\varphi(x(t_0),\,y(t_0)) = u_0 = u(t_0)$$

implies that $k = 0$. Therefore

$$u(t) = \varphi(x(t),\,y(t))$$

and the characteristic lies on Σ.

Observation Two If we begin with an arbitrary (but noncharacteristic) curve Γ and construct the family of characteristics passing through points of Γ, as in Figure 1.2, then the resulting surface Σ is the graph of a solution of the partial differential equation.

To see why this is true, assume that Σ is the graph of $u = \varphi(x, y)$. We want to show that φ is a solution of equation 1.11.

Suppose Γ is parameterized by

$$x = x(s),\ y = y(s),\ z = z(s).$$

At any (x, y, u) on Σ,

$$\frac{dx}{ds} = f(x,\,y,\,u),\quad \frac{dy}{ds} = g(x,\,y,\,u),\quad \frac{du}{ds} = h(x,\,y,\,u)$$

because the surface is made up of characteristics. Then

$$\frac{du}{ds} = h(x,\,y,\,u) = \varphi_x\,\frac{dx}{ds} + \varphi_y\,\frac{dy}{ds} = f\varphi_x + g\varphi_y$$

so φ is a solution.

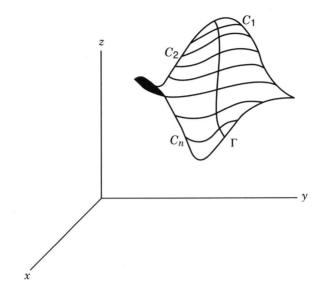

FIGURE 1.2   A solution surface formed by constructing characteristics through points of Γ.

These observations suggest the *method of characteristics* for solving the Cauchy problem when the partial differential equation is quasi-linear. Suppose we want the solution of equation 1.11 assuming prescribed values on a given curve Γ that is not characteristic. Construct the characteristic through each point of Γ. This defines a surface in three space, and this surface is the graph of the solution of the Cauchy problem.

This strategy also suggests why we do not want to specify data along a characteristic C. If we did so, then the characteristic through each point of C would be just C itself, not a surface.

Here are two illustrations of the method of characteristics.

Example 5   We want the solution of

$$yu_x - xu_y = e^u$$

that passes through the curve Γ given by $y = \sin(x)$, $u = 0$. This means that we require

$$u(x, \sin(x)) = 0.$$

The characteristics of this partial differential equation are specified by

$$\frac{dx}{dt} = y, \quad \frac{dy}{dt} = -x, \quad \frac{du}{dt} = e^u.$$

From the first two of these equations we can write

$$\frac{dy}{dx} = -\frac{x}{y},$$

or

$$ydy + xdx = 0,$$

with general solution (in terms of t)

$$x = a\ \cos(t) + b\ \sin(t),\ y = b\ \cos(t) - a\ \sin(t),$$

with a and b constant. From $du/dt = e^u$ we obtain

$$-e^{-u} = t + c.$$

The characteristics therefore have parametric representation

$$x = a\ \cos(t) + b\ \sin(t),\ y = b\ \cos(t) - a\ \sin(t),\ e^{-u} = c - t.$$

Parameterize Γ as

$$x = s,\quad y = \sin(s),\ u = 0.$$

We use s as parameter on Γ to distinguish between points on Γ and points on characteristics.

We want to construct a characteristic through each point of Γ (as in Figure 1.2). Let $P:(s, \sin(s), 0)$ be a point of Γ and suppose a characteristic passes through P when $t = 0$ (this is just a scaling of the parameter). Then at $t = 0$,

$$x = a = s,\ y = b = \sin(s),\ \text{and}\ e^0 = 1 = 0 + c,$$

giving us $a = s$, $b = \sin(s)$, and $c = 1$ at this point of intersection of Γ with P. Therefore, the characteristic intersecting Γ at P has parametric equations

$$x = s\ \cos(t) + \sin(s)\sin(t),\ y = \sin(s)\cos(t) - s\ \sin(t),\ e^{-u} = 1 - t.$$

Now eliminate t and s from these equations. From the first two equations,

$$s = x\ \cos(t) - y\ \sin(t)$$

and

$$\sin(s) = y\ \cos(t) + x\ \sin(t).$$

Therefore

$$\sin(x \cos(t) - y \sin(t)) = y \cos(t) + x \sin(t). \tag{1.13}$$

But $e^{-u} = 1 - t$ implies that $t = 1 - e^{-u}$. Substitute this into equation (1.13) to get

$$\sin(x \cos(1 - e^{-u}) - y \sin(1 - e^{-u})) = y \cos(1 - e^{-u}) + x \sin(1 - e^{-u})).$$

This equation implicitly defines the solution of the Cauchy problem. It is easy to check that $y = \sin(x)$, $u = 0$ satisfies this equation.

Figure 1.3(a) shows a graph of part of the surface (solution), and Figure 1.3(b) shows the same surface from a different perspective. ∎

Example 6 Consider the quasi-linear equation

$$xu_x + yu_y = \sec(u).$$

We would like the solution passing through

$$\Gamma : x = s^2, \ y = \sin(s), \ u = 0.$$

The characteristics satisfy

$$\frac{dx}{dt} = x, \ \frac{dy}{dt} = y, \ \frac{du}{dt} = \sec(u),$$

which have solutions defined by

$$x = Ae^t, \ y = Be^t, \ \sin(u) = t + c.$$

As in Example 5, we want to construct the characteristic through each point

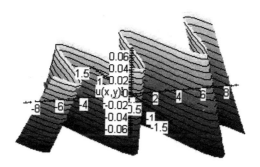

FIGURE 1.3(a) Solution of $yu_x - xu_y = e^u$ containing the curve $y = \sin(x)$, $u = 0$ (Example 5).

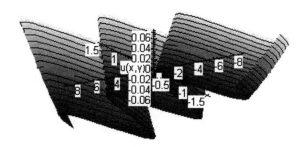

FIGURE 1.3(b) Another perspective of the solution surface shown in Figure 1.3(a) (Example 5).

of Γ. Suppose a characteristic passes through Γ at $P:(s^2, \sin(s), 0)$ at $t = 0$. Then

$$x = A = s^2, \; y = B = \sin(s), \; \sin(0) = 0 = 0 + c,$$

so

$$A = s^2, \; B = \sin(s) \text{ and } c = 0$$

at this point. Then

$$x = s^2 e^t, \; y = \sin(s)e^t \text{ and } \sin(u) = t.$$

We want to eliminate s and t from these equations. Since $t = \sin(u)$, then

$$y = \sin(s)e^{\sin(u)}$$

so

$$\sin(s) = ye^{-\sin(u)}$$

and

$$s = \arcsin(ye^{-\sin(u)}).$$

Then

$$x = e^{\sin(u)}[\arcsin(ye^{-\sin(u)})]^2.$$

This equation implicitly defines $u(x, y)$ such that $u(s^2, \sin(s)) = 0$—that is, the solution surface contains the data curve Γ. Figure 1.4(a) shows a graph of Γ, and Figure 1.4(b) part of the solution surface. ∎

EXERCISE 12   For each of the following, use the method of characteristics to find a solution of the partial differential equation that passes through the given curve Γ. The solution may be implicitly defined. Graph part of the solution surface.

1. $xu_x + yu_y = \sec(u)$; Γ is the curve defined by $y = x^3$, $u = 0$.
2. $u_x - xu_y = 4$; Γ is given by $y = 4x$, $u = 0$.
3. $u_x - y^2u_y = 1$; Γ is given by $y = x^2 + 2$; $u = 0$.
4. $u_x - y^3u_y = \sec(u)$; Γ is given by $y = x^2$, $u = 0$.
5. $u_x + yu_y = u$; Γ is given by $y = 1 - x$, $u = 1$.
6. $u_x + y^2u_y = \cos(u)$; Γ is given by $x = y^2$, $u = 0$.
7. $u_x - u_y = u^2$; Γ is given by $y = 2x - 1$, $u = 4$.
8. $x^3u_x - yu_y = u$; Γ is given by $x = y^2 - 1$, $u = 1$.
9. $u_x - y^2u_y = u$; Γ is given by $y = 1 - x^2$, $u = 2$.
10. $xu_x + u_y = e^u$; Γ is given by $y = x - 1$, $u = 0$.

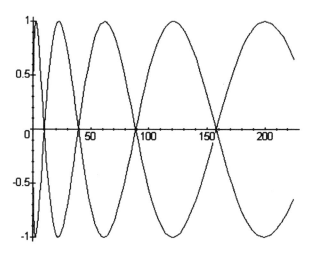

FIGURE 1.4(a)   Graph of the curve $x = t^2$, $y = \sin(t)$, $u = 0$ (Example 6).

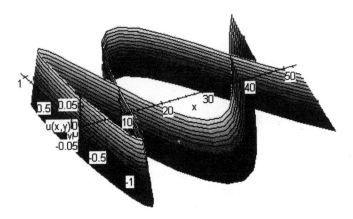

FIGURE 1.4(b)   Solution of $xu_x + yu_y = \sec(u)$ containing the curve shown in Figure 1.4(a).

EXERCISE 13   Use the method of characteristics to show that the solution of the problem

$$uu_x + u_y = 0$$

$$u(x, 0) = f(x)$$

is defined implicitly by the equation

$$u(x, y) = f(x - u(x, y)y).$$

Hint: Think of Γ as the curve $x = s, y = 0, u = f(s)$.

   Next use the idea of Exercise 4 to solve this problem. Compare these solutions.

2

LINEAR SECOND ORDER PARTIAL DIFFERENTIAL EQUATIONS

2.1 CLASSIFICATION

Consider the linear second order partial differential equation

$$Au_{xx} + 2Bu_{xy} + Cu_{yy} + Du_x + Eu_y + Fu + G = 0 \qquad (2.1)$$

in which the coefficients are functions of x and y that are continuous in some region of the plane. The factor of 2 in the coefficient of u_{xy} simplifies an expression we will derive shortly. We assume that A, B, and C do not vanish at the same point.

We were able to solve the linear first order partial differential equation by a change of variables. We will explore the idea of transforming equation 2.1 into a form more accessible for obtaining solutions, or at least information about solutions.

Begin with a general transformation

$$\xi = \xi(x, y), \ \eta = \eta(x, y)$$

whose Jacobian does not vanish in some region $\mathcal{D}$ of the plane:

$$J = \begin{vmatrix} \xi_x & \xi_y \\ \eta_x & \eta_y \end{vmatrix} \neq 0 \text{ for } (x, y) \text{ in } \mathcal{D}.$$

Let $w(\xi, \eta) = u(x(\xi, \eta), y(\xi, \eta))$. Then

$$u_x = w_\xi \xi_x + w_\eta \eta_x,$$

$$u_{xx} = \xi_{xx} w_\xi + \xi_x(w_{\xi\xi}\xi_x + w_{\eta\xi}\eta_x) + \eta_{xx}w_\eta + \eta_x(w_{\xi\eta}\xi_x + w_{\eta\eta}\eta_x)$$

$$= w_{\xi\xi}\xi_x^2 + 2w_{\eta\xi}\xi_x\eta_x + w_{\eta\eta}\eta_x^2 + w_\xi \xi_{xx} + w_\eta \eta_{xx},$$

and so on for the other derivatives occurring in equation 2.1. After a routine calculation the transformed equation is

$$aw_{\xi\xi} + 2bw_{\xi\eta} + cw_{\eta\eta} + dw_\xi + ew_\eta + fw + g = 0 \qquad (2.2)$$

in which

$$a = A\xi_x^2 + 2B\xi_x\xi_y + C\xi_y^2, \ c = A\eta_x^2 + 2B\eta_x\eta_y + C\eta_y^2$$

and

$$b = A\xi_x\eta_x + B(\xi_x\eta_y + \eta_x\xi_y) + C\xi_y\eta_y.$$

The coefficients $d, \ldots, g$ are not difficult to obtain, but we will not need them for this discussion.

Equation 2.2 is of the same form as the original equation 2.1. This should not be surprising, since we have not yet placed any restrictions on the transformation, other than that it be invertible ($J \neq 0$). We now seek ways to choose the transformation to simplify equation 2.2.

A routine calculation shows that $J = \text{Jacobian}$

$$b^2 - ac = (B^2 - AC)J^2. \qquad (2.3)$$

EXERCISE 14   Verify equation 2.3.

The quantity $B^2 - AC$ is called the *discriminant* of equation 2.1, and equation 2.3 implies that the original partial differential equation and the transformed equation 2.2 have discriminants of the same sign. The sign of the discriminant is called an *invariant* of the transformation, and the partial differential equation is defined to be

hyperbolic in $\mathcal{D}$ if $B^2 - AC > 0$,
parabolic in $\mathcal{D}$ if $B^2 - AC = 0$,
and
elliptic in $\mathcal{D}$ if $B^2 - AC < 0$,
for all (x, y) in $\mathcal{D}$.

This scheme for assigning names to equation 2.1 defines a classification of the linear second order partial differential equation which remains fixed through transformations having nonvanishing Jacobian. We will now show that it is possible to transform equation 2.1 to a relatively simple form, called its *canonical form*, which varies according to whether the equation is hyperbolic, parabolic, or elliptic. The type of the equation determines what properties might be expected of solutions and what kinds of information must be supplied with the equation to have a unique solution, and also influences the kinds of numerical techniques used to approximate solutions.

Consider each type of equation in turn.

2.2 | CANONICAL FORM OF THE HYPERBOLIC EQUATION

Suppose first that $B^2 - AC > 0$ in some region $\mathcal{D}$ of the plane. We will assume that $A(x, y) \neq 0$ for (x, y) in $\mathcal{D}$. A similar discussion follows if $C(x, y) \neq 0$. (If $A = C = 0$ throughout the region of interest, we will see that the partial differential equation is already in one of the canonical forms we will define.)

We can make $a = 0$ in equation 2.2 if we choose ξ so that

$$A\xi_x^2 + 2B\xi_x\xi_y + C\xi_y^2 = 0.$$

Since $A \neq 0$, we can divide this equation by $A\xi_y^2$ to get

$$\left(\frac{\xi_x}{\xi_y}\right)^2 + 2\frac{B}{A}\left(\frac{\xi_x}{\xi_y}\right) + \frac{C}{A} = 0.$$

Then

$$\frac{\xi_x}{\xi_y} = \frac{-B \pm \sqrt{B^2 - AC}}{A}.$$

Along any curve $\xi(x, y) = k$, the slope at any point is given by

$$\frac{dy}{dx} = -\frac{\xi_x}{\xi_y} = \frac{B \pm \sqrt{B^2 - AC}}{A}. \tag{2.4}$$

We can make $c = 0$ in equation 2.2 if we choose η so that

$\eta = u$ in notes.

$$A\eta_x^2 + 2B\eta_x\eta_y + C\eta_y^2 = 0.$$

This is the same form of the equation we had involving ξ_x and ξ_y, and we obtain

$$\frac{\eta_x}{\eta_y} = \frac{-B \pm \sqrt{B^2 - AC}}{A}.$$

Further, the slope along any curve $\eta(x, y) = k$ is

$$\frac{dy}{dx} = -\frac{\eta_x}{\eta_y} = \frac{B \pm \sqrt{B^2 - AC}}{A}. \tag{2.5}$$

We can obtain a transformation with nonzero Jacobian, and making both $a = c = 0$, by choosing different signs before the radical in equations 2.4 and 2.5. Let $\xi(x, y) = k$ be an integral of

$$\frac{dy}{dx} = \frac{B + \sqrt{B^2 - AC}}{A} \tag{2.6}$$

char

and let $\eta(x, y) = K$ be an integral of

$$\frac{dy}{dx} = \frac{B - \sqrt{B^2 - AC}}{A}. \tag{2.7}$$

Equations 2.6 and 2.7 are the *characteristic equations* for equation 2.1 in the hyperbolic case, and their integrals $\xi(x, y) = k$, $\eta(x, y) = K$ define two classes of curves called *characteristics* for this equation. The transformation

$$\xi = \xi(x, y), \ \eta = \eta(x, y)$$

has been chosen so that $a = c = 0$ in the transformed equation 2.2, which becomes

$$2bw_{\xi\eta} + dw_\xi + ew_\eta + fw + g = 0.$$

Upon dividing by $2b$ we obtain

$$w_{\xi\eta} + d^*w_\xi + e^*w_\eta + f^*w + g^* = 0. \tag{2.8}$$

This is the *canonical form for the hyperbolic equation.* We can write this canonical form as

$$w_{\xi\eta} + \Phi(\xi, \eta, w, w_\xi, w_\eta) = 0,$$

with the only second derivative term being a mixed second partial derivative with coefficient 1.

Example 7 Consider

$$u_{xx} + 2\cos(x)u_{xy} - \sin^2(x)u_{yy} - \sin(x)u_y = 0. \tag{2.9}$$

Here $A = 1$, $B = \cos(x)$ and $C = -\sin^2(x)$. The discriminant is

$$B^2 - AC = \cos^2(x) + \sin^2(x) = 1 > 0.$$

The partial differential equation 2.9 is hyperbolic. From equations 2.6 and 2.7, the characteristic equations are

$$\frac{dy}{dx} = \cos(x) + 1 \text{ and } \frac{dy}{dx} = \cos(x) - 1.$$

Integrate the first of these equations to get $y = \sin(x) + x + k$; hence let

$$\xi(x, y) = y - \sin(x) - x.$$

Curves in the family $\xi(x, y) = k$, obtained by choosing different values of k, are characteristics.

Solve the other characteristic equation to get $y = \sin(x) - x + K$, so let

$$\eta(x, y) = y - \sin(x) + x.$$

Curves in the family $\eta(x, y) = K$ are also characteristics. Some curves in these families of characteristics are shown in Figure 2.1.

Now transform equation 2.9. Let

$$\xi = y - \sin(x) - x, \; \eta = y - \sin(x) + x$$

and

$$w(\xi, \eta) = u(x, y).$$

Compute

$$u_x = w_\xi(-\cos(x) - 1) + w_\eta(-\cos(x) + 1),$$

$$u_y = w_\xi + w_\eta,$$

$$u_{xx} = \sin(x)w_\xi + \sin(x)w_\eta$$
$$- (\cos(x) + 1)[w_{\xi\xi}(-\cos(x) - 1) + w_{\xi\eta}(-\cos(x) + 1)]$$
$$+ (1 - \cos(x))[w_{\eta\xi}(-\cos(x) - 1) + w_{\eta\eta}(-\cos(x) + 1)],$$

$$u_{yy} = w_{\xi\xi} + 2w_{\xi\eta} + w_{\eta\eta},$$

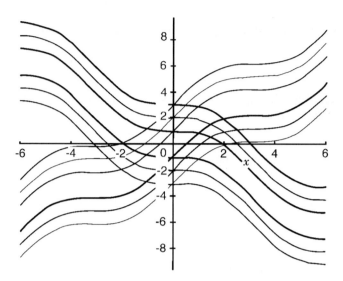

FIGURE 2.1   Families of characteristics of the hyperbolic equation $u_{xx} + 2\cos(x)u_{xy} - \sin^2(x)u_{yy} - \sin(x)u_y = 0$ of Example 7.

and

$$u_{xy} = w_{\xi\xi}(-\cos(x) - 1) + w_{\xi\eta}(-\cos(x) + 1)$$
$$+ w_{\eta\xi}(-\cos(x) - 1) + w_{\eta\eta}(-\cos(x) + 1).$$

Substitute these into equation 2.9 to obtain

$$w_{\xi\eta} = 0.$$

This is the canonical form of equation 2.9.

This canonical form is so simple that we can immediately write solutions. Any function of the form

$$w(\xi, \eta) = f(\xi) + g(\eta)$$

will certainly satisfy $w_{\xi\eta} = 0$, if f and g are twice differentiable functions of a single variable. Thus, for any such f and g,

$$u(x, y) = f(y - \sin(x) - x) + g(y - \sin(x) + x)$$

is a solution to equation 2.9. The fact that we were able to extract this nonobvious solution of equation 2.9 from its canonical form suggests one value of canonical forms.

As a specific example, if we choose $f(t) = t^2$ and $g(t) = e^t$, we obtain the solution

$$u(x, y) = (y - \sin(x) - x)^2 + e^{y - \sin(x) + x}. \qquad \blacksquare$$

The canonical form of a hyperbolic equation may contain first derivative and other terms, and so need not be as simple as that of Example 7.

If $C \neq 0$ in some region $\mathcal{D}$, we can again begin with the equation

$$A\xi_x^2 + 2B\xi_x\xi_y + C\xi_y^2 = 0,$$

but this time divide by $C\xi_x^2$ to obtain

$$\left(\frac{\xi_y}{\xi_x}\right)^2 + 2\frac{B}{C}\left(\frac{\xi_y}{\xi_x}\right) + \frac{A}{C} = 0.$$

Now we solve for ξ_y/ξ_x instead of ξ_x/ξ_y and the discussion proceeds parallel to the preceding one, resulting again in a transformation to the canonical form 2.8.

EXERCISE 15   Carry out the derivation of the canonical form for this case that $C \neq 0$ throughout $\mathcal{D}$.

EXERCISE 16   In each of the following, find a solution involving two arbitrary, twice differentiable functions, following the idea of Example 7.

1. $u_{xx} + 2u_{xy} - 3u_{yy} = 0$
2. $u_{xx} + 4u_{xy} + u_{yy} = 0$
3. $2u_{xx} + 2u_{xy} - 5u_{yy} = 0$
4. $3u_{xx} + 6u_{xy} + 2u_{yy} = 0$

2.3   CANONICAL FORM OF THE PARABOLIC EQUATION

Suppose $B^2 - AC = 0$ in $\mathcal{D}$. Now A and C cannot both vanish at any point of $\mathcal{D}$ because then we would also have $B = 0$, contrary to the assumption made beginning in Section 2.1. Suppose $A(x, y) \neq 0$ throughout $\mathcal{D}$. We can begin as we did in the hyperbolic case, except now equations 2.6 and 2.7 are the same, namely

$$\frac{dy}{dx} = \frac{B}{A}.$$

This is the characteristic equation for equation 2.1 in the parabolic case. Now the characteristics are graphs of $\xi(x, y) = k$, where this equation implicitly defines an integral of the characteristic equation. We can choose η as *any* continuous function with continuous first and second partial derivatives, and such that the Jacobian $J \neq 0$ throughout $\mathcal{D}$.

Make the transformation defined by

$$\xi = \xi(x, y), \quad \eta = \eta(x, y).$$

In the transformed equation 2.2 we will have $a = 0$ by the way ξ was chosen. As a bonus, we claim that $b = 0$ also. The reason is that

$$b = (A\xi_x + B\xi_y)\eta_x + (B\xi_x + C\xi_y)\eta_y.$$

But $A\xi_x + B\xi_y = 0$ because $\xi_x/\xi_y = -B/A$. And $B^2 - AC = 0$ implies that $B/A = C/B$, so $\xi_x/\xi_y = -C/B$, hence $B\xi_x + C\xi_y = 0$ also.

With $a = b = 0$ the transformed equation 2.2 is

$$cw_{\eta\eta} + dw_\xi + ew_\eta + fw + g = 0.$$

Upon dividing by c we obtain the *canonical form of the parabolic equation*:

$$w_{\eta\eta} + d^*w_\xi + e^*w_\eta + f^*w + g^* = 0.$$

We can write this canonical form as

$$w_{\eta\eta} + \Phi(\xi, \eta, w, w_\xi, w_\eta) = 0.$$

Example 8   Consider the constant coefficient equation

$$9u_{xx} + 12u_{xy} + 4u_{yy} + u_x = 0.$$

This equation is parabolic because $B^2 - AC = 6^2 - 4 \cdot 9 = 0$. The characteristic equation is

$$\frac{dy}{dx} = \frac{6}{9}$$

with general solution defined by $y - (2/3)x = k$. The characteristics are straight lines corresponding to different choices of k. Choose

$$\xi = y - \frac{2}{3}x.$$

We can choose η as any continuous function with continuous first and second partial derivatives, such that the Jacobian of ξ and η is not zero in the region under consideration. If we choose $\eta = x$ then

$$J = \begin{vmatrix} -\dfrac{2}{3} & 1 \\ 1 & 0 \end{vmatrix} = -1$$

and this is nonzero in the entire plane. Let $w(\xi, \eta) = u(x, y)$ and compute

$$u_x = -\frac{2}{3} w_\xi + w_\eta,$$

$$u_y = w_\xi,$$

$$u_{xx} = -\frac{2}{3}\left(-\frac{2}{3} w_{\xi\xi} + w_{\xi\eta}\right) - \frac{2}{3} w_{\eta\xi} + w_{\eta\eta},$$

$$u_{yy} = w_{\xi\xi},$$

and

$$u_{xy} = -\frac{2}{3} w_{\xi\xi} + w_{\xi\eta}.$$

Substitute these into the partial differential equation to obtain

$$9w_{\eta\eta} - \frac{2}{3} w_\xi + w_\eta = 0.$$

The canonical form is

$$w_{\eta\eta} - \frac{2}{27} w_\xi + \frac{1}{9} w_\eta = 0. \qquad \blacksquare$$

Example 9   Suppose we want solutions of

$$u_{xx} + 4u_{xy} + 4u_{yy} = 0. \tag{2.10}$$

Here $A = 1$, $B = 2$ and $C = 4$, so $B^2 - AC = 0$ and equation 2.10 is parabolic. Solve the characteristic equation

$$\frac{dy}{dx} = \frac{B}{A} = 2$$

to get

$$y - 2x = k.$$

Let $\xi = y - 2x$ and choose η so that $J \neq 0$, say $\eta = x$. We find that

$$u_{xx} = 4w_{\xi\xi} - 4w_{\xi\eta} + w_{\eta\eta},$$

$$u_{yy} = w_{\xi\xi}$$

and

$$u_{xy} = -2w_{\xi\xi} + w_{\xi\eta}.$$

Upon substituting these into equation 2.10 we obtain the canonical form

$$w_{\eta\eta} = 0.$$

This equation is simple enough that we can write solutions. Write the canonical form as

$$(w_\eta)_\eta = 0,$$

which implies that w_η is independent of η, say

$$w_\eta = F(\xi).$$

Integrate this equation with respect to η to obtain

$$w(\xi, \eta) = \int F(\xi) \, d\eta = \eta F(\xi) + G(\xi).$$

Therefore

$$u(x, y) = xF(y - 2x) + G(y - 2x)$$

is a solution for any twice differentiable functions F and G of a single variable. The student can verify by substitution that $u(x, y)$ is indeed a solution of equation 2.10. ∎

EXERCISE 17   In Example 8 the choice $\eta = x$ was arbitrary, requiring only that $J \neq 0$ in $\mathcal{D}$. Determine the canonical form that would result from choosing $\eta = y$. What would result if we chose $\eta = x^2$ and agreed that $\mathcal{D}$ contains no points on the y-axis?

EXERCISE 18   For each of the following, find solutions involving two arbitrary functions.

1. $u_{xx} + 6u_{xy} + 9u_{yy} = 0$
2. $u_{xx} - 4u_{xy} + 4u_{yy} = 0$
3. $25u_{xx} + 20u_{xy} + 4u_{yy} = 0$
4. $9u_{xx} + 12u_{xy} + 4u_{yy} = 0$

2.4   CANONICAL FORM OF THE ELLIPTIC EQUATION

Suppose $B^2 - AC < 0$ in $\mathcal{D}$. Now $B^2 - AC$ has no real square root, so the equations 2.6 and 2.7 used to generate characteristics in the hyperbolic and parabolic cases have no real solutions. The elliptic equation has no characteristics. Nevertheless, we can seek a transformation $\xi = \xi(x, y)$, $\eta = \eta(x, y)$ which simplifies equation 2.2.

We will begin by attempting to make the coefficient of $w_{\xi\eta}$ zero in the transformed equation. This will require that

$$\xi_x(A\eta_x + B\eta_y) + \xi_y(B\eta_x + C\eta_y) = 0.$$

Choose any "reasonable" function $\eta(x, y)$. Then along any curve defined by $\xi(x, y) = k$ the last equation tells us that

$$\frac{dy}{dx} = -\frac{\xi_x}{\xi_y} = \frac{B\eta_x + C\eta_y}{A\eta_x + B\eta_y}.$$

We attempt to solve this equation to determine curves $\xi(x, y) = k$. Assuming that we can do this, we can transform the elliptic partial differential equation to the form

$$aw_{\xi\xi} + cw_{\eta\eta} + \Phi(\xi, \eta, w, w_\xi, w_\eta) = 0. \qquad (2.11)$$

Thus far we will have succeeded only in eliminating the mixed derivative term. We do not choose this as the canonical form for the parabolic case because equation 2.11 can be further simplified by a second change of variables. In order to avoid having to assign new symbols to another transformation, we will rewrite equation 2.11 in terms of x and y and imagine that we are starting over with this equation. Thus, we want to simplify

$$Au_{xx} + Cu_{yy} + \Phi(x, y, u, u_x, u_y) = 0. \qquad (2.12)$$

We have also written the coefficient functions in uppercase to use previous results in which this notation was used.

We will determine a change of variables that transforms equation 2.12 to

$$w_{\xi\xi} + w_{\eta\eta} + \Psi(\xi, \eta, w, w_\xi, w_\eta) = 0.$$

In order to do this, we must obtain $a = c$ and $b = 0$ in the transformed equation 2.2. Thus (keeping in mind that we already have $B = 0$ in the new starting equation 2.12), we need

$$A\xi_x^2 + C\xi_y^2 = A\eta_x^2 + C\eta_y^2$$

and

$$A\xi_x\eta_x + C\xi_y\eta_y = 0.$$

One way to make headway on the first equation is to try to find ξ and η such that

$$\sqrt{A}\,\xi_x = \sqrt{C}\,\eta_y \text{ and } \sqrt{C}\,\xi_y = -\sqrt{A}\,\eta_x. \tag{2.13}$$

If we are able to find η, then we can put

$$\xi(x, y) = k + \int_{(x_0, y_0)}^{(x,y)} \sqrt{\frac{C}{A}}\,\eta_y \, dx - \sqrt{\frac{A}{C}}\,\eta_x \, dy \tag{2.14}$$

in which k is a constant and (x_0, y_0) some conveniently chosen point. Such a function will satisfy equations 2.13 exactly when the integral defining $\xi(x, y)$ is independent of path, and the condition for this is

$$\left(-\sqrt{\frac{A}{C}}\,\eta_x\right)_x = \left(\sqrt{\frac{C}{A}}\,\eta_y\right)_y. \tag{2.15}$$

The strategy is therefore to first attempt to solve equation 2.15 for $\eta(x, y)$ (keeping in mind that A and C are in general functions of x and y), and then use this η and equation 2.14 to obtain ξ.

Assuming success in solving these equations to find a suitable transformation, then equation 2.12 transforms to an equation of the form

$$aw_{\xi\xi} + aw_{\eta\eta} + H(\xi, \eta, w, w_\xi, w_\eta) = 0.$$

Finally, upon dividing by a (in a region in which this function is nonzero), we obtain the canonical form

$$w_{\xi\xi} + w_{\eta\eta} + \Psi(\xi, \eta, w, w_\xi, w_\eta) = 0.$$

This is the *canonical form of the elliptic equation.*

Example 10 Consider the elliptic equation

$$u_{xx} + 2u_{xy} + 3u_{yy} + 4u = 0. \tag{2.16}$$

Since $B^2 - AC = 1 - (1)(3) = -2 < 0$, this equation is elliptic and so has no characteristics. We have to start somewhere, so try $\eta = x + y$ and consider

$$\frac{dy}{dx} = \frac{B\eta_x + C\eta_y}{A\eta_x + B\eta_y} = 2.$$

This has general solution defined by $y - 2x = k$, so let

$$\xi = y - 2x, \ \eta = x + y.$$

The Jacobian is $J = -3$. Compute

$$u_x = -2w_\xi + w_\eta,$$

$$u_{xx} = 4w_{\xi\xi} - 4w_{\xi\eta} + w_{\eta\eta},$$

$$u_y = w_\xi + w_\eta,$$

$$u_{yy} = w_{\xi\xi} + 2w_{\xi\eta} + w_{\eta\eta},$$

and

$$u_{xy} = -2w_{\xi\xi} - w_{\xi\eta} + w_{\eta\eta}.$$

Upon substituting these into the partial differential equation, we obtain

$$3w_{\xi\xi} + 6w_{\eta\eta} + 4w = 0$$

which we write as

$$w_{\xi\xi} + 2w_{\eta\eta} + \frac{4}{3}w = 0.$$

Thus far we have obtained an intermediate form in which there is no mixed derivative term. To simplify further and obtain the canonical form, write this equation in terms of u, x, and y again to avoid having to use new letters for the second transformation. Thus, consider

$$u_{xx} + 2u_{yy} + \frac{4}{3}u = 0 \tag{2.17}$$

in which $A = 1$ and $C = 2$ so $A/C = 1/2$ and $C/A = 2$. Equation 2.17 is again elliptic. Equation 2.15 is

$$-\frac{1}{\sqrt{2}} \eta_{xx} = \sqrt{2} \eta_{yy}.$$

Of course there are many solutions of this equation. We will choose a simple one, say

$$\eta = x + y.$$

(Keep in mind here that we are reusing symbols—this is a new η, and is only coincidentally the same as that used in the first reduction.) Now use equation 2.14 to write

$$\xi(x, y) = k + \int_{(x_0, y_0)}^{(x,y)} \sqrt{2}(1) \, dx - \frac{1}{\sqrt{2}} (1) \, dy = k + \sqrt{2} \int_{(x_0, y_0)}^{(x,y)} dx - \frac{1}{2} \, dy.$$

A potential function for this line integral (which is independent of path) is $\varphi(x, y) = x - (1/2)y$ and

$$\xi(x, y) = k + \sqrt{2}[\varphi(x, y) - \varphi(x_0, y_0)] = \sqrt{2} \left(x - \frac{1}{2} y \right)$$

in which we choose the arbitrary constant k and the arbitrary point (x_0, y_0) so that $k - \varphi(x_0, y_0) = 0$ (for example, choose $k = 0$ and (x_0, y_0) to be the origin). Finally, we have the transformation

$$\xi = \sqrt{2}x - \frac{1}{\sqrt{2}} y$$

$$\eta = x + y.$$

It is routine to compute the partial derivatives and substitute into the partial differential equation 2.17 to obtain

$$3w_{\xi\xi} + 3w_{\eta\eta} + \frac{4}{3} w = 0.$$

The canonical form of the partial differential equation 2.16 is

$$w_{\xi\xi} + w_{\eta\eta} + \frac{4}{9} w = 0. \qquad \blacksquare$$

EXERCISE 19   In each of 1 through 10, (a) classify the partial differential equation, (b) sketch some of the characteristics (if the equation is not elliptic), and (c) determine the canonical form of the partial differential equation.

1. $u_{xx} - 8u_{xy} + 2u_{yy} + xu_x - yu_y = 0$
2. $4u_{xx} + 2u_{xy} + u_{yy} - u_x + xyu_y + u = 0$
3. $3u_{xx} + 2u_{xy} - u_{yy} + yu_x - u_y = 0$
4. $2u_{xx} - 4u_{xy} + 2u_{yy} - y^2u_x + u_y - xu = 0$
5. $3u_{xx} - 8u_{xy} + 2u_{yy} + (x + y)u_y = 0$
6. $3u_{xx} + 6u_{xy} + 4u_{yy} - u_x - xu_y + xyu = 0$
7. $u_{xx} - 4u_{xy} + 4u_{yy} + u_x + u_y = 0$
8. $6u_{xy} + 5u_{yy} - x^2u_x + yu_y - xu = 0$
9. $2u_{xx} - 10u_{xy} + 8u_{yy} + u_x - u_y = 0$
10. $2u_{xx} - 2u_{xy} - 3u_{yy} + y^2u_x - u = 0$

EXERCISE 20   Consider the transformation to canonical form in terms of variables ξ, η in the hyperbolic case. Let $\hat{\xi} = \xi + \eta$ and $\hat{\eta} = \xi - \eta$ and $\hat{w}(\hat{\xi}, \hat{\eta}) = w(\xi, \eta)$ to obtain

$$\hat{w}_{\hat{\xi}\hat{\xi}} - \hat{w}_{\hat{\eta}\hat{\eta}} + \Psi(\hat{\xi}, \hat{\eta}, \hat{w}, \hat{w}_{\hat{\xi}}, \hat{w}_{\hat{\eta}}) = 0.$$

This is another canonical form commonly used for the hyperbolic second order linear partial differential equation. Obtain this canonical form for the following hyperbolic equations:

1. $u_{xx} - 8u_{xy} + 7u_{yy} + u_x = 0$
2. $u_{xx} - 10u_{xy} + 16u_{yy} - xu_x = 0$
3. $2u_{xx} + 6u_{xy} - 8u_{yy} + yu = 0$
4. $u_{xx} + 4u_{xy} - 5u_{yy} + y^2u_y = 0$

EXERCISE 21   Let a, b, and c be constants and suppose that

$$u_{xx} - u_{yy} + au_x + bu_y + cu = 0.$$

Let $v(x, y) = e^{\alpha x + \beta y}u(x, y)$. Determine constants α, β, and h so that

$$v_{xx} - v_{yy} = -hv.$$

EXERCISE 22   Let a, b, and c be constants and suppose that

$$u_y - au_{xx} - bu_x - cu = 0.$$

Let $v(x, y) = e^{\alpha x + \beta y}u(x, y)$. Determine constants α and β so that

$$v_y - av_{xx} = 0.$$

EXERCISE 23   Determine the characteristics of the canonical form of the hyperbolic equation.

EXERCISE 24   Determine the characteristics of the canonical form of the parabolic equation.

EXERCISE 25   Tricomi's equation is

$$u_{xx} + xu_{yy} = 0.$$

(a) Determine the region of the plane in which Tricomi's equation is hyperbolic, the region in which it is parabolic, and the region in which it is elliptic.
(b) Determine the characteristics of the Tricomi equation in the part of the plane where the equation is hyperbolic.
(c) Can you determine the characteristics of the Tricomi equation in the part of the plane where the equation is parabolic?

EXERCISE 26   Consider equation 2.1 in the case that each coefficient is constant. For each of the cases (hyperbolic, parabolic, elliptic), go through the discussion in the text and determine a transformation of equation 2.1 to canonical form. You should find that the transformation and the canonical form in each case are simplified by the assumption of constant coefficients.

EXERCISE 27   Show that the hyperbolic canonical form

$$U_{\xi\xi} - U_{\eta\eta} + \Phi(\xi, \eta, U, U_\xi, U_\eta) = 0$$

can be obtained from the alternate hyperbolic canonical form

$$u_{xy} + \Psi(x, y, u, u_x, u_y) = 0$$

(see Exercise 20) by a rotation of coordinate axes through $\pi/4$ radians.

EXERCISE 28   Let u satisfy the hyperbolic equation $u_{xx} - u_{yy} = 0$. Show that u_x, u_y, u_{xx}, u_{yy}, and u_{xy} are also solutions of this equation. Verify this conclusion by direct calculation for the solution

$$u(x, y) = \sin(nx)\cos(ncy)$$

of the equation

$$c^2 u_{xx} - u_{yy} = 0.$$

Here c is a positive number and n is a positive integer.

EXERCISE 29 Let u satisfy the parabolic equation $u_y - ku_{xx} = 0$, in which k is a positive number. Show that u_x, u_y, u_{xx}, u_{yy}, and u_{xy} are also solutions. Verify this conclusion by direct calculation for the solution

$$u(x, y) = e^{-\alpha^2\pi^2 y} \cos(\alpha\pi x)$$

in which α can be any nonzero real number.

EXERCISE 30 Let u satisfy the elliptic equation $u_{xx} + u_{yy} = 0$. Show that u_x, u_y, u_{xx}, u_{yy}, and u_{xy} are also solutions. Verify this conclusion by direct calculation for the solution

$$u(x, y) = \sin(x)\cosh(y).$$

2.5 CANONICAL FORMS AND EQUATIONS OF MATHEMATICAL PHYSICS

We now have canonical forms for hyperbolic, parabolic, and elliptic second order linear partial differential equations in two independent variables. We next want to develop the Cauchy problem and information which, together with the partial differential equation, determines a unique solution. But first we will relate the canonical forms to equations describing physical phenomena. This will afford us additional insight into properties we might expect solutions to have, and provide another point of view from which these partial differential equations can be approached.

2.5.1 The Wave Equation

Consider an elastic string stretched between two pegs, as on a guitar or harp. We want to describe the motion if the string is displaced and released to vibrate in a plane.

Place the string along the x-axis from 0 to L and let the plane of motion be the x, y-plane. We seek a function $u(x, t)$ such that at any time $t \geq 0$, the graph of the function $y = u(x, t)$ is the shape of the string at the time (Figure 2.2).

To begin with a simple case, neglect damping forces such as air resistance and the weight of the string and assume that the tension $T(x, t)$ in the string acts tangentially to the string. Also assume that the mass ρ per unit length is constant. Apply Newton's second law of motion to the segment of string between x and $x + \Delta x$:

net force of this segment due to the tension

= acceleration of the center of mass of the segment × mass of the segment.

This is a vector equation. For small Δx, consideration of the vertical component of this equation (Figure 2.3) gives approximately

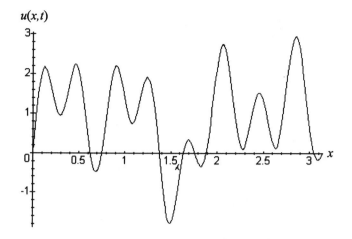

FIGURE 2.2 Typical profile of a vibrating string at some time.

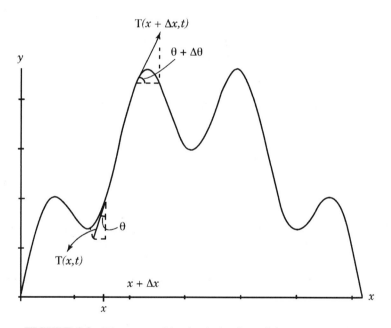

FIGURE 2.3 Vectors used in the derivation of the wave equation.

$$T(x + \Delta x, t)\sin(\theta + \Delta\theta) - T(x, t)\sin(\theta) = \rho(\Delta x)u_{tt}(\bar{x}, t)$$

in which $\bar{x}$ is the center of mass of the segment. Then

$$\frac{T(x + \Delta x, t)\sin(\theta + \Delta\theta) - T(x, t)\sin(\theta)}{\Delta x} = \rho u_{tt}(\bar{x}, t).$$

Now $T(x, t)\sin(\theta)$ is the vertical component of the tension. Write this as $v(x, t)$. The last equation becomes

$$\frac{v(x + \Delta x, t) - v(x, t)}{\Delta x} = \rho u_{tt}(\bar{x}, t).$$

In the limit as $\Delta x \to 0$, $\bar{x} \to x$ and this equation yields

$$v_x = \rho u_{tt}.$$

The horizontal component of the tension is $h(x, t) = T(x, t)\cos(\theta)$. Now

$$v(x, t) = h(x, t)\tan(\theta) = h(x, t)u_x$$

so

$$(hu_x)_x = \rho u_{tt}.$$

Since the horizontal component of the tension on the segment is zero,

$$h(x + \Delta x, t) - h(x, t) = 0.$$

Therefore h is independent of x and $(hu_x)_x = hu_{xx}$, leading to

$$hu_{xx} = \rho u_{tt}$$

or

$$u_{tt} = c^2 u_{xx} \tag{2.18}$$

where $c^2 = h/\rho$. Equation 2.18 is the *one-dimensional wave equation* (one space dimension), and is one of the canonical forms of the hyperbolic equation (see Exercise 20), with x and t as the independent variables instead of x and y in the above discussion. We may therefore associate hyperbolic partial differential equations with wave motion, providing some intuition about how solutions might behave under various conditions.

We would expect that the motion of the string is influenced by whatever is done to the ends of the string. Thus we impose conditions of the form

$$u(0, t) = \alpha(t), u(L, t) = \beta(t) \text{ for } t > 0.$$

These are called *boundary conditions*. We also specify *initial conditions*

$$u(x, 0) = \varphi(x), u_t(x, 0) = \psi(x) \text{ for } 0 < x < L.$$

These give, respectively, the initial position from which the string is released, and the velocity with which it is released. Under reasonable assumptions on the functions α, β, φ, and ψ, the wave equation, together with these initial and boundary conditions, has a unique solution. This is consistent with the intuition that, in the absence of external driving forces acting on the string, the motion is completely determined by the string's initial configuration and velocity, and any subsequent action at the ends of the string.

It is a remarkable fact that, despite the severe assumptions made in the derivation, the wave equation, together with boundary and initial conditions, has been found to accurately model certain kinds of wave motion, such as we see in variations of an elastic string, thus forging a link between a partial differential equation and a phenomenon of importance in physics, engineering, music, and other areas.

For a string which we imagine covers the entire real line, we have initial, but no boundary conditions, since now $-\infty = x < \infty$ (the string has no ends).

If the string is thought of as extending over the half-line ($x \geq 0$), then we seek solutions satisfying an initial condition, and a boundary condition at the left end $x = 0$. While we cannot actually produce an infinitely long string, such models are of practical importance in many contexts. For example, an astronomer sending a signal from a radio telescope on earth may for all practical purposes assume that the signal extends indefinitely from its source.

The derivation of the wave equation can be adapted to include external forces acting on the string. For example, if there is at each point an external force of magnitude F per unit length, acting parallel to the y-axis, then the profile $u(x, t)$ of the string at time t satisfies

$$u_{tt} = c^2 u_{xx} + \frac{1}{\rho} F.$$

In two space dimensions the wave equation is

$$u_{tt} = c^2(u_{xx} + u_{yy}).$$

This equation models oscillations of an elastic membrane whose shape at time t is the graph of the surface $z = u(x, y, t)$ in 3-space. Again, we would seek solutions specifying a given initial position, and boundary conditions describing the status of the membrane along its boundary (for example, a drum is usually fastened along a frame, and these boundary points remain motionless).

EXERCISE 31   Show that the function

$$u(x, t) = k \sin(n\pi x/L)\cos(n\pi ct/L)$$

with k constant and n any integer, satisfies the one-dimensional wave equation. L is a positive constant.

EXERCISE 32   Show that, for any integers m and n, the function defined by

$$u(x, y, t) = k \sin(nx)\cos(my)\cos(\sqrt{n^2 + m^2}ct),$$

with k any constant, satisfies the two-dimensional wave equation.

EXERCISE 33   Obtain the one-dimensional wave equation under the added assumption that the motion of the string is opposed by air resistance, which has a force at each point of magnitude proportional to the square of the velocity at that point.

EXERCISE 34   Obtain the one-dimensional wave equation if the motion of the string is being driven by an external force whose magnitude at x and time t is proportional to the product of the velocity at that point and time, and the vertical distance from the x-axis to the string at that time.

2.5.2.   The Heat Equation

We would like to derive an equation describing the flow of heat energy.

We will focus on a specific setting to clarify the assumptions made in the derivation. Consider a bar of constant density ρ and length L. Assume that the material of the bar conducts heat, and that the bar has uniform cross-sectional area A. The lateral surface of the bar is insulated, so there is no heat loss across this surface.

Draw the x-axis along the length of the bar, as in Figure 2.4. We assume that, at a given time, the temperature is the same throughout a given cross section, though it will in general vary from one cross section to another. We will derive an equation for $u(x, t)$, the temperature in the cross section of the bar at x, at time t.

Let c be the specific heat of the material of the bar. This is the amount of heat energy that must be supplied to a unit mass of the material to raise its temperature one degree. Consider the segment of bar between x and $x + \Delta x$. This segment has mass $\rho A\Delta x$, and it will take approximately $\rho cAu(x, t)\Delta x$ units of heat energy to change the temperature of this segment from zero to $u(x, t)$, its temperature at time t. The total heat energy in this segment at time t is

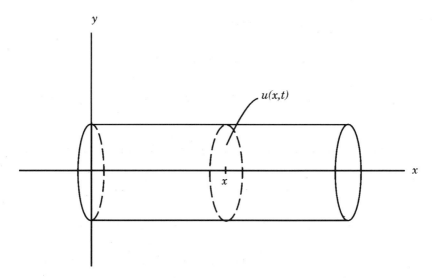

FIGURE 2.4 $u(x, t)$ is the temperature in the cross section at x at time t.

$$E(x, \Delta x, t) = \int_x^{x+\Delta x} \rho c A u(\xi, t) \, d\xi$$

for $t > 0$.

The amount of heat energy within this segment at time t can increase in two ways: heat energy may flow into the segment across its ends (this increase is called the flux of the heat energy), or there may be a source of heat energy within the segment (for example, a chemical reaction). The rate of change of the temperature within the segment, with respect to time, is therefore

$$\frac{\partial E}{\partial t} = \text{flux} + \text{source} = \int_x^{x+\Delta x} \rho c A \, \frac{\partial u}{\partial t} (\xi, t) \, d\xi.$$

Assuming for now that there is no energy source within the bar, then

$$\text{flux} = \int_x^{x+\Delta x} \rho c A \, \frac{\partial u}{\partial t} (\xi, t) \, d\xi. \tag{2.19}$$

Now let $F(x, t)$ be the amount of heat energy per unit area flowing across the cross section at x at time t, in the direction of increasing x. The rate of flow of heat energy across the section at x into the part of the bar to the right of x is $AF(x, t)$, while the rate of flow of heat energy across the section at $x + \Delta x$ into the part of the bar to the right of $x + \Delta x$ is $AF(x + \Delta x, t)$. The rate of flow of heat energy into the segment of bar between x and $x + \Delta x$ at time

t is the rate of flow into this segment to the right across the section at x, minus the rate of flow out of this segment to the right across the section at $x + \Delta x$ (Figure 2.5), giving us

$$\text{flux} = AF(x, t) - AF(x + \Delta x, t),$$

or

$$\text{flux} = -A(F(x + \Delta x, t) - F(x, t)). \tag{2.20}$$

Next recall Newton's law of cooling, which states that heat energy flows from the warmer to the cooler segment, and the amount of heat energy is proportional to the temperature gradient. That is,

$$F(x, t) = -K \frac{\partial u}{\partial x} (x, t),$$

with the positive constant of proportionality K called the heat conductivity of the bar. The negative sign in the equation indicates that heat energy flows from the warmer to the cooler segment. Substituting this into equation 2.20 gives us

$$\text{flux} = -A \left(-K \frac{\partial u}{\partial x} (x + \Delta x, t) + K \frac{\partial u}{\partial x} (x, t) \right),$$

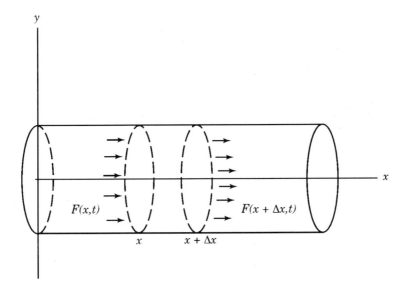

FIGURE 2.5   Heat flow across typical cross sections, used in the derivation of the heat equation.

which we write as

$$\text{flux} = \int_x^{x+\Delta x} \frac{\partial}{\partial x}\left(KA\,\frac{\partial u}{\partial x}\,(\xi, t)\right) d\xi. \tag{2.21}$$

Equating the two expressions for the flux from equations 2.19 and 2.21, we obtain

$$\int_x^{x+\Delta x} \rho c A\,\frac{\partial u}{\partial t}\,(\xi, t)\,d\xi = \int_x^{x+\Delta x} \frac{\partial}{\partial x}\left(KA\,\frac{\partial u}{\partial x}\,(\xi, t)\right) d\xi,$$

or

$$\int_x^{x+\Delta x}\left(\rho c\,\frac{\partial u}{\partial t}\,(\xi, t) - \frac{\partial}{\partial x}\left(K\,\frac{\partial u}{\partial x}\,(\xi, t)\right)\right) d\xi = 0.$$

This equation must hold for $0 < x < x + \Delta x < L$. If the integrand were nonzero at any x_0, then, assuming continuity of this integrand, we could choose Δx so small that the integrand is nonzero throughout $[x_0, x_0 + \Delta x]$. In this case the integral of this nonzero continuous function over this interval would be nonzero, a contradiction. We conclude that

$$\rho c\,\frac{\partial u}{\partial t} = K\,\frac{\partial^2 u}{\partial x^2}$$

for $0 < x < L$. This equation is usually written

$$u_t = ku_{xx} \tag{2.22}$$

where $k = K/c\rho$ is the *diffusivity* of the material of the bar. Equation 2.22 is the one-dimensional heat equation. With t in place of y, it is a parabolic equation in canonical form.

We may therefore think of the canonical parabolic equation in two independent variables as describing heat flow through a homogeneous bar.

As with the wave equation, we usually solve the heat equation subject to additional information needed to determine a unique solution having certain properties. Boundary conditions

$$u(0, t) = \alpha(t),\ u(L, t) = \beta(t) \text{ for } t > 0$$

specify the amount of heat energy introduced at the ends of the bar at time t, and the initial condition

$$u(x, 0) = f(x) \text{ for } 0 < x < L$$

specifies the temperature of the bar at time zero. We would expect to have to supply this kind of information to completely determine the temperature function, and in Chapter 5 we will see that this is indeed the case.

If there is a source of heat energy within the bar, then the heat equation takes the form

$$u_t = ku_{xx} + Q(x, t).$$

In two space dimensions, the heat equation (with no sources) has the form

$$u_t = k(u_{xx} + u_{yy}),$$

while in three space dimensions it is

$$u_t = k(u_{xx} + u_{yy} + u_{zz}).$$

This is often written

$$u_t = k\nabla^2 u,$$

where

$$\nabla^2 u = u_{xx} + u_{yy} + u_{zz}$$

is the Laplacian of u (in two dimensions, omit the u_{zz} term).

EXERCISE 35   Show that, for any real number α, the function

$$u(x, t) = e^{-\alpha^2 \pi^2 t} \cos(\alpha \pi x)$$

satisfies $u_t = u_{xx}$.

EXERCISE 36   Show that

$$u(x, t) = t^{-3/2} e^{-x^2/4kt}$$

satisfies $u_t = ku_{xx}$ for $x > 0$, $t > 0$.

2.5.3   Laplace's Equation, or the Potential Equation

The *steady state* case of the heat equation occurs when $u_t = 0$. In this case the heat equation becomes

$$\nabla^2 u = 0.$$

This is *Laplace's equation*. In two dimensions this is the canonical elliptic partial differential equation

$$u_{xx} + u_{yy} = 0.$$

In the context of heat conduction, Laplace's equation models the steady state temperature distribution in a medium. This is the limit of the temperature function as $t \to \infty$, and is therefore independent of t. Laplace's equation is important in other areas as well. If $Q(x, y, z)$ is the electrostatic potential in a region of space that has no electric charges, then the negative of the gradient of Q gives the electrostatic intensity at each point (the electrical force exerted on a unit charge at that point). The potential Q is found to satisfy Laplace's equation.

The real and imaginary parts of an analytic complex function also satisfy Laplace's equation in the plane, so harmonic functions bridge the gap between real and complex analysis.

Laplace's equation, and conditions under which solutions are uniquely determined, are the subject of Chapter 6.

To sum up this section, we relate hyperbolic equations to wave motion, parabolic equations to heat flow or diffusion, and elliptic equations to phenomena involving potential functions.

2.6 THE SECOND ORDER CAUCHY PROBLEM *(omit all of 2.6)*

For a first order ordinary differential equation, a condition for the uniqueness of a solution is the prescription of the function value at a given point. The analogue of this for the quasi-linear (hence also for the linear) first order partial differential equation is to prescribe function values along a noncharacteristic curve Γ.

For the second order ordinary differential equation, two pieces of information are needed to determine a unique solution. They are the value of the solution at a point, and the value of its derivative at that point.

Drawing from this experience with ordinary differential equations and first order partial differential equations, we might conjecture that conditions for uniqueness of the solution of the second order linear partial differential equation 2.1 might consist of specification of values the solution is to have along a given curve Γ, together with some appropriately formulated second condition. In this section we will explore a rationale for selecting this second condition.

In doing this it is instructive to consider a simple example. Suppose $u(x, y) = \varphi(x, y)$ is a solution of equation 2.1, so

$$A\varphi_{xx} + 2B\varphi_{xy} + C\varphi_{yy} + D\varphi_x + E\varphi_y + F\varphi + G = 0 \qquad (2.23)$$

with coefficients defined over some region $\mathcal{D}$ of the x, y-plane and $A(x, y)$, $B(x, y)$, and $C(x, y)$ not simultaneously zero. In the absence of constraining conditions this problem may have many solutions.

To simplify calculations and illustrate a point, we will choose Γ to be the y-axis and specify that φ is to assume given values on Γ, say

$$\varphi(0, y) = f(y).$$

Assuming that f can be differentiated repeatedly, then we can compute partial derivatives of φ at points $(0, y)$:

$$\varphi_y(0, y) = f'(y), \ \varphi_{yy}(0, y) = f''(y), \ \varphi_{yyy}(0, y) = f'''(y), \ \cdots.$$

There, is however, insufficient information as yet to determine partial derivatives of φ with respect to x at points $(0, y)$. Let us therefore provide such information by specifying that

$$\varphi_x(0, y) = g(y).$$

Assuming that we can repeatedly differentiate g, then

$$\varphi_{xy}(0, y) = g'(y), \ \varphi_{xyy}(0, y) = g''(y), \ \cdots.$$

Now we can obtain $\varphi_{xx}(0, y)$ by solving for it in equation 2.23, assuming that $A(0, y) \neq 0$. Further, once we know $\varphi_{xx}(0, y)$, we can obtain y-derivatives of φ_{xx} at $(0, y)$:

$$\varphi_{xxy}(0, y) = \frac{\partial}{\partial y} \varphi_{xx}(0, y), \ \varphi_{xxyy}(0, y) = \frac{\partial^2}{\partial y^2} \varphi_{xx}(0, y), \ \cdots.$$

There are still some partial derivatives we have not computed. But we can obtain these as well at points $(0, y)$. If we differentiate equation 2.23 with respect to x (keeping in mind that the coefficients are functions of x and y), we obtain an equation we can solve for $\varphi_{xxx}(0, y)$ in terms of quantities previously calculated. We can then differentiate again to get $\varphi_{xxxx}(0, y)$, and so on to higher order x-derivatives.

Now here is the point to obtaining these partial derivatives. For a given y, we can expand $\varphi(x, y)$ as a function of x in a Taylor series about $x = 0$:

$$\varphi(x, y) = \varphi(0, y) + \varphi_x(0, y)x + \frac{1}{2!} \varphi_{xx}(0, y)x^2 + \frac{1}{3!} \varphi_{xxx}(0, y)x^3 + \cdots,$$

and we can in theory compute the coefficients in this expansion because the given information (on $\varphi(0, y)$ and $\varphi_x(0, y)$) was chosen to enable us to do this.

In Section 2.9, when we discuss the Cauchy-Kowalevski Theorem, we will see that this series expansion defines a unique analytical solution of equation 2.1, in some region of the plane containing the origin, if the functions B/A, $C/A, \ldots, G/A$ have power series expansions as functions of x and y in this region, and f and g have Taylor expansions in y for the part of the y-axis in this region.

Example 11 Consider

$$u_{xx} + 4u_{xy} - u_{yy} + yu_x + u = 0.$$

Suppose $u = \varphi(x, y)$ is a solution satisfying

$$\varphi(0, y) = e^{3y} \text{ and } \varphi_x(0, y) = y^4.$$

We will obtain the first few terms of the Taylor expansion of $\varphi(x, y)$ about $x = 0$. This series has the form

$$\varphi(x, y) = \varphi(0, y) + \varphi_x(0, y)x + \frac{1}{2} \varphi_{xx}(0, y)x^2$$

$$+ \frac{1}{6} \varphi_{xxx}(0, y)x^3 + \frac{1}{24} \varphi_{xxxx}(0, y)x^4 + \cdots.$$

We know $\varphi(0, y)$ and $\varphi_x(0, y)$. To compute $\varphi_{xx}(0, y)$, use the partial differential equation to write

$$\varphi_{xx} = -4\varphi_{xy} + \varphi_{yy} - y\varphi_x - \varphi. \tag{2.24}$$

Then

$$\varphi_{xx}(0, y) = -4\varphi_{xy}(0, y) + \varphi_{yy}(0, y) - y\varphi_x(0, y) - \varphi(0, y).$$

But,

$$\varphi_{xy}(0, y) = \frac{\partial}{\partial y} \varphi_x(0, y) = \frac{\partial}{\partial y} y^4 = 4y^3$$

and

$$\varphi_{yy}(0, y) = \frac{\partial^2}{\partial y^2} \varphi(0, y) = \frac{\partial^2}{\partial y^2} e^{3y} = 9e^{3y}.$$

Then

$$\varphi_{xx}(0, y) = -4(4y^3) + 9e^{3y} - y^5 - e^{3y}$$

$$= -16y^3 - y^5 + 8e^{3y}.$$

Next, from equation 2.24,

$$\varphi_{xxx} = -4\varphi_{xxy} + \varphi_{xyy} - y\varphi_{xx} - \varphi_x,$$

in which we interchanged the order in some of the mixed partial derivatives (for example, $\varphi_{xyx} = \varphi_{xxy}$). Now

$$\varphi_{xxy}(0, y) = \frac{\partial}{\partial y}\, \varphi_{xx}(0, y) = -48y^2 - 5y^4 + 24e^{3y}$$

and

$$\varphi_{xyy}(0, y) = \frac{\partial^2}{\partial y^2}\, \varphi_x(0, y) = 12y^2.$$

Therefore

$$\varphi_{xxx}(0, y) = -4(-48y^2 - 5y^4 + 24e^{3y}) + 12y^2 - y(-16y^3 - y^5 + 8e^{3y}) - y^4$$

$$= y^6 + 35y^4 + 204y^2 - 96e^{3y} - 8ye^{3y}.$$

Thus far we have

$$\varphi(x, y) = e^{3y} + y^4x + \frac{1}{2}\,(-16y^3 - y^5 + 8e^{3y})x^2$$

$$+ \frac{1}{6}\,(y^6 + 35y^4 + 204y^2 - 96e^{3y} - 8ye^{3y})x^3 + \cdots. \quad \blacksquare$$

What is a reasonable generalization of this discussion? Think of the y-axis as a curve in the x, y-plane. Then φ_x is the derivative of φ in the direction normal to this curve. We therefore conjecture that a reasonable problem (in the sense of having a unique solution) might be to show that equation 2.1 has a unique solution if a curve Γ is specified, along with information about the solution and its derivative in the direction normal to Γ. Such information is called *Cauchy data*. The problem of obtaining a solution to equation 2.1 which satisfies Cauchy data is called the *Cauchy problem* for this partial differential equation.

Thus far we have sought to motivate the formulation of a reasonable and interesting problem. This does not imply that such a problem always has a solution. Indeed, in the first order case we saw that it might not if the curve is a characteristic. We will now explore the connection between characteristics and the Cauchy problem in the second order case.

EXERCISE 37   In each of 1 through 5, use the partial differential equation and the Cauchy data on the y-axis to calculate the first four terms of the Taylor expansion about $x = 0$ of the solution $\varphi(x, y)$.

1. $u_{xx} - 4u_{xy} + yu_{yy} + u_x + yu_y = 0;\ u(0, y) = y^3,\ u_x(0, y) = 4y$
2. $u_{xx} - x^2u_{xy} + u_x - 3u_y = 0;\ u(0, y) = 4;\ u_x(0, y) = y^3$
3. $u_{xx} + xyu_{xy} - xu_{yy} + 8u_x = 0;\ u(0, y) = -y^2 + y,\ u_x(0, y) = \cos(y)$
4. $e^{-y}u_{xx} + xu_{xy} + 2u_x - u_y + u = 0;\ u(0, y) = y^3 - 2y^2,\ u_x(0, y) = e^y$
5. $u_{xx} - 2u_{xy} + y^2u_{yy} - 3u_x + xu = 0;\ u(0, y) = \sin(2y),\ u_x(0, y) = 2y^2$

EXERCISE 38   In each of 1 through 5, use the partial differential equation and the Cauchy data on the x-axis to calculate the first four terms of the Taylor expansion about $y = 0$ of the solution $\varphi(x, y)$.

1. $u_{xx} - u_{yy} + 2u_x - \sin(y)u = 0$; $u(x, 0) = x^2$, $u_y(x, 0) = xe^{-x}$
2. $u_{xx} + xu_{xy} + u_{yy} - u_x + y^2u = 0$; $u(x, 0) = x^2 - x$, $u_y(x, 0) = \sin(2x)$
3. $u_{xx} - xu_{xy} + u_{yy} - 4xu_y + 2u = 0$; $u(x, 0) = 1 - x^3$, $u_y(x, 0) = x^4$
4. $u_{xx} + xu_{yy} + xu_y - yu = 0$; $u(x, 0) = x \sin(x)$, $u_y(x, 0) = x$
5. $u_{xx} - u_{yy} + xyu_x - \cos(x)u = 0$; $u(x, 0) = e^{-x}$, $u_y(x, 0) = x^3$

2.7   CHARACTERISTICS AND THE CAUCHY PROBLEM

Suppose we are given a curve Γ in the plane, specified by an equation $\xi(x, y) = 0$. Assume that Γ is smooth and that a unit normal vector $\mathbf{n}(x, y)$ is specified at each point, as in Figure 2.6. This defines a "preferred side" of Γ into which normal vectors drawn from points of Γ are directed.

Let φ be a solution of equation 2.1, so equation 2.23 is valid in some region of the plane containing Γ. Suppose φ satisfies Cauchy data that is prescribed on Γ. This means that values $\varphi(x, y)$ and $\varphi_n(x, y)$ are specified at points (x, y)

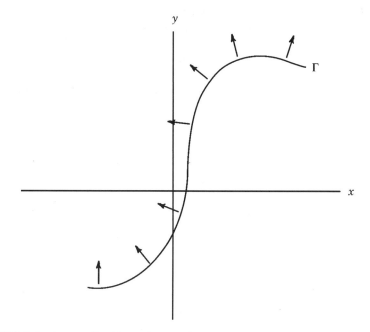

FIGURE 2.6   A curve Γ with a unit normal chosen at each point, defining a "preferred side" of Γ.

of Γ, with φ_n the derivative of φ at (x, y) in the direction of the selected normal to Γ at that point. Our objective is to generalize the discussion of Section 2.6, where Γ was the y-axis, to determine the partial derivatives of φ on Γ. In theory this determines a Taylor expansion of the solution about any point on Γ.

The equation $\xi(x, y) = k$ determines a family of smooth curves in the plane, of which Γ is one. Define another family $\eta(x, y) = k$. This can be done somewhat arbitrarily, but we want these curves to be smooth as well, and no curve from one family should be tangent to a curve from the other family at points of intersection. Such a tangency occurs at a point if and only if

$$\frac{\xi_x}{\xi_y} = \frac{\eta_x}{\eta_y}$$

at this point. We therefore prevent such tangency by requiring that

$$\xi_x\eta_y - \xi_y\eta_x \neq 0.$$

Since $\xi_x\eta_y - \xi_y\eta_x$ is the Jacobian J of the transformation

$$\xi = \xi(x, y)$$

$$\eta = \eta(x, y)$$

then we are really back to the familiar condition that this Jacobian is not zero. This assumption allows us in principle to invert the transformation and write $x = x(\xi, \eta)$, $y = y(\xi, \eta)$. Figure 2.7 shows some typical members of the family of curves defined by the transformation.

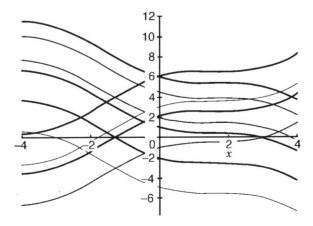

FIGURE 2.7 Typical families of curves defined by a transformation.

Applying this transformation to the partial differential equation, we can think of φ as a function of (ξ, η). The transformed equation is equation 2.2.

We will now show that the Cauchy data is sufficient to determine φ_ξ and φ_η along Γ. Imagine that Γ is parameterized in terms of arc length:

$$x = x(s), \, y = y(s)$$

on Γ. This means that $\xi(x(s), y(s)) \equiv 0$. Let

$$\varphi(x(s), y(s)) = f(s) \tag{2.25}$$

and

$$\varphi_\mathbf{n}(x(s), y(s)) = g(s) \tag{2.26}$$

with f and g given as the Cauchy data.

Now

$$\varphi_x = \varphi_\xi \xi_x + \varphi_\eta \eta_x$$

and

$$\varphi_y = \varphi_\xi \xi_y + \varphi_\eta \eta_y.$$

Solve these equations to obtain

$$\varphi_\xi = \frac{\begin{vmatrix} \varphi_x & \eta_x \\ \varphi_y & \eta_y \end{vmatrix}}{\begin{vmatrix} \xi_x & \eta_x \\ \xi_y & \eta_y \end{vmatrix}} = \frac{1}{J} (\varphi_x \eta_y - \varphi_y \eta_x). \tag{2.27}$$

Differentiate equation 2.25 to obtain

$$\varphi_x x'(s) + \varphi_y y'(s) = f'(s). \tag{2.28}$$

The unit tangent to Γ is $x'(s)\mathbf{i} + y'(s)\mathbf{j}$, and the unit normal into the region bounded by Γ is $\mathbf{n} = -y'(s)\mathbf{i} + x'(s)\mathbf{j}$, so, using equation 2.26,

$$\varphi_\mathbf{n}(x(s), y(s)) = \nabla\varphi \cdot \mathbf{n} = (\varphi_x \mathbf{i} + \varphi_y \mathbf{j}) \cdot (-y'\mathbf{i} + x'\mathbf{j})$$

$$= -y'\varphi_x + x'\varphi_y = g(s). \tag{2.29}$$

Solve equations 2.28 and 2.29 to obtain

$$\varphi_x = \frac{\begin{vmatrix} f' & y' \\ g & x' \end{vmatrix}}{\begin{vmatrix} x' & y' \\ -y' & x' \end{vmatrix}} = \frac{x'f' - y'g}{(x')^2 + (y')^2} = x'f' - y'g \tag{2.30}$$

and

$$\varphi_y = \frac{\begin{vmatrix} x' & f' \\ -y' & g \end{vmatrix}}{\begin{vmatrix} x' & y' \\ -y' & x' \end{vmatrix}} = x'g + y'f'. \tag{2.31}$$

Here we have used the fact that the tangent and normal vectors are of length 1, so $(x')^2 + (y')^2 = 1$.

Substitute the results from equations 2.30 and 2.31 into equation 2.27 to obtain

$$\varphi_\xi = \frac{1}{J} [(x'f' - y'g)\eta_y - (x'g + y'f')\eta_x]. \tag{2.32}$$

By similar reasoning we obtain

$$\varphi_\eta = \frac{1}{J} [(x'g + y'f')\xi_x - (x'f' - y'g)\xi_y]. \tag{2.33}$$

Equations 2.32 and 2.33 mean that the Cauchy data is sufficient to compute φ_ξ and φ_η along Γ in terms of the information provided. Since $\xi = 0$ on Γ, φ_ξ and φ_η are functions of just η along Γ, and we can use the expansions just obtained for φ_ξ and φ_η on Γ to compute $\varphi_{\xi\eta}$, $\varphi_{\eta\eta}$, $\varphi_{\xi\eta\eta}$, $\varphi_{\eta\eta\eta}$, $\ldots$ along this curve.

However, we must use the transformed partial differential equation 2.2 to solve for $\varphi_{\xi\xi}$. This requires that the coefficient of $\varphi_{\xi\xi}$ in equation 2.2 does not vanish. But this coefficient is $A\xi_x^2 + 2B\xi_x\xi_y + C\xi_y^2$; hence we require that

$$A\xi_x^2 + B\xi_x\xi_y + C\xi_y^2 \neq 0.$$

Therefore Γ cannot be a characteristic of the partial differential equation.

Once we have $\varphi_{\xi\xi}$ on Γ, as a function of η, we can compute $\varphi_{\xi\xi\eta}$, $\varphi_{\xi\xi\eta\eta}$, and so on. Continuing in this way, we can compute all partial derivatives of φ on Γ, just as we did in the preceding section when Γ was the y-axis. The requirement, again, is that Γ is not a characteristic of the partial differential equation.

When we know these partial derivatives, we can expand $\varphi(\xi, \eta)$ in a Taylor series about any point of Γ. If this series converges, it provides a solution of the partial differential equation in a neighborhood of a point of Γ. Convergence of this series depends on properties of the coefficient functions in the partial differential equation. We will pursue this issue in more detail in Section 2.9.

We conclude informally that Cauchy data is sufficient to determine a unique solution of the partial differential equation provided that this data is specified along a curve that is not a characteristic. The hyperbolic equation has two families of characteristics, and Cauchy data is insufficient if specified along a curve of either family. The parabolic equation has one family of characteristics, and Cauchy data is also insufficient along any curve of this family. The elliptic equation has no characteristics, which means that Cauchy data may be specified along any smooth curve in the plane (in a region throughout which the functions B/A, C/A, ..., G/A are analytic).

Example 12   Consider

$$u_{xx} + 4u_{xy} - u_{yy} + xu = 0.$$

Let Γ be the line $y = 3x$. Define the family of curves $\xi(x, y) = y - 3x = k$, so Γ is the curve $\xi = 0$.

Suppose we are given Cauchy data on Γ:

$$u(x, 3x) = \sin(x)$$

and

$$-\frac{3}{\sqrt{10}} u_x(x, 3x) + \frac{1}{\sqrt{10}} u_y(x, 3x) = -7x.$$

Here

$$\mathbf{n} = -\frac{3}{\sqrt{10}} \mathbf{i} + \frac{1}{\sqrt{10}} \mathbf{j}$$

is a unit normal vector to Γ, and

$$u_{\mathbf{n}} = \nabla u \cdot \mathbf{n} = -\frac{3}{\sqrt{10}} u_x + \frac{1}{\sqrt{10}} u_y$$

is the derivative of u in the direction of this normal.

Define $\eta = x + y$, so $\eta(x, y) = c$ defines a second family of curves. This choice of η is somewhat arbitrary, but no curve from one family is tangent to a curve from the other at a point of intersection. Note that

$$J = \begin{vmatrix} \xi_x & \xi_y \\ \eta_x & \eta_y \end{vmatrix} = -4.$$

We will first obtain the transformed partial differential equation. Let $u(x, y) = \varphi(\xi, \eta)$ and compute

$$u_x = -3\varphi_\xi + \varphi_\eta,$$

$$u_{xx} = 9\varphi_{\xi\xi} - 6\varphi_{\xi\eta} + \varphi_{\eta\eta},$$

$$u_y = \varphi_\xi + \varphi_\eta,$$

$$u_{yy} = \varphi_{\xi\xi} + 2\varphi_{\xi\eta} + \varphi_{\eta\eta},$$

and

$$u_{xy} = -3\varphi_{\xi\xi} - 2\varphi_{\xi\eta} + \varphi_{\eta\eta}.$$

Substitution into the partial differential equation yields the transformed equation

$$\varphi_{\xi\xi} + 4\varphi_{\xi\eta} - \varphi_{\eta\eta} + \frac{1}{4}(\xi - \eta)\varphi = 0.$$

On Γ, using arc length as parameter,

$$x(s) = \frac{s}{\sqrt{10}}, \quad y(s) = \frac{3s}{\sqrt{10}}.$$

Then $(x'(s))^2 + (y'(s))^2 = 1$, as required to use equations 2.32 and 2.33. The Cauchy data in terms of s is

$$\varphi(x(s), y(s)) = \sin(s/\sqrt{10}) = f(s)$$

and

$$\varphi_n(x(s), y(s)) = -\frac{7s}{\sqrt{10}} = g(s).$$

By equation 2.32,

$$\varphi_\xi(x(s), y(s)) = -\frac{1}{4}\left[\frac{1}{\sqrt{10}}\frac{1}{\sqrt{10}}\cos\left(\frac{s}{\sqrt{10}}\right) - \frac{3}{\sqrt{10}}\left(\frac{-7s}{\sqrt{10}}\right)\right]$$

$$+ \frac{1}{4}\left[\frac{1}{\sqrt{10}}\left(\frac{-7s}{\sqrt{10}}\right) + \frac{3}{\sqrt{10}}\frac{1}{\sqrt{10}}\cos\left(\frac{s}{\sqrt{10}}\right)\right]$$

$$= \frac{1}{20}\cos\left(\frac{s}{\sqrt{10}}\right) - \frac{7}{10}s.$$

And, by equation 2.33,

$$\varphi_\eta(x(s), y(s)) = -\frac{1}{4}\left[\frac{1}{\sqrt{10}}\left(\frac{-7s}{\sqrt{10}}\right) + \frac{3}{\sqrt{10}}\frac{1}{\sqrt{10}}\cos\left(\frac{s}{\sqrt{10}}\right)\right](-3)$$

$$+ \frac{1}{4}\left[\frac{1}{\sqrt{10}}\frac{1}{\sqrt{10}}\cos\left(\frac{s}{\sqrt{10}}\right) - \frac{3}{\sqrt{10}}\left(\frac{-7s}{\sqrt{10}}\right)\right]$$

$$= \frac{1}{4}\cos\left(\frac{s}{\sqrt{10}}\right).$$

These partial derivatives can be written as functions of η on Γ, where $y = 3x$ and $\eta = 4x$; hence

$$s = \sqrt{10}x = \frac{\sqrt{10}}{4}\eta.$$

Therefore on Γ,

$$\varphi_\xi = \frac{1}{20}\cos(\eta/4) - \frac{7}{4\sqrt{10}}\eta$$

and

$$\varphi_\eta = \frac{1}{4}\cos(\eta/4).$$

From these we can compute $\varphi_{\xi\eta}$, $\varphi_{\xi\eta\eta}$, . . . , $\varphi_{\eta\eta}$, $\varphi_{\eta\eta\eta}$, . . . on Γ. In particular, on Γ,

$$\varphi_{\xi\eta} = \frac{\partial}{\partial\eta}\varphi_\xi = -\frac{1}{80}\sin(\eta/4) - \frac{7}{4\sqrt{10}}$$

and

$$\varphi_{\eta\eta} = -\frac{1}{16} \sin(\eta/4).$$

To compute $\varphi_{\xi\xi}$ on Γ, we must use the transformed partial differential equation to write

$$\varphi_{\xi\xi} = -4\varphi_{\xi\eta} + \varphi_{\eta\eta} + \frac{1}{4}(\eta - \xi)\varphi.$$

Since $\xi \equiv 0$ on Γ, this equation becomes

$$\varphi_{\xi\xi} = -4\varphi_{\xi\eta} + \varphi_{\eta\eta} + \frac{1}{4}\eta\varphi.$$

Therefore, on Γ,

$$\varphi_{\xi\xi} = -4\left[-\frac{1}{80}\sin(\eta/4) - \frac{7}{4\sqrt{10}}\right] - \frac{1}{16}\sin(\eta/4) + \frac{1}{4}\eta\sin(\eta/4)$$

$$= -\frac{1}{80}\sin(\eta/4) + \frac{1}{4}\eta\sin(\eta/4) + \frac{7}{\sqrt{10}}.$$

Using these partial derivatives, we can compute the first few terms of the Taylor series $\varphi(\xi, \eta)$ about the origin, which occurs when $\eta = 0$. This series is

$$\varphi(\xi, \eta) = \varphi(0, 0) + \varphi_{\xi}(0, 0)\xi + \varphi_{\eta}(0, 0)\eta$$

$$+ \frac{1}{2}(\varphi_{\xi\xi}(0, 0)\xi^2 + 2\varphi_{\xi\eta}(0, 0)\xi\eta + \varphi_{\eta\eta}(0, 0)\eta^2) + \cdots. \quad (2.34)$$

We have

$$\varphi(0, 0) = 0, \quad \varphi_{\xi}(0, 0) = \frac{1}{20}, \quad \varphi_{\eta}(0, 0) = \frac{1}{4},$$

$$\varphi_{\xi\eta}(0, 0) = -\frac{7}{4\sqrt{10}}, \quad \varphi_{\eta\eta}(0, 0) = 0 \text{ and } \varphi_{\xi\xi}(0, 0) = \frac{7}{\sqrt{10}},$$

so

$$\varphi(\xi, \eta) = \frac{1}{20}\xi + \frac{1}{4}\eta + \frac{7}{2\sqrt{10}}\xi^2 - \frac{7}{4\sqrt{10}}\xi\eta + \cdots,$$

and we can compute more terms by repeatedly differentiating the partial differential equation and using partial derivatives previously calculated. The resulting series is a solution of the transformed equation in some disk about the origin, if this series converges in such a disk.   ∎

Of course, nothing in Example 12 shows that the resulting Taylor series actually converges in some disk about the origin. This will follow from the Cauchy-Kowalevski Theorem following the next section.

EXERCISE 39   Derive equation 2.33.

EXERCISE 40   In the spirit of equations 2.32 and 2.33, derive expressions for $\varphi_{\xi\xi}$, $\varphi_{\xi\eta}$, and $\varphi_{\eta\eta}$. *Hint*: Begin with the chain rule derivatives for φ_x and φ_y and obtain φ_{xx}, φ_{yy} and φ_{xy}. Solve these equations for $\varphi_{\xi\xi}$, $\varphi_{\eta\eta}$, and $\varphi_{\xi\eta}$. This involves some algebra, and a program such as MAPLE will simplify the task.

EXERCISE 41   Carry out the details of the discussion and Example 12 for the following problem. Consider the partial differential equation

$$u_{xx} + 4u_{xy} - 3u_{yy} + u_x = 0.$$

Let Γ be the line $y = x$. Check that Γ is not a characteristic of the differential equation. Choose

$$\mathbf{n} = \frac{1}{\sqrt{2}}(-\mathbf{i} + \mathbf{j})$$

as the unit normal to Γ and specify values of u and the normal derivative of u along $\mathbf{n}$ by

$$u(x, x) = x^2,$$

$$-u_x(x, x) + u_y(x, x) = \cos(x).$$

Choosing a suitable $\eta(x, y)$, as allowed in the discussion, compute φ, φ_ξ, φ_η, $\varphi_{\xi\xi}$, $\varphi_{\xi\eta}$, and $\varphi_{\eta\eta}$ along Γ. Use these to compute the terms listed in equation 2.34 of the Taylor series for φ about the origin (which is a point of Γ).

EXERCISE 42   Carry out the program of Exercise 41 with the partial differential equation

$$u_{xx} + 5u_{yy} = 0.$$

Let Γ be the line $y = 2x$ and let the Cauchy data on this line be given by

$$u(x, 2x) = 1 - x^2,$$

$$2u_x(x, 2x) - u_y(x, 2x) = 2 + x.$$

EXERCISE 43 Carry out the program of Exercise 41 with the partial differential equation

$$u_{xx} - 2u_x - 6u_{yy} = 0.$$

Let Γ be the line $y = 5x$ and let the Cauchy data on this line be given by

$$u(x, 5x) = \sin(x)$$

$$5u_x(x, 5x) - u_y(x, 5x) = x^2.$$

EXERCISE 44 Carry out the program of Exercise 41 with the partial differential equation

$$u_{xx} - u_{yy} = 0.$$

Let Γ be the line $y = -2x$ and let Cauchy data on Γ be given by

$$u(x, -2x) = 1 + x$$

$$2u_x(x, -2x) + u_y(x, -2x) = x^2 - x.$$

EXERCISE 45 Let φ be a solution of the linear second order partial differential equation 2.1. Suppose Cauchy data are prescribed along a curve Γ, which is parameterized in terms of arc length by

$$x = \alpha(s), \ y = \beta(s).$$

Along this curve we can think of φ and its partial derivatives as functions of s. Show that, along Γ,

$$\varphi_{xx}\alpha' + \varphi_{xy}\beta' = \frac{d}{ds}\,\varphi_x,$$

$$\varphi_{xy}\alpha' + \varphi_{yy}\beta' = \frac{d}{ds}\,\varphi_y,$$

and

$$A\varphi_{xx} + 2B\varphi_{xy} + C\varphi_{yy} = -D\varphi_x - E\varphi_y - F\varphi - G.$$

Think of these as three linear algebraic equations to be solved for φ_{xx}, φ_{xy}, and φ_{yy} along Γ, in terms of known quantities along Γ. From algebra, these equations have a unique solution only if the determinant of the coefficients is nonzero. Show that this condition holds if Γ is not characteristic. Further show that, if Γ is characteristic, then these three equations have a solution only if the terms on the right sides of the equations satisfy a certain condition. Obtain this condition.

2.8 CHARACTERISTICS AS CARRIERS OF DISCONTINUITIES

Characteristics may also be regarded as possible carriers of discontinuities of the second derivative. We will discuss what this means.

Suppose as usual that we have made the transformation $\xi = \xi(x, y)$, $\eta = \eta(x, y)$, obtaining the transformed equation 2.2. Suppose also that $\Gamma : \xi(x, y) = \xi_0$ has been chosen, and we want to consider the possibility that $w_{\xi\xi}$ is discontinuous at points of Γ. What does this imply about Γ?

Consider a particular point (ξ_0, η) on Γ at which $w_{\xi\xi}$ may be discontinuous, but the other terms appearing in equation 2.2 are continuous. Take the limit of the terms in equation 2.2 as $(\xi, \eta) \to (\xi_0, \eta)$, first from one side of Γ, then from the other. Denote the limit from one side with a superscript $+$, and that from the other side with a superscript $-$. This will yield the equations

$$a^+(\xi_0, \eta)w_{\xi\xi}^+ + 2b^+(\xi_0, \eta)w_{\xi\eta}^+ + c^+(\xi_0, \eta)w_{\eta\eta}^+ + d^+(\xi_0, \eta)w_\xi^+$$
$$+ e^+(\xi_0, \eta)w_\eta^+ + f^+(\xi_0, \eta)w^+ + g^+(\xi_0, \eta) = 0$$

and

$$a^-(\xi_0, \eta)w_{\xi\xi}^- + 2b^-(\xi_0, \eta)w_{\xi\eta}^- + c^-(\xi_0, \eta)w_{\eta\eta}^- + d^-(\xi_0, \eta)w_\xi^-$$
$$+ e^-(\xi_0, \eta)w_\eta^- + f^-(\xi_0, \eta)w^- + g^-(\xi_0, \eta) = 0.$$

Subtract these equations. Assuming that all terms except possibly $w_{\xi\xi}$ are continuous, most terms cancel because the limits from either side of Γ are equal. We obtain

$$a(\xi_0, \eta)(w_{\xi\xi}^+(\xi_0, \eta) - w^-(\xi_0, \eta)) = 0.$$

This equation can be satisfied in either of two ways. Either $a(\xi_0, \eta) = 0$, or $w_{\xi\xi}^+(\xi_0, \eta) = w_{\xi\xi}^-(\xi_0, \eta)$. In the latter case the second derivative $w_{\xi\xi}$ is continuous at (ξ_0, η). In the former case, which must hold if $w_{\xi\xi}$ is discontinuous at (ξ_0, η), we have

$$A\xi_x^2 + 2B\xi_x\xi_y + C\xi_y^2]_{\xi=\xi_0} = 0$$

and this is exactly the condition for the curve $\xi(x, y) = \xi_0$ to be a characteristic.

We conclude that the characteristics $\xi(x, y) = k$ are curves along which $w_{\xi\xi}$ may be discontinuous. By a similar analysis, characteristics $\eta(x, y) = k$ are curves along which $w_{\eta\eta}$ may have discontinuities.

Geometrically, this means that there may be two solutions of the partial differential equation which branch away from each other at Γ. Graphs of such solutions could have the appearance of the surfaces shown in Figure 2.8. One solution consists of the surfaces Σ_1 and Σ_2, and the other of surfaces Σ_1 and Σ_3, with Σ_1 branching onto the distinct surface elements Σ_2 and Σ_3 at Γ.

We now have three ways of thinking of characteristics of the second order linear partial differential equation 2.1:

1. As curves defining a transformation of the partial differential equation to canonical form;
2. As curves along which Cauchy data do not determine a solution, or perhaps not a unique solution; and
3. As curves along which a second derivative of the solution may have a discontinuity.

EXERCISE 46   Let a, b, and c be real numbers with not both a and b zero. Let Γ be the straight line determined by $ax + by + c = 0$. Define

$$\varphi(x, y) = |ax + by + c|$$

for all (x, y) in R^2. Show that φ satisfies Laplace's equation in two dimensions.

$$u_{xx} + u_{yy} = 0.$$

However, show that the normal derivative $\partial\varphi/\partial n$ is discontinuous across this line.

Laplace's equation is elliptic and hence has no characteristic. However, while we have discussed how characteristics can be carriers of discontinuities

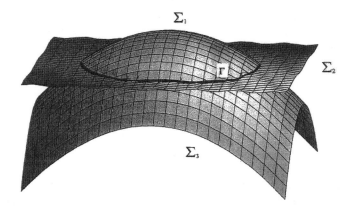

FIGURE 2.8   Branching of a solution surface into two distinct surfaces at a characteristic.

of second derivative terms, this example suggests that we cannot approach characteristics by thinking of discontinuities of first derivative terms.

2.9   THE CAUCHY-KOWALEVSKI THEOREM $\left(\text{omit all of } 2.9\right)$

In this section we will state the Cauchy-Kowalevski Theorem, which is the fundamental existence theorem for systems of partial differential equations with Cauchy data. We will prove the theorem in the linear case, which applies to the partial differential equations we will study.

The theorem is named after Augustin-Louis Cauchy (1789–1857), one of the leading mathematicians of the nineteenth century, and Sophie Kowalevski (1850–1891), a contemporary of Weierstrass in the last half of that century. They are jointly honored for their contributions to the theorem's formulation and proof. There are many transliterations of Kowalevski's name in the literature, including Kowalewski, Kowalewskaya and Kovalevski. We have chosen the spelling she attached to papers she authored in German and French.

We will approach the theorem and its proof in stages.

2.9.1   The General Cauchy Problem

We will deal with a system of N partial differential equations in functions $u_j(t, x_1, \ldots, x_n)$ of $n + 1$ independent variables. The equations are assumed to have the form

$$\frac{\partial^{n_i} u_i}{\partial t^{n_i}} = F_i \left(t, x_1, \ldots, x_n, u_1(x_1, \ldots, x_n), \ldots, u_N(x_1, \ldots, x_n), \right.$$

$$\left. \cdots \frac{\partial^k u_j}{\partial t^{k_0} \partial x_1^{k_1} \cdots \partial x_n^{k_n}}, \cdots \right) \quad (2.35)$$

in which i and j range independently from 1 through N inclusive. The ith equation has a derivative order n_i of u_i with respect to t isolated on one side of the equation, while the right side is a function F_i of the variables $t, x_1, \ldots, x_n$ and of the functions $u_1, \ldots, u_N$ and their partial derivatives of orders k, where

$$k = k_0 + k_1 + \cdots + k_n \le n_j$$

and

$$k_0 < n_j.$$

Thus in each partial derivative on the right in the ith equation, k can be any sum of nonnegative integers not exceeding n_i. No partial derivative appearing on the right side of the ith equation can be of order higher than n_i, and all derivatives with respect to t on the right must be of order strictly less than n_i.

We call the system of partial differential equations 2.35, a *Cauchy system*.

Example 13   The system consisting of the single partial differential equation

$$\frac{\partial^2 u}{\partial t^2} = t + x_1^2 - x_2^2 + x_1 \frac{\partial u}{\partial x_1} + 2 \frac{\partial^2 u}{\partial t \partial x_1} - \frac{\partial^2 u}{\partial x_1 \partial x_2} + \frac{\partial^2 u}{\partial x_1^2} + \frac{\partial^2 u}{\partial x_2^2}$$

is a Cauchy system with $N = 1$ (number of equations), $n = 2$, and independent variables t, x_1, x_2. Here $n_1 = 2$ (order of the isolated highest partial derivative with respect to t). Note that no derivative on the right is of order higher than 2, and no t derivative on the right is of order greater than 1. ∎

Example 14   The system

$$\frac{\partial^3 u_1}{\partial t^3} = \frac{\partial^2 u_1}{\partial x_1^2} - \frac{\partial^3 u_2}{\partial x_1^2 \partial x_2} + x_1 x_2 \frac{\partial^2 u_1}{\partial x_1 \partial x_2} + t^2 x_1$$

$$\frac{\partial^2 u_2}{\partial t^2} = \frac{\partial^2 u_2}{\partial t \partial x_1} + \frac{\partial^2 u_2}{\partial x_2^2} - \frac{\partial u_1}{\partial t} - \frac{\partial u_2}{\partial t}$$

is a Cauchy system. Here $N = 2$ and $n = 2$, with independent variables t, x_1, x_2, and $n_1 = 3$ and $n_2 = 2$. ∎

The equations in a Cauchy system need not be linear. If they are, the system is called linear.

As we have already seen informally, we usually attempt to solve a partial differential equation subject to certain given information. The information we will specify is that, at some $t = t_0$, each u_i and its partial derivatives with respect to t, up to and including order $n_i - 1$, must equal functions of $x_1, \ldots, x_n$ for $(x_1, \ldots, x_n)$ in some specified region of n-dimensional space R^n. Thus, we assume that

$$\frac{\partial^k u_i}{\partial t^k} (t_0, x_1, \ldots, x_n) = \varphi_{i,k}(x_1, \ldots, x_n) \tag{2.36}$$

for $i = 1, \ldots, N$ and, for each i, for $k = 0, 1, \ldots, n_i - 1$, with each of the given functions $\varphi_{i,k}$ defined in some common region of n-dimensional space. Of course, when $k = 0$ we have the zero order partial derivative

$$\partial^0 u_i / \partial t^0 (t_0, x_1, \ldots, x_n),$$

which we define to be the function u_i itself evaluated at $(t_0, x_1, \ldots, x_n)$.

We call equations 2.36 *Cauchy data* for the system 2.35. These conditions are also called *initial conditions*, thinking of t as time and of equations 2.36

as specifying the functions and their derivatives up to and including orders n_i − 1 at the initial time t_0.

The *Cauchy problem* consists of finding a solution of the system 2.35 satisfying the initial conditions 2.36.

The concepts we have just developed are consistent with the analysis in Section 2.6 of a simple Cauchy problem for the linear, second order partial differential equation. Write equation 2.1 as

$$u_{xx} = -2\frac{B}{A}u_{xy} - \frac{C}{A}u_{yy} - \frac{D}{A}u_x - \frac{E}{A}u_y - \frac{F}{A}u - G.$$

This is in the form of a Cauchy system with one equation and with x in place of t, and y in place of x_1. In Section 2.5 we specified $\varphi(0, y)$ and $\varphi_x(0, y)$, and in the current context these are $u(0, x_1)$ and $\partial u/\partial t(0, x_1)$, which are the Cauchy data for that problem.

2.9.2   The Cauchy-Kowalevski Theorem

We are almost ready to state the main theorem. Recall that a subset S of n-dimensional space R^n is open if each point of S is the center of an n-ball wholly contained in S. An n-ball about $(x_1^0, \ldots, x_n^0)$ consists of all points satisfying

$$(x_1 - x_1^0)^2 + \cdots + (x_n - x_n^0)^2 < r^2$$

for some positive number r. In the plane this inequality determines the set of points interior to a circle, and in 3-space, within a (standard) sphere.

An n-ball centered at a point in R^n is often referred to as a *neighborhood* of that point.

A function of f of n real variables $x_1, \ldots, x_n$ is *analytic* at a point $(x_1^0, x_2^0, \ldots, x_n^0)$ if it has a power series representation in some neighborhood of this point. This power series in n variables has the form

$$f(x_1, \ldots, x_n) = \sum_{N=0}^{\infty} \sum_{k_1+k_2+\cdots+k_n=N}^{\infty} A_{k_1 k_2 \cdots k_n}(x_1 - x_1^0)^{k_1}(x_2 - x_2^0)^{k_2} \cdots (x_n - x_n^0)^{k_n},$$

in which

$$\sum_{N=0}^{\infty} \sum_{k_1+k_2+\cdots+k_n=N}$$

denotes a sum over all n-tuples $(k_1, k_2, \ldots, k_n)$ of nonnegative integers whose sum is N, for $N = 0, 1, 2, \ldots$. When $N = 0$, then $k_1 = k_2 = \cdots = k_n = 0$ and we obtain the constant term $A_{0,0,\ldots,0}$ in the series. When $N = 1$, each k_j in turn can equal 1 while the other $k_i's$ are zero. This yields terms in the series of the form constant times $(x_j - x_j^0)$. There are n such terms (though some may have zero coefficients). When $N = 2$ then two of k_i and k_j can be 1 and the others zero, yielding terms $(x_i - x_i^0)(x_j - x_j^0)$, or one k_i can equal 2 and the other $k_j's$ zero, yielding terms $(x_i - x_i^0)^2$. And so on for higher order terms. Thus the series is

$$f(x_1, \ldots, x_n) = A_{0,0,\ldots,0} + A_{1,0,\ldots,0}(x_1 - x_1^0) + A_{0,1,0,\ldots,0}(x_2 - x_2^0) + \cdots$$
$$+ A_{0,0,\ldots,1}(x_n - x_n^0) + A_{1,1,0,\ldots,0}(x_1 - x_1^0)(x_2 - x_2^0) + \cdots$$
$$+ A_{0,0,\ldots,1,1}(x_{n-1} - x_{n-1}^0)(x_n - x_n^0) + A_{2,0,\ldots,0}(x_1 - x_1^0)^2 + \cdots$$
$$+ A_{0,0,\ldots,2}(x_n - x_n^0)^2 + \text{higher order terms.}$$

By definition, f is analytic at $(x_1^0, \ldots, x_n^0)$ if and only if this series expansion converges to $f(x_1, \ldots, x_n)$ at each $(x_1, \ldots, x_n)$ in some neighborhood of $(x_1^0, \ldots, x_n^0)$. When $n = 1$, this expression reduces to the usual Taylor expansion in an interval about x_1^0 on the real line.

As with Taylor series in one variable, there is a formula for the coefficients in this n-dimensional expansion:

$$A_{k_1 k_2 \cdots k_n} = \frac{1}{k_1! k_2! \cdots k_n!} \left[\frac{\partial^{k_1 + k_2 + \cdots + k_n} f}{\partial x_1^{k_1} \partial x_2^{k_2} \cdots \partial x_n^{k_n}} \right]_{x_1 = x_1^0, x_2 = x_2^0, \ldots, x_n = x_n^0}.$$

We are now ready to state the theorem.

Theorem 1 (Cauchy-Kowalevski) Suppose each F_i in the system 2.35 is analytic in some open set containing

$$\left(t_0, x_1^0, x_2^0, \ldots, x_n^0, \ldots, \left[\frac{\partial^{k - k_0} \varphi_{i,k_0}}{\partial x_1^{k_1} \partial x_2^{k_2} \ldots \partial x_n^{k_n}} \right]_{x_1 = x_1^0, \ldots, x_n = x_n^0}, \ldots \right).$$

Suppose also that each $\varphi_{i,k}$ is analytic in some open set containing $(x_1^0, \ldots, x_n^0)$. Then the Cauchy problem consisting of the system 2.35 and the initial conditions 2.36 has exactly one solution that is analytic in a neighborhood of $(t_0, x_1^0, x_2^0, \ldots, x_n^0)$. ∎

The Cauchy-Kowalevski Theorem gives sufficient conditions for a Cauchy problem to have a unique analytic solution in some neighborhood of the initial point. This uniqueness is only in the class of analytic functions, however, and there can be nonanalytic solutions of a Cauchy problem.

2.9.3 An Example

Before proving the Cauchy-Kowalevski Theorem, it is instructive to see the idea of it in a specific case. Consider the Cauchy problem

$$u_{tt} = u_{xx} - tu_{xt} + xu_x$$

$$u(0, x) = -4 \cos(x), \; u_t(0, x) = e^{3x}.$$

We will derive a few terms of a series "solution" about $(0, 0)$. The process mirrors the discussions of Sections 2.6 and 2.7, except that now the variables are called t and x.

We want to compute the coefficients in the expansion

$$u(t, x) = u(0, 0) + u_t(0, 0)t + u_x(0, 0)x + \frac{1}{2} u_{tt}(0, 0)t^2$$

$$+ \; u_{xt}(0, 0)xt + \frac{1}{2} u_{xx}(0, 0)x^2 + \frac{1}{6} u_{ttt}(0, 0)t^3 + \frac{1}{2} u_{ttx}(0, 0)t^2x$$

$$+ \frac{1}{2} u_{txx}(0, 0)tx^2 + \frac{1}{6} u_{xxx}(0, 0)x^3 + \cdots.$$

Now,

$$u(0, 0) = -4 \cos(0) = -4$$

and

$$u_t(0, 0) = 1$$

are given. Compute

$$u_x(0, x) = \frac{d}{dx}(-4 \cos(x)) = 4 \sin(x),$$

so

$$u_x(0, 0) = 0.$$

Next, from the partial differential equation,

$$u_{tt} = u_{xx} - tu_{xt} + xu_x$$

so

$$u_{tt}(0, 0) = u_{xx}(0, 0).$$

But $u_x(0, x) = 4 \sin(x)$, so $u_{xx}(0, x) = 4 \cos(x)$ and

$$u_{tt}(0, 0) = 4.$$

Next, by differentiating the partial differential equation,

$$u_{ttt} = u_{xxt} - u_{xt} - tu_{xtt} + xu_{xt}.$$

Since $u_t(0, x) = e^{3x}$, then $u_{tx}(0, x) = 3e^{3x}$ and

$$u_{xt}(0, 0) = 3.$$

Similarly,

$$u_{txx}(0, x) = 9e^{3x}$$

so

$$u_{txx}(0, 0) = 9.$$

Then, from the t-derivative of the partial differential equation,

$$u_{ttt}(0, 0) = u_{xxt}(0, 0) - u_{xt}(0, 0) = 9 - 3 = 6.$$

Differentiating the partial differential equation with respect to x gives

$$u_{ttx} = u_{xxx} - tu_{xtx} + u_x + xu_{xx}$$

so

$$u_{ttx}(0, 0) = u_{xxx}(0, 0) + u_x(0, 0).$$

Now $u_{xxx}(0, x) = -4 \sin(x)$, so $u_{xxx}(0, 0) = 0$. Therefore

$$u_{ttx}(0, 0) = 0.$$

Thus far, we have

$$u(x, t) = -4 + t + 0 \cdot x + \frac{1}{2} 4t^2 + 3xt + \frac{4}{2} x^2$$

$$+ \frac{1}{6} 6t^3 + \frac{1}{2} (0)t^2 x + \frac{1}{2} 9tx^2 + \frac{1}{6} (0)x^3 + \cdots$$

$$= -4 + t + 2t^2 + 3xt + 2x^2 + t^3 + \frac{9}{2} x^2 t + \cdots.$$

We can continue to calculate as many terms as we like. This is a mechanical process. The issue is whether this series converges. In the next subsection we will prove the Cauchy-Kowalevski Theorem in three stages: first, a reduction of the general problem to a simpler standard form; then a proof of uniqueness of the solution, following the idea of this example; and finally a proof of convergence of the series defining the proposed solution.

EXERCISE 47   In each of 1 through 10, check that the given problem is a Cauchy problem and then determine terms through third order in the series expansion of the solution about the origin, as in the above example.

1. $u_t = u - 2u_x$; $u(0, x) = 3 + x$

2. $u_t = u + u_x$; $u(0, x) = x$

3. $u_t = xu + tu_x$; $u(0, x) = 1 - x$

4. $u_{tt} = u_x + 2u_{xt}$; $u(0, x) = 1$, $u_t(0, x) = -x$

5. $u_{tt} = u_x - u_{xx}$; $u(0, x) = x^2$, $u_t(0, x) = 2 + x$

6. $u_{tt} = -u_x + u_t - u_{xx}$; $u(0, x) = 1$, $u_t(0, x) = x^2$

7. $u_{tt} = u + u_x - u_{xt}$; $u(0, x) = \cos(x)$, $u_t(0, x) = -x$

8. $u_{ttt} = u_x - u_{tx} + 2u_{xx}$; $u(0, x) = 4 + x$, $u_t(0, x) = 2$, $u_{tt}(0, x) = e^x$

9. $u_{tt} = u - 3u_t - u_x + 2u_{xx}$; $u(0, x) = x - \cos(x)$, $u_t(0, x) = 1 - x^2$

10.
$$\frac{\partial u_1}{\partial t} = u_2 - \frac{\partial u_1}{\partial x}$$

$$\frac{\partial^2 u_2}{\partial t^2} = u_1 + u_2 + \frac{\partial u_1}{\partial t} - 3\frac{\partial^2 u_2}{\partial x^2}$$

$$u_1(0, x) = x^2; \quad u_2(0, x) = x; \quad \frac{\partial u_2}{\partial t}(0, x) = 2$$

2.9.4   Proof of the Cauchy-Kowalevski Theorem for Linear Systems

We will outline a proof of the theorem for linear systems, leaving some of the details as exercises along the way.

Reduction to Initial Conditions at the Origin and First Order Equations
First, we may assume in the Cauchy problem that the initial conditions are specified at time zero, and that the point at which the data is specified is the origin in R^n. That is, we may assume that

$$t_0 = x_1^0 = \cdots = x_n^0 = 0.$$

EXERCISE 48   Verify this assertion. *Hint:* Make the change of variables

$$t = T + t_0, \; x_1 = X_1 + x_1^0, \; \ldots, \; x_n = X_n + x_n^0.$$

Show that this yields a new Cauchy problem whose solution in a neighborhood of

$$T = X_1 = \cdots = X_n = 0$$

yields a solution of the original system in a neighborhood of $(t_0, x_1^0, \ldots, x_n^0)$.

Next, we can replace a system of linear partial differential equations by a perhaps larger system (that is, more equations) of first order partial differential equations, in such a way that a solution of either system yields a solution of the other.

To see why this is true, suppose, for example, we have just a single partial differential equation, and it is of second order in the time derivative:

$$\frac{\partial^2 u}{\partial t^2} = \sum_{j=1}^{n} \sum_{i=1}^{n} a_{ij} \frac{\partial^2 u}{\partial x_i \partial x_j} + \sum_{i=1}^{n} b_i \frac{\partial^2 u}{\partial t \partial x_i} + \sum_{i=1}^{n} c_i \frac{\partial u}{\partial x_i} + d \frac{\partial u}{\partial t} + eu + f \quad (2.37)$$

in which $a_{ij} = a_{ji}$ because of equality of the mixed partial derivatives, and each of the coefficients is a function of $(t, x_1, \ldots, x_n)$. By the preceding discussion, initial conditions are assumed to be given by

$$u(0, x_1, \ldots, x_n) = \varphi(x_1, \ldots, x_n)$$

$$u_t(0, x_1, \ldots, x_n) = \psi(x_1, \ldots, x_n) \qquad (2.38)$$

with φ and ψ analytic in some neighborhood of $(0, , \ldots, 0)$.

Produce a system of linear first order partial differential equations from equation 2.37 as follows. Define

$$u_0 = \frac{\partial u}{\partial t}$$

and, for $j = 1, 2, \ldots, n$,

$$u_j = \frac{\partial u}{\partial x_j}.$$

These functions satisfy the linear, first order system

$$\frac{\partial u_0}{\partial t} = \sum_{j=1}^{n} \sum_{i=1}^{n} a_{ij} \frac{\partial u_i}{\partial x_j} + \sum_{i=1}^{n} b_i \frac{\partial u_0}{\partial x_i} + \sum_{i=1}^{n} c_i u_i + eu + f$$

$$\frac{\partial u_j}{\partial t} = \frac{\partial u_0}{\partial x_j} \text{ for } j = 1, 2, \ldots, n$$

$$\frac{\partial u}{\partial t} = u_0 \tag{2.39}$$

and initial conditions

$$u(0, x_1, \ldots, x_n) = \varphi(x_1, \ldots, x_n)$$

$$u_0(0, x_1, \ldots, x_n) = \psi(x_1, \ldots, x_n)$$

$$u_1(0, x_1, \ldots, x_n) = \frac{\partial \varphi}{\partial x_1} (x_1, \ldots, x_n)$$

$$\vdots$$

$$u_n(0, x_1, \ldots, x_n) = \frac{\partial \varphi}{\partial x_n} (x_1, \ldots, x_n) \tag{2.40}$$

in some neighborhood of the origin of R^n.

EXERCISE 49   Verify that any solution of the equation 2.37, satisfying the initial data 2.38, yields a solution of the system 2.39 satisfying the conditions 2.40, and conversely.

We may therefore restrict our attention to the linear first order system

$$\frac{\partial u_i}{\partial t} = \sum_{j=1}^{N} \sum_{s=1}^{N} a_{ijs} \frac{\partial u_j}{\partial x_s} + \sum_{j=1}^{N} b_{ij} u_j + c_i \tag{2.41}$$

and initial conditions

$$u_i(0, x_i, \ldots, x_n) = \varphi_i(x_1, \ldots, x_n) \tag{2.42}$$

with each φ_i analytic in some neighborhood of the origin in R^n, for $i = 1, \ldots, N$.

We can make one final simplification. It is no loss of generality to assume that each φ_i is identically zero in some n-ball about the origin. To prove this, define

$$w_i(t, x_1, \ldots, x_n) = u_i(t, x_1, \ldots, x_n) - \varphi_i(x_1, \ldots, x_n)$$

for $i = 1, \ldots, N$.

EXERCISE 50   Verify that the functions w_i satisfy a new system of N linear partial differential equations having the same general form as the system 2.41. Further, any solution of this new system satisfying initial conditions

$$w_i(0, x_1, \ldots, x_n) = 0$$

in some neighborhood of the origin in R^n is also a solution of system 2.41 satisfying the initial conditions 2.42.

We may therefore assume, for purposes of proving an existence and uniqueness theorem, that each φ_i is identically zero in some neighborhood S of the origin in R^n:

$$u_i(0, x_1, \ldots, x_n) \equiv 0 \tag{2.43}$$

for $(x_1, \ldots, x_n)$ in S.

Uniqueness   We will now prove the uniqueness part of the conclusion of the Cauchy-Kowalevski Theorem. We may restrict our attention to system 2.41 with homogeneous initial conditions 2.43. We want to prove that there is some neighborhood S of the origin in R^{n+1} in which this system can have at most one analytic solution. The argument is like that carried out in Section 2.6 for a simple case when we were discussing characteristics.

Suppose $u_1, \ldots, u_N$ are analytic and satisfy the system and the initial conditions in some neighborhood S of the origin O in R^{n+1}. By the analyticity assumption, each of these functions has a power series expansion about O. A typical coefficient of a term in one of these expansions is

$$\frac{1}{k_0! k_1! \cdots k_n!} \left[\frac{\partial^{k_0 + k_1 + \cdots + k_n} u_i}{\partial t^{k_0} \partial x_1^{k_1} \cdots \partial x_n^{k_n}} \right]_{t = x_1 = \cdots = x_n = 0}.$$

It is enough to show that these partial derivatives are uniquely determined by the initial conditions 2.43 for solutions of 2.41.

First, all derivatives with respect to just $x_1, \ldots, x_n$ (that is, $k_0 = 0$) are zero at O:

$$\left[\frac{\partial^{k_1 + \cdots + k_n} u_i}{\partial x_1^{k_1} \cdots \partial x_n^{k_n}} \right]_{t = x_1 = \cdots = x_n = 0} = 0. \tag{2.44}$$

This conclusion is immediate by simply differentiating the initial condition $u_i(0,$

$x_1, \ldots, x_n) \equiv 0$ the requisite number of times with respect to each of x_1, $\ldots, x_n$ and setting each of these variables equal to zero.

For derivatives involving t we must turn to the partial differential equations of system 2.41. First differentiate the ith equation k_1 times with respect to x_1, and so on through k_n times with respect to x_n. We get

$$\frac{\partial^{1+k_1+\cdots+k_n}u_i}{\partial t \partial x_1^{k_1} \cdots \partial x_n^{k_n}} = \sum_{j=1}^{N} \sum_{s=1}^{N} \left(\frac{\partial^{k_1+\cdots+k_n}a_{ijs}}{\partial x_1^{k_1} \cdots \partial x_n^{k_n}} \frac{\partial u_j}{\partial x_s} + a_{ijs} \frac{\partial^{k_1+\cdots+k_n+1}u_j}{\partial x_1^{k_1} \cdots \partial x_s^{k_s+1} \cdots \partial x_n^{k_n}} \right)$$

$$+ \sum_{j=1}^{N} \left(\frac{\partial^{k_1+\cdots+k_n}b_{ij}}{\partial x_1^{k_1} \cdots \partial x_n^{k_n}} u_j + b_{ij} \frac{\partial^{k_1+\cdots+k_n}u_j}{\partial x_1^{k_1} \cdots \partial x_n^{k_n}} \right) + \frac{\partial^{k_1+\cdots+k_n}c_i}{\partial x_1^{k_1} \cdots \partial x_n^{k_n}}. \quad (2.45)$$

We want to evaluate the derivative on the left at $t = x_1 = \cdots = x_n = 0$. But each of the terms on the right is either known from equation 2.44, or involves derivatives of known coefficient functions from the system, and these can in principle be evaluated. Thus the derivative on the left side of equation 2.45 is uniquely determined from the system and the initial conditions at O.

Now we know the derivatives

$$\left[\frac{\partial^{1+k_1+k_2+\cdots+k_n}u_i}{\partial t \partial x_1^{k_1} \cdots \partial x_n^{k_n}} \right]_{t=x_1=\cdots=x_n=0}. \quad (2.46)$$

We can continue to systematically carry out higher differentiations with respect to t. Differentiate equation 2.45 with respect to t. Upon setting $t = x_1 = \cdots = x_n = 0$, this yields an equation whose right side contains derivatives of known coefficient functions of the system, together with terms given by equations 2.44 and 2.46. In this way we evaluate

$$\left[\frac{\partial^{2+k_1+\cdots+k_n}u_i}{\partial t^2 \partial x_1^{k_1} \cdots \partial x_n^{k_n}} \right]_{t=x_1=\cdots=x_n=0}. \quad (2.47)$$

Continuing in this way, the system and initial conditions determine uniquely all of the coefficients in the power series expansion of u_i about O, for $i = 1$, $\ldots, N$, and hence uniquely determine the $u_i's$ in some neighborhood of O.

Notice that the argument exploits the assumption that each u_i is analytic, and this is the reason that uniqueness is claimed only in the class of analytic solutions. We have not proved that there cannot be another solution of the Cauchy problem that is not analytic.

This uniqueness argument has proceeded from the assumption that an analytic solution exists. We will now prove the existence of an analytic solution of the linear Cauchy problem, following the line of argument developed by Petrovsky [12]. The idea is to carry out a formal calculation yielding a power series representation of each u_i to form a solution. The task is then to show that these power series converge in some neighborhood of the origin. For this we will develop a method used to prove convergence of power series.

Majorants   Let S be a neighborhood of the origin in R^{n+1} and let φ and ψ have power series expansions about the origin in S, say

$$\varphi(t, x_1, \ldots, x_n) = \sum_{N=0}^{\infty} \sum_{k_0+k_1+\cdots+k_n=N}^{\infty} a_{k_0 k_1 \cdots k_n} t^{k_0} x_1^{k_1} \cdots x_n^{k_n}$$

and

$$\psi(t, x_1, \ldots, x_n) = \sum_{N=0}^{\infty} \sum_{k_0+k_1+\cdots+k_n=N}^{\infty} b_{k_0 k_1 \cdots k_n} t^{k_0} x_1^{k_1} \cdots x_n^{k_n},$$

for $(t, x_1, \ldots, x_n)$ in S.

We say that ψ is a *majorant* of φ if

$$\left| a_{k_0 k_1 \cdots k_n} \right| \leq b_{k_0 k_1 \cdots k_n}$$

for each such $(n + 1)$-tuple $(k_0, k_1, \ldots, k_n)$ of nonnegative integers. Of course, this implies that the coefficients in the expansion of ψ are nonnegative.

We also say that ψ *majorizes* φ.

Lemma 1   Any function that is analytic in a neighborhood of the origin has a majorant.   ∎

Proof: Let φ be analytic in a neighborhood of the origin in R^{n+1}. We may assume that φ is not identically zero, since in this case the lemma is obviously true. By the assumption of analyticity, we can write

$$\varphi(t, x_1, \ldots, x_n) = \sum_{N=0}^{\infty} \sum_{k_0+k_1+\cdots+k_n=N}^{\infty} a_{k_0 k_1 \cdots k_n} t^{k_0} x_1^{k_1} \cdots x_n^{k_n} \qquad (2.48)$$

in some neighborhood S of the origin. Choose a point $(\xi_0, \xi_1, \ldots, \xi_n)$ in S, with each $\xi_i \neq 0$. Since

$$\sum_{k_0+k_1+\cdots+k_n=0}^{\infty} a_{k_0 k_1 \cdots k_n} \xi_0^{k_0} \xi_1^{k_1} \cdots \xi_n^{k_n}$$

converges, its terms are bounded and there is a positive number M such that

$$\left| a_{k_0 k_1 \cdots k_n} \xi_0^{k_0} \xi_1^{k_1} \cdots \xi_n^{k_n} \right| \leq M$$

for every $(n + 1)$-tuple of nonnegative integers $k_0, \ldots k_n$. Thus

$$\left| a_{k_0 k_1 \cdots k_n} \right| \leq \frac{M}{\left| \xi_0 \right|^{k_0} \left| \xi_1 \right|^{k_1} \cdots \left| \xi_n \right|^{k_n}}. \qquad (2.49)$$

Now define the function

$$\psi(t, x_0, \ldots, x_n) = \frac{M}{\left(1 - \dfrac{t}{|\xi_0|}\right)\left(1 - \dfrac{x_1}{|\xi_1|}\right) \cdots \left(1 - \dfrac{x_n}{|\xi_n|}\right)}$$

for $(t, x_1, \ldots, x_n)$ within the sphere S_0 of radius $\sqrt{\xi_0^2 + \xi_1^2 + \cdots + \xi_n^2}$ about the origin in R^{n+1}. We claim that ψ is analytic in S_0, and majorizes φ in this neighborhood. To prove this, recall the geometric series

$$\frac{1}{1 - r} = \sum_{k=0}^{\infty} r^k$$

for $|r| < 1$. Hence, within S_0,

$$\psi(t, x_1, \ldots, x_n) = M \left[\sum_{k_0=0}^{\infty} \left(\frac{t}{|\xi_0|}\right)^{k_0} \sum_{k_1=0}^{\infty} \left(\frac{x_1}{|\xi_1|}\right)^{k_1} \cdots \sum_{k_n=0}^{\infty} \left(\frac{x_n}{|\xi_n|}\right)^{k_n} \right]$$

$$= \sum_{N=0}^{\infty} \sum_{k_0+k_1+\cdots+k_n=N} \frac{M}{|\xi_0|^{k_0}|\xi_1|^{k_1} \cdots |\xi_n|^{k_n}} t^{k_0} x_1^{k_1} \cdots x_n^{k_n}$$

so ψ is analytic in S_0. By inequality 2.49, ψ majorizes φ in S_0. ∎

EXERCISE 51   Other majorants of φ could have been produced to prove the lemma. For example, choose

$$\alpha = \min\{|\xi_0|, |\xi_1|, \ldots, |\xi_n|\}.$$

Prove that, for $0 < \rho < 1$,

$$\psi(t, x_1, \ldots, x_n) = \frac{M}{1 - \dfrac{1}{\alpha}((t/\rho) + x_1 + \cdots + x_n)}$$

$$= \sum_{k=0}^{\infty} \frac{M}{\alpha^k} \left(\frac{t}{\rho} + x_1 + \cdots + x_n\right)^k$$

majorizes φ in a neighborhood of the origin.

Existence   We are now ready to prove the existence of an analytic solution of the linear Cauchy problem, under the conditions stated in the Cauchy-Kowalevski Theorem.

We will recycle some of the calculations done in the uniqueness argument. We showed there that we can in principle compute the numbers

$$\frac{1}{k_0!k_1! \, \cdots \, k_n!} \left[\frac{\partial^{k_0+k_1+\,\cdots\,+k_n} u_i}{\partial t^{k_0} \partial x_1^{k_1} \, \cdots \, \partial x_n^{k_n}} \right]_{t=x_1=\,\cdots\,=x_n=0}$$

from the system 2.41 of partial differential equations and the initial conditions 2.43 and hence formally obtain power series representations of functions u_1, ..., u_N satisfying this system of partial differential equations and these initial conditions. There remains to prove that these series converge in a neighborhood of the origin. This is done using majorants.

Refer to system 2.41 with initial data 2.43 as CP1, for Cauchy Problem 1. The idea is to construct a second Cauchy problem, CP2, such that (1) we can find an analytic solution of CP2, and (2) this solution majorizes the series candidate for a solution to CP1. This will prove that the power series defining u_1, ..., u_N converge in a neighborhood of the origin and therefore that CP1 has an analytic solution.

To begin, choose positive numbers M, α, and ρ so that

$$\frac{M}{1 - \dfrac{1}{\alpha} ((t/\rho) + x_1 + \cdots + x_n)}$$

is larger in magnitude than all of the coefficients a_{ijs} of equation 2.41. Choose a positive number $\tilde{M}$ so that

$$\frac{\tilde{M}}{1 - \dfrac{1}{\alpha} ((t/\rho) + x_1 + \cdots + x_n)}$$

majorizes the coefficients b_{ij} and c_i. Let $m = \tilde{M}/M$ and define the new system

$$\frac{\partial w_i}{\partial t} = \frac{M}{1 - \dfrac{1}{\alpha} ((t/\rho) + x_1 + \cdots + x_n)} \left[\sum_{j=1}^{N} \sum_{k=1}^{N} \frac{\partial w_j}{\partial x_k} + \sum_{j=1}^{N} w_j + m \right] \quad (2.50)$$

for $i = 1, \ldots, N$. We will seek a solution of this system of the form

$$w_i(t, x_1, \ldots, x_n) = W(y)$$

for $i = 1, \ldots, N$, where

$$y = \frac{t}{\rho} + x_1 + \cdots + x_n.$$

Upon substituting $W(y)$ into equation 2.50 we find that W must satisfy

$$\frac{1}{\rho} W'(y) = \frac{M}{1 - (y/\alpha)} (NnW'(y) + NW(y) + m). \tag{2.51}$$

Upon writing

$$A(y) = \frac{M}{1 - (y/\alpha)}, \quad B(y) = \frac{mA(y)}{\dfrac{1}{\rho} - NnA(y)}$$

we find after routine manipulation that equation 2.51 is a separable equation which can be written

$$\frac{1}{\dfrac{N}{m} W + 1} dW = B(y) \, dy. \tag{2.52}$$

Adjust ρ so that

$$\frac{1}{\rho} - Nn|A(y)| > 0$$

in some neighborhood of the origin in R^{n+1}. This makes B analytic in this neighborhood. A solution of equation 2.52 is

$$W(y) = \frac{m}{N} \left[\exp\left(\frac{N}{m} \int_0^y B(s) \, ds \right) - 1 \right].$$

Now keep in mind that $w_i(t, x_1, \ldots, x_n) = W(y)$ are solutions of system 2.50 which was constructed to majorize system 2.41. Thus, to complete the proof, it is enough to show that, at $t = 0$, $W(y)$ can be expanded in a power series with positive coefficients. But

$$B(y) = \frac{mA(y)}{\dfrac{1}{\rho} - NnA(y)} = \frac{m\rho A(y)}{1 - Nn\rho A(y)} = m\rho A(y) \sum_{k=0}^{\infty} (Nn\rho A(y))^k$$

has nonnegative coefficients; hence so does the expansion of

$$\exp\left(\frac{N}{m} \int_0^y B(s) \, ds \right) - 1.$$

Thus the series expansion of $W(x_1 + \cdots + x_n)$ has nonnegative coefficients.

We conclude that the functions w_i defined by

$$w_i(t, x_1, \ldots, x_n) = W((t/\rho) + x_1 + \cdots + x_n)$$

are solutions of CP2 in a neighborhood of the origin, and that these solutions have power series expansions about the origin, and hence are analytic. Since these functions majorize the power series u_i for $i = 1, \ldots, N$, the power series constructed for the $u_i's$ from the Cauchy system and Cauchy data converge in a neighborhood of the origin, and hence determine an analytic solution of CP1.

EXERCISE 52   The following argument, from a simple case in ordinary differential equations, may help in understanding the more general argument used to prove the Cauchy-Kowalevski Theorem. Consider the ordinary Cauchy problem

$$u'(t) = f(t, u); \ u(t_0) = u_0,$$

assuming that f is analytic in a neighborhood of (t_0, u_0). In the spirit of the Cauchy-Kowalevski Theorem, attempt a solution

$$u(t) = \sum_{n=0}^{\infty} a_n(t - t_0)^n.$$

As in the proof of uniqueness, it is easy to check that the coefficients a_j are computable from information given in the problem. They key is to show that this power series converges in some interval about t_0. To do this by the method of majorants, consider the Cauchy problem

$$v'(t) = \frac{M}{\left(1 - \dfrac{t - t_0}{r}\right)\left(1 - \dfrac{v - u_0}{r}\right)}; \ v(t_0) = u_0.$$

Solve this problem explicitly (use separation of variables). Then show that it is possible to choose M large enough and r small enough that the series expansion $\sum_{n=0}^{\infty} b_n(t - t_0)^n$ for $v(t)$ majorizes the above series for $u(t)$ in some open interval about t_0. This proves that the series for u converges in some open interval about t_0, and hence establishes existence of a solution of the problem for u.

EXERCISE 53   Consider the Cauchy system consisting of the single second order partial differential equation

$$\frac{\partial^2 u}{\partial t^2} = a_{11} \frac{\partial^2 u}{\partial x_1^2} + 2a_{12} \frac{\partial^2 u}{\partial x_1 \partial x_2} + a_{22} \frac{\partial^2 u}{\partial x_2^2} + b_1 \frac{\partial^2 u}{\partial t \partial x_1}$$

$$+ b_2 \frac{\partial^2 u}{\partial t \partial x_2} + c_1 \frac{\partial u}{\partial x_1} + c_2 \frac{\partial u}{\partial x_2} + eu + f$$

for (t, x_1, x_2) in some neighborhood (sphere) N_1 about $(0, x_1^0, x_2^0)$, and Cauchy data

$$u(0, x_1, x_2) = \varphi(x_1, x_2), \quad \frac{\partial u}{\partial t}(0, x_1, x_2) = \psi(x_1, x_2)$$

in some neighborhood (disk) N_2 about (x_1^0, x_2^0). Assume that the functions a_{ij}, b_i, c_i, e, and f are analytic in N_1, and that φ and ψ are analytic in N_2. Adapting details from the above proof of the Cauchy-Kowalevski Theorem for linear systems, write a complete proof of the theorem for this case.

EXERCISE 54   Consider the Cauchy problem

$$u_t = u_{xx} \text{ for } -1 < x < 1, \, t > 0$$

$$u(x, 0) = \frac{1}{1 - x} \text{ for } -1 < x < 1.$$

Assume that this problem has an analytic solution and derive a contradiction. Why does this not violate the conclusion of the Cauchy-Kowalevski Theorem?

2.9.5 Hörmander's Theorem

This century has seen important advances in the development of new methods to probe existence questions. We will mention one result due to the contemporary Swedish mathematician Lars Hörmander. First we need some standard notation. If

$$\alpha = (\alpha_1, \ldots, \alpha_n)$$

and each α_j is a nonnegative integer, then $D^\alpha u$ denotes the derivative defined by

$$D^\alpha u = D_1^{\alpha_1} D_2^{\alpha_2} \cdots D_n^{\alpha_n} u = \frac{\partial^{\alpha_1 + \alpha_2 + \cdots + \alpha_n} u}{\partial x_1^{\alpha_1} \partial x_2^{\alpha_2} \cdots \partial x_n^{\alpha_n}}.$$

Also denote

$$|\alpha| = \alpha_1 + \alpha_2 + \cdots + \alpha_n.$$

If $\beta = (\beta_1, \ldots, \beta_n)$ and each β_j is a positive integer, consider the problem

$$D^\beta u = \sum_{|\alpha| \leq |\beta|} g_\alpha D^\alpha u + f$$

$$D_i^j u = h_{ij} \text{ for } 0 \leq j < \beta_i \text{ and } i = 1, \ldots, n, \tag{2.53}$$

in which the functions h_{ij} are given in some neighborhood of the origin in R^n. Here the summation

$$\sum_{|\alpha| \leq |\beta|}$$

is over all n-tuples α with each α_j a nonnegative integer and

$$\alpha_1 + \cdots + \alpha_n \leq \beta_1 + \cdots \beta_n.$$

We can now state the following.

Theorem 2 (Hörmander) Let β_i be a positive integer for $i = 1, 2, \ldots, n$. Assume in problem 2.53 that the functions g_α, h_{ij}, and f are analytic in some neighborhood of the origin. Suppose also that, for some positive number K depending only on $|\beta|$,

$$\sum_{|\alpha|=|\beta|} |g_\alpha(0)| \leq K.$$

Then the problem 2.53 has a unique solution that is analytic in some neighborhood of the origin in R^n. ∎

3

ELEMENTS OF FOURIER ANALYSIS

Up to this point we have discussed some aspects of first and second order partial differential equations. For the remainder of this book we will study constant coefficient, linear second order partial differential equations, concentrating on their canonical forms. These equations have a rich and important theory, and, as we have seen, also occur in models of interesting physical phenomena. We will seek to determine information about solutions, and, where possible, develop explicit formulas for solutions.

The last objective will employ a variety of techniques, including a set of tools and methods from Fourier analysis. This chapter is devoted to a development of these tools at a level of rigor sufficient to give us confidence in their use. The student who is familiar with Fourier series, integrals, and transforms may skip over this chapter and refer to it later as needed.

3.1 WHY FOURIER SERIES?

In his seminal 1807 paper on heat conduction (which was not published until 1822), Joseph Fourier (1768–1830) developed and solved partial differential equations governing heat flow under a variety of conditions. This led him to attempt to expand the function specifying the initial temperature of the medium in a series of trigonometric functions. We will explore how he was led to this position.

Imagine a homogeneous bar of some material, with uniform cross section and lateral sides insulated, the ends kept at temperature zero, and lying along the x-axis from 0 to π. Suppose the diffusivity of the material is k and the temperature initially along the bar in the cross section at x is $f(x)$. Let $u(x, t)$

be the temperature in the cross section at x at time t (Figure 2.4). In Section 2.5.2 (with $L = \pi$) we argued that u satisfies the initial-boundary value problem

$$u_t = ku_{xx} \text{ for } 0 < x < \pi, \quad t > 0$$

$$u(0, t) = u(\pi, t) = 0$$

$$u(x, 0) = f(x) \text{ for } 0 < x < \pi.$$

We can solve this problem by a technique called separation of variables, or the Fourier method. Let $u(x, t) = X(x)T(t)$. Upon substitution into the heat equation we obtain

$$XT' = kX''T$$

or, after dividing by kXT,

$$\frac{T'(t)}{kT(t)} = \frac{X''(x)}{X(x)}.$$

This must hold for $0 < x < \pi$ and $t > 0$. Since x and t are independent, we can substitute any positive t into $T'(t)/kT(t)$, and then $X''(x)/X(x)$ must equal this number for all x in $(0, \pi)$. Thus $X''(x)/X(x)$, and hence also $T'(t)/kT(t)$, must equal the same constant:

$$\frac{X''(x)}{X(x)} = -\lambda \text{ for } 0 < x < \pi$$

and

$$\frac{T'(t)}{kT(t)} = -\lambda \text{ for } t > 0.$$

(Calling the constant $-\lambda$ instead of λ is just a convention, and we would eventually get the same result either way). Now

$$X'' + \lambda X = 0 \text{ and } T' + \lambda kT = 0.$$

Since

$$u(0, t) = X(0)T(t) = 0$$

then $X(0) = 0$. Assume that $T(t)$ must be nonzero at some time, a necessary assumption if $u(x, t)$ is not identically zero (as must occur if there is any heat flow in the bar). Similarly,

$$u(\pi, t) = X(\pi)T(t) = 0$$

implies that $X(\pi) = 0$. Now we have an ordinary boundary value problem for X:

$$X'' + \lambda X = 0; \quad X(0) = X(\pi) = 0. \tag{3.1}$$

We must find numerical values for λ for which this problem has nontrivial solutions for X. Consider cases on λ.

If $\lambda = 0$ then $X'' = 0$ implies that $X(x) = ax + b$ for some constants a and b. But then $X(0) = b = 0$ and $X(\pi) = a\pi = 0$ implies that $a = 0$. We get only the trivial solution for X if $\lambda = 0$, so $\lambda = 0$ is not useful for problem 3.1.

Suppose $\lambda < 0$, say $\lambda = -\alpha^2$, with $\alpha > 0$. Then $X'' - \alpha^2 X = 0$, with general solution

$$X(x) = ae^{\alpha x} + be^{-\alpha x}.$$

Now

$$X(0) = 0 = a + b$$

implies that $b = -a$ and $X(x) = 2a \sinh(\alpha x)$. Further,

$$X(\pi) = 2a \sinh(\alpha \pi) = 0$$

implies that $a = 0$, since $\alpha\pi > 0$ forces $\sinh(\alpha\pi) > 0$. This case gives no nontrivial solutions for X. Thus we will not use any negative values of λ for this problem.

Finally, try $\lambda > 0$, say $\lambda = \alpha^2$ with $\alpha > 0$. Now the differential equation is $X'' + \alpha^2 X = 0$, with general solution

$$X(x) = a \cos(\alpha x) + b \sin(\alpha x).$$

Since $X(0) = a = 0$, then $X(x) = b \sin(\alpha x)$. We need

$$X(\pi) = b \sin(\alpha \pi) = 0.$$

This will certainly occur if $b = 0$, but then we have the trivial solution again. To obtain a nontrivial solution for X, observe that $b \sin(\alpha\pi) = 0$ is also satisfied if $\sin(\alpha\pi) = 0$, which occurs if α is a positive integer. Thus λ can be chosen as the square of a positive integer:

$$\lambda = n^2 \text{ for } n = 1, 2, \ldots.$$

Corresponding to each positive integer n,

$$X_n(x) = \sin(nx),$$

or any constant multiple of this function, is a solution of problem 3.1 corresponding to $\lambda = n^2$.

With $\lambda = n^2$, the equation for T is

$$T' + n^2 kT(t) = 0$$

with general solution any constant multiple of

$$T_n(t) = e^{-n^2 kt}.$$

For each positive integer n, we now have a function

$$u_n(x, t) = b_n \sin(nx)e^{-n^2 kt}$$

which satisfies the heat equation and the boundary conditions $u(0, t) = u(\pi, t) = 0$, for any constant b_n.

To satisfy the initial condition with one such function, we need n and b_n so that

$$u_n(x, 0) = f(x) = b_n \sin(nx) \text{ for } 0 < x < \pi.$$

This condition is impossible to satisfy unless $f(x)$ is a constant multiple of $\sin(nx)$ for some n.

A finite sum of such functions, say

$$u(x, t) = \sum_{n=1}^{N} b_n \sin(nx)e^{-n^2 kt}$$

likewise satisfies the heat equation and the boundary conditions. But again, it is in general impossible to choose the $b_n's$ to satisfy

$$u(x, 0) = f(x) = \sum_{n=1}^{N} b_n \sin(nx) \text{ for } 0 < x < \pi.$$

For example, if $f(x)$ is a nontrivial polynomial, then $f(x)$ cannot be written as a sum of constants times these sine functions for all x in any interval.

Here is where Fourier had a brilliant insight. He attempted an infinite superposition

$$u(x, t) = \sum_{n=1}^{\infty} b_n \sin(nx)e^{-n^2 kt}.$$

This function certainly satisfies the boundary conditions. It will satisfy the initial condition if it is possible to choose the $b'_n s$ so that

$$u(x, 0) = f(x) = \sum_{n=1}^{\infty} b_n \sin(nx) \text{ for } 0 < x < L.$$

In this way, an attempt to solve a problem in heat conduction has led to the problem of writing the initial temperature function as a series of constant multiples of sine functions. In view of the latitude available in choosing an initial temperature function, and the fact that we expect such heat conduction problems to have solutions, Fourier's reasoning suggests the necessity of being able to expand "arbitrary" functions in series of sines. This was an astonishing prospect for the natural philosophers of Fourier's day, and led to sometimes heated debate and a period of intense research into trigonometric series and representations of functions.

Other problems can lead to expansions of functions in series of cosines, or in series involving both sine and cosine functions. We will now develop such series, or Fourier series, and explore the possibility of expanding arbitrary functions in trigonometric series. We will also develop other tools of Fourier analysis which we will use, including Fourier integrals and transforms.

EXERCISE 55 Solve the initial-boundary value problem for u if $f(x) = \sqrt{3} \sin(2x)$.

EXERCISE 56 Solve the initial-boundary value problem for u if $f(x) = 5 \sin(t) - 12 \sin(4t)$.

EXERCISE 57 Prove that $f(x) = x$ cannot be written as a finite sum

$$\sum_{n=1}^{N} b_n \sin(nx)$$

for $0 < x < \pi$, for any choice of real numbers $b_1, \ldots, b_N$.

3.2   THE FOURIER SERIES OF A FUNCTION

Begin with the trigonometric series

$$\frac{1}{2} a_0 + \sum_{n=1}^{\infty} (a_n \cos(nx) + b_n \sin(nx)),$$

whose terms are constant multiples of sines and cosines. Calling the constant term $a_0/2$ is a convention which will have a notational advantage soon.

The idea that an arbitrary function might be represented by such a series was not original with Fourier. Euler, Lagrange, and others had obtained trigonometric expansions of particular functions. These were, however, usually regarded as special cases or curiosities and were not well understood. Fourier was the first to require the expansion of an arbitrary function in a trigonometric series, as part of the solution of an important problem.

Here is how a person not too concerned about convergence questions might approach such an expansion. Suppose we want to write a given function f in a trigonometric series:

$$f(x) = \frac{1}{2} a_0 + \sum_{n=1}^{\infty} (a_n \cos(nx) + b_n \sin(nx)) \tag{3.2}$$

for $-\pi \le x \le \pi$. Integrate both sides of equation 3.2 and interchange the summation and the integral to obtain

$$\int_{-\pi}^{\pi} f(x) \, dx = \frac{1}{2} a_0 \int_{-\pi}^{\pi} dx + \sum_{n=1}^{\infty} \left(a_n \int_{-\pi}^{\pi} \cos(nx) \, dx + b_n \int_{-\pi}^{\pi} \sin(nx) \, dx \right) = \pi a_0,$$

because all of the integrals in the series are zero. Thus

$$a_0 = \frac{1}{\pi} \int_{-\pi}^{\pi} f(x) \, dx. \tag{3.3}$$

Now solve for the $a_n's$ for $n = 1, 2, \ldots$ as follows. Multiply equation 3.2 by $\cos(mx)$, with m any positive integer, and integrate the resulting equation, interchanging the summation and the integral. We obtain:

$$\int_{-\pi}^{\pi} f(x)\cos(mx) \, dx = \frac{1}{2} a_0 \int_{-\pi}^{\pi} \cos(mx) \, dx$$

$$+ \sum_{n=1}^{\infty} \left(a_n \int_{-\pi}^{\pi} \cos(nx)\cos(mx) \, dx + b_n \int_{-\pi}^{\pi} \sin(nx)\cos(mx) \, dx \right).$$

By a straightforward integration, all integrals on the right side of this equation are zero except $\int_{-\pi}^{\pi} \cos^2(mx) \, dx$, which occurs when $n = m$ in the summation. This integral equals π. Thus the last equation reduces to

$$\int_{-\pi}^{\pi} f(x)\cos(mx) \, dx = a_m \int_{-\pi}^{\pi} \cos^2(mx) \, dx = \pi a_m.$$

This enables us to solve for

$$a_m = \frac{1}{\pi} \int_{-\pi}^{\pi} f(x)\cos(mx)\ dx \text{ for } m = 1, 2, \ldots.. \tag{3.4}$$

Finally, multiply equation 3.2 by $\sin(mx)$ and integrate both sides of the resulting equation, interchanging the sum and the integrals to obtain:

$$\int_{-\pi}^{\pi} f(x)\sin(mx)\ dx = \frac{1}{2} a_0 \int_{-\pi}^{\pi} \sin(mx)\ dx$$

$$+ \sum_{n=1}^{\infty} \left(a_n \int_{-\pi}^{\pi} \cos(nx)\sin(mx)\ dx + b_n \int_{-\pi}^{\pi} \sin(nx)\sin(mx)\ dx \right).$$

Again, all integrals on the right are zero except $\int_{-\pi}^{\pi} \sin^2(mx)\ dx$, which occurs in the summation when $n = m$. This integral also equals π. The last equation simplifies to

$$\int_{-\pi}^{\pi} f(x)\sin(mx)\ dx = b_m \int_{-\pi}^{\pi} \sin^2(mx)\ dx = b_m \pi$$

and therefore

$$b_m = \frac{1}{\pi} \int_{-\pi}^{\pi} f(x)\sin(mx)\ dx \text{ for } m = 1, 2, \ldots.. \tag{3.5}$$

The reasoning we have just pursued is flawed because the interchange of an integral and an infinite series can lead to incorrect results. Nevertheless, the fact that series like 3.2, with the coefficients 3.3, 3.4, and 3.5, were occurring in the solution of important problems suggested that a closer look was warranted, and some of the greatest mathematicians (Dirichlet, Riemann, and others) took up the challenge. There followed a period of intense activity which inspired advances in real function theory and integration theory and spawned such important areas as harmonic analysis and transfinite arithmetic. Georg Cantor, who is credited with much of the early work on infinite ordinals and cardinals, was led to these ideas through investigations of the convergence of Fourier series.

The numbers given by equations 3.3, 3.4, and 3.5 are called the *Fourier coefficients* of f on $[-\pi, \pi]$. With these coefficients, the series

$$\frac{1}{2} a_0 + \sum_{n=1}^{\infty} (a_n \cos(nx) + b_n \sin(nx)) \tag{3.6}$$

is the *Fourier series* of f on $[-\pi, \pi]$.

We are interested in the following question. What can be said about the convergence of the Fourier series 3.6? Fourier believed that for any function the Fourier series always converges to the function. It is not difficult to find examples in which this is not true, even for very simple and well-behaved functions.

Example 15   Let $f(x) = x$ for $-\pi \leq x \leq \pi$. The Fourier coefficients are

$$a_0 = \frac{1}{\pi} \int_{-\pi}^{\pi} x \, dx = 0;$$

$$a_n = \frac{1}{\pi} \int_{-\pi}^{\pi} x \cos(nx) \, dx = 0;$$

and

$$b_n = \frac{1}{\pi} \int_{-\pi}^{\pi} x \sin(nx) \, dx = -2 \frac{\cos(n\pi)}{n} = 2 \frac{(-1)^{n+1}}{n}.$$

The Fourier series of f is

$$\sum_{n=1}^{\infty} 2 \frac{(-1)^{n+1}}{n} \sin(nx).$$

But this series equals 0 at $x = \pm\pi$, whereas $f(-\pi) = -\pi$ and $f(\pi) = \pi$. At least at $-\pi$ and at π, then, the Fourier series of this function does not converge to the function.   ∎

EXERCISE 58

(a) Following in Euler's footsteps, use a geometric series to derive the series

$$\sum_{n=1}^{\infty} a^n \cos(nx) = \frac{a \cos(x) - a^2}{1 - 2a \cos(x) + a^2}$$

and

$$\sum_{n=1}^{\infty} a^n \sin(nx) = \frac{a \sin(x)}{1 - 2a \cos(x) + a^2}$$

for $|a| < 1$. *Hint:* Recall that

$$\sum_{n=1}^{\infty} r^n = \frac{r}{1 - r} \quad \text{for} \quad |r| < 1$$

and let

$$r = ae^{ix} = a\,\cos(x) + ia\,\sin(x).$$

(b) Let $a = 1$ in the first series in part (a) to conclude that

$$\sum_{n=1}^{\infty} \cos(nx) = -\frac{1}{2}.$$

This "equation," though incorrect, also appeared in a study of sound waves by Joseph-Louis Lagrange.

(c) Integrate the divergent series of part (b) term by term from x to π to obtain

$$\sum_{n=1}^{\infty} \frac{1}{n} \sin(nx) = \frac{1}{2}(\pi - x).$$

When we discuss Fourier sine expansions, we will see that this equation is correct for $0 < x \leq \pi$. Thus the adage "garbage in, garbage out" does not seem to hold in all cases!

3.3 CONVERGENCE OF FOURIER SERIES

We will determine the sum of the Fourier series 3.6 for functions satisfying certain conditions. This will involve some preliminaries, culminating in the proof of a convergence theorem.

3.3.1 Periodic Functions

First observe that the terms in the Fourier series are all periodic of period 2π. We will therefore assume that f is periodic of period 2π. This means that $f(x)$ is defined for $-\infty < x < \infty$ and $f(x + 2\pi) = f(x)$ for all x. Examples of such functions are $\sin(x)$ and $\cos(4x)$.

This assumption of periodicity is not overly restrictive. Suppose we are given a function g defined just on $(-\pi, \pi]$. We have only to replicate the graph for $-\pi < x \leq \pi$ over successive intervals of length 2π. Figure 3.1(b) illustrates this process for the function graphed in Figure 3.1(a). This replication defines a new function f over the entire real line, and f is periodic with period 2π. Further, $f(x) = g(x)$ for $-\pi < x \leq \pi$. We call f the *periodic extension* of g to the entire real line.

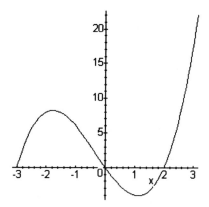

FIGURE 3.1(a)   Graph of a typical function defined on $(-\pi, \pi]$.

Any interval of length 2π carries a complete copy of the graph of a function of period 2π, and hence contains enough information to evaluate the function at any number. In particular, if $f(x)$ is given explicitly for $-\pi < x \le \pi$, then we know $f(x)$ for all x. For example, $f(-\pi) = f(\pi)$, explaining why we need only specify values of the function on the half-closed interval $(-\pi, \pi]$. As another example,

$$f(3\pi/2) = f(-\pi/2 + 2\pi) = f(-\pi/2)$$

and $f(-\pi/2)$ is known because $-\pi/2$ is in the interval $(-\pi, \pi]$ in which function values have been given.

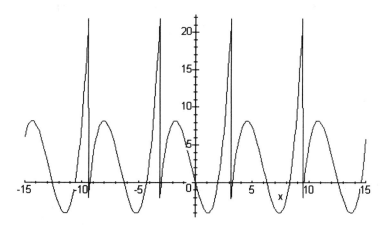

FIGURE 3.1(b)   Periodic extension of the function shown in Figure 3.1(a).

It will be useful to observe that, if f is integrable and has period 2π, then

$$\int_a^{a+2\pi} f(x)\ dx = \int_{-\pi}^{\pi} f(x)\ dx \tag{3.7}$$

for any number a. In particular, in the formulas for the Fourier coefficients of f, the integration can be carried out over any interval of length 2π.

3.3.2   Dirichlet's Formula

We want to understand the relationship between values of the Fourier series 3.6 and values $f(x)$ of the periodic function f. As with any series, convergence of the Fourier series 3.6 hinges on the behavior of its partial sums. Denote the Nth partial sum as $S_N(x)$:

$$S_N(x) = \frac{1}{2} a_0 + \sum_{n=1}^{N} (a_n \cos(nx) + b_n \sin(nx)).$$

We will rewrite this partial sum in a way that will help us determine its limit as $N \to \infty$. Insert the Fourier coefficients to obtain

$$S_N(x) = \frac{1}{2\pi} \int_{-\pi}^{\pi} f(\xi)\ d\xi + \sum_{n=1}^{N} \left[\frac{1}{\pi} \int_{-\pi}^{\pi} f(\xi)\cos(n\xi)\ d\xi \cos(nx) \right.$$

$$+ \frac{1}{\pi} \int_{-\pi}^{\pi} f(\xi)\sin(n\xi)\ d\xi \sin(nx) \left.\right] = \frac{1}{\pi} \int_{-\pi}^{\pi} f(\xi) \left[\frac{1}{2} + \sum_{n=1}^{N} [\cos(n\xi)\cos(nx) \right.$$

$$+ \sin(n\xi)\sin(nx) \left.\right]\ d\xi = \frac{1}{\pi} \int_{-\pi}^{\pi} f(\xi) \left[\frac{1}{2} + \sum_{n=1}^{N} \cos(n(\xi - x)) \right]\ d\xi. \tag{3.8}$$

To simplify the quantity in square brackets in the last integral, let $y = \xi - x$ and let

$$\sigma = \frac{1}{2} + \sum_{n=1}^{N} \cos(ny).$$

Upon multiplying both sides of this equation by $2 \sin(y/2)$ we obtain

$$2\sigma \sin(y/2) = \sin(y/2) + \sum_{n=1}^{N} 2 \cos(ny)\sin(y/2) = \sin(y/2)$$

$$+ \sum_{n=1}^{N} [\sin((n + 1/2)y) - \sin((n - 1/2)y)] = \sin(y/2) + [\sin(3y/2) - \sin(y/2)]$$

$$+ [\sin(5y/2) - \sin(3y/2)] \cdots + [\sin((N - 1/2)y) - \sin((N - 3/2)y)]$$

$$+ [\sin((N + 1/2)y) - \sin((N - 1/2)y)].$$

This series is telescoping—the first term in each square bracket cancels the second term in the next bracket. Further, the second term in the first bracket is canceled by the $\sin(y/2)$ term. All that remains is the first term in the last bracket, and we have

$$2\sigma \sin(y/2) = \sin((N + 1/2)y).$$

Therefore

$$\sigma = \frac{\sin((N + 1/2)y)}{2 \sin(y/2)} = \frac{\sin((N + 1/2)(\xi - x))}{2 \sin((\xi - x)/2)}$$

provided that $\sin((\xi - x)/2) \neq 0$. Upon inserting this result into equation 3.8 we obtain

$$S_N(x) = \frac{1}{\pi} \int_{-\pi}^{\pi} f(\xi) \frac{\sin((N + 1/2)(\xi - x))}{2 \sin((\xi - x)/2)} \, d\xi. \qquad (3.9)$$

Now put $t = \xi - x$ in equation 3.9 to obtain

$$S_N(x) = \frac{1}{\pi} \int_{-\pi-x}^{\pi-x} f(x + t) \frac{\sin((N + 1/2)t)}{2 \sin(t/2)} \, dt.$$

Since f is periodic of period 2π, this integrand also has period 2π. This means that we can carry out the integral over any interval of length 2π. In particular,

$$S_N(x) = \frac{1}{\pi} \int_{-\pi}^{\pi} f(x + t) \frac{\sin((N + 1/2)t)}{2 \sin(t/2)} \, dt. \qquad (3.10)$$

This is *Dirichlet's formula*, and it is the special form we sought for $S_N(x)$. The function

$$\frac{\sin((N + 1/2)t)}{2 \sin(t/2)}$$

is called the *Dirichlet kernel*.

Using Dirichlet's formula, we can derive another result we will use shortly.

Lemma 2

$$\frac{1}{\pi} \int_{-\pi}^{0} \frac{\sin((N + 1/2)t)}{2 \sin(t/2)} \, dt = \frac{1}{\pi} \int_{0}^{\pi} \frac{\sin((N + 1/2)t)}{2 \sin(t/2)} \, dt = \frac{1}{2}. \qquad \blacksquare$$

Proof of the Lemma: Let $f(x) = 1$ in Dirichlet's formula. The Fourier coefficients of this function are

$$a_0 = \frac{1}{\pi} \int_{-\pi}^{\pi} d\xi = 2$$

and for $n = 1, 2, \ldots$,

$$a_n = \frac{1}{\pi} \int_{-\pi}^{\pi} \cos(n\xi) \, d\xi = 0, \quad b_n = \frac{1}{\pi} \int_{-\pi}^{\pi} \sin(n\xi) \, d\xi = 0.$$

Therefore $S_N(x) = 1$ and, since $f(x + t) = 1$ for all x and t, Dirichlet's formula gives

$$\frac{1}{\pi} \int_{-\pi}^{0} \frac{\sin((N + 1/2)t)}{2 \sin(t/2)} \, dt + \frac{1}{\pi} \int_{0}^{\pi} \frac{\sin((N + 1/2)t)}{2 \sin(t/2)} \, dt = 1. \qquad (3.11)$$

Let $t = -w$ in the first integral on the left to write

$$\frac{1}{\pi} \int_{-\pi}^{0} \frac{\sin((N + 1/2)t)}{2 \sin(t/2)} \, dt = \frac{1}{\pi} \int_{\pi}^{0} \frac{\sin((N + 1/2)w)}{2 \sin(w/2)} (-1) \, dw$$

$$= \frac{1}{\pi} \int_{0}^{\pi} \frac{\sin((N + 1/2)w)}{2 \sin(w/2)} \, dw$$

Therefore the integrals on the left side of equation 3.11 are equal. Since their sum is 1, both integrals equal 1/2. ∎

3.3.3 The Riemann-Lebesgue Lemma

We will now prove a result which will enable us to compute the limit of $S_N(x)$ as $N \to \infty$.

Lemma 3 (Riemann-Lebesgue Lemma) If g is piecewise continuous on $[a, b]$, then

$$\lim_{\omega \to \infty} \int_{a}^{b} g(t)\sin(\omega t) \, dt = 0. \qquad \blacksquare$$

Proof of the Riemann-Lebesgue Lemma: Suppose first that g is continuous on $[a, b]$ and let $I = \int_{a}^{b} g(t)\sin(\omega t) \, dt$. Let $t = \xi + \pi/\omega$, with ω chosen large enough that $b - \pi/\omega \geq a$. Then

$$I = \int_{a-\pi/\omega}^{b-\pi/\omega} g(\xi + \pi/\omega)\sin(\omega\xi + \pi) \, d\xi = -\int_{a-\pi/\omega}^{b-\pi/\omega} g(\xi + \pi/\omega)\sin(\omega\xi) \, d\xi.$$

In order to maintain t as the variable of integration, replace ξ with t in the last integral:

$$I = -\int_{a-\pi/\omega}^{b-\pi/\omega} g(t + \pi/\omega)\sin(\omega t) \, dt.$$

Now add this expression for I to the definition of I to write

$$2I = \int_a^b g(t)\sin(\omega t) \, dt - \int_{a-\pi/\omega}^{b-\pi/\omega} g(t + \pi/\omega)\sin(\omega t) \, dt$$

$$= \int_a^{b-\pi/\omega} [g(t) - g(t + \pi/\omega)]\sin(\omega t) \, dt + \int_{b-\pi/\omega}^b g(t)\sin(\omega t) \, dt$$

$$- \int_{a-\pi/\omega}^a g(t + \pi/\omega)\sin(\omega t) \, dt. \tag{3.12}$$

Since g is continuous on $[a, b]$, then for some M, $|g(t)| \leq M$ for $a \leq t \leq b$. Therefore

$$\left| \int_{b-\pi/\omega}^b g(t)\sin(\omega t) \, dt \right| \leq M \frac{\pi}{\omega}$$

and also

$$\left| \int_{a-\pi/\omega}^a g(t + \pi/\omega)\sin(\omega t) \, dt \right| \leq M \frac{\pi}{\omega}.$$

For the other two integrals in equation 3.12, use the fact that g is uniformly continuous on $[a, b]$. Let $\epsilon > 0$. There is some $\delta > 0$ such that

$$|g(x) - g(y)| < \epsilon/3 \text{ if } |x - y| < \delta.$$

Then

$$|g(t) - g(t + \pi/\omega)| < \epsilon/3 \text{ if } \frac{\pi}{\omega} < \delta.$$

Therefore, for $\omega > \pi/\delta$, and also ω large enough that $b - \pi/\omega \geq a$ and $M\pi/\omega < \epsilon/3$, we have from equation 3.12 and the bounds we have just made that

$$|2I| < M\frac{\pi}{\omega} + M\frac{\pi}{\omega} + \frac{\epsilon}{3} < \frac{\epsilon}{3} + \frac{\epsilon}{3} + \frac{\epsilon}{3} = \epsilon.$$

But then

$$|I| < \frac{\epsilon}{2} < \epsilon \text{ if } \omega > \pi/\delta,$$

proving that $\lim_{\omega\to\infty} I = 0$.

This leaves the case that g is piecewise continuous but not continuous. In this event g is continuous except for jump discontinuities at finitely many points $t_1, \ldots, t_{k_s}$. Now write I as a sum of integrals from a to t_1, t_1 to t_2, $\ldots$, and finally from t_n to b. By redefining $g(t)$ at the end point of each of these intervals, if necessary, we can write I as a finite sum of integrals, each having the same form as I, but with each having a continuous integrand. These integrals all have limit 0 as $\omega \to \infty$ by the case just proved, hence $I \to 0$ as $\omega \to \infty$. This completes the proof of the Riemann-Lebesgue lemma. ■

3.3.4   Convergence of the Fourier Series

We can now give what is essentially Dirichlet's proof of a convergence theorem for the Fourier series 3.6. We will use the following notation and terminology.

For any real number x, denote the left limit of f at x by $f(x-)$, and the right limit by $f(x+)$:

$$f(x-) = \lim_{\xi\to x-} f(\xi) \text{ and } f(x+) = \lim_{\xi\to x+} f(\xi).$$

A function is *piecewise smooth* on an interval if the function and its derivative are piecewise continuous on the interval. This means that f and f' are both continuous at all but finitely many points of the interval, and both f and f' have finite right and left limits at each point at which they are discontinuous. Thus, even if f is not continuous at x, we know that $f(x+)$ and $f(x-)$ exist, and if f' is not defined or continuous at x, then $f'(x+)$ and $f'(x-)$ exist.

EXERCISE 59   Let f be periodic and differentiable. Prove that f' is also periodic with the same period as f.

Theorem 3 (Convergence of Fourier Series)   Let f be piecewise smooth on $[-\pi, \pi]$ and periodic of period 2π. Then at each x the Fourier series 3.6 converges to

$$\frac{1}{2}(f(x+) - f(x-)).$$ ■

If f is actually continuous at x, then $f(x+) = f(x-) = f(x)$ and the Fourier series of f on $[-\pi, \pi]$ converges to $f(x)$. Thus, over any interval on which f is continuous and satisfies the hypotheses of the theorem, the Fourier series is an exact representation of the function.

If f has a jump discontinuity at x, then the Fourier series converges to the average of the left and right limits at x. Geometrically, this number is midway between the "ends" of the graph at the jump discontinuity (Fig. 3.2).

Proof of the Convergence Theorem: Use Dirichlet's formula and Lemma 2 to write

$$S_N(x) = \frac{1}{\pi} \int_{-\pi}^{0} f(x + t) \frac{\sin((N + 1/2)t)}{2 \sin(t/2)} dt + \frac{1}{\pi} \int_{0}^{\pi} f(x + t) \frac{\sin((N + 1/2)t)}{2\sin(t/2)} dt$$

$$= \frac{1}{\pi} \int_{-\pi}^{0} f(x + t) \frac{\sin((N + 1/2)t)}{2 \sin(t/2)} dt - \frac{1}{2} f(x-) + \frac{1}{2} f(x-)$$

$$+ \frac{1}{\pi} \int_{0}^{\pi} f(x + t) \frac{\sin((N + 1/2)t)}{2 \sin(t/2)} dt - \frac{1}{2} f(x+) + \frac{1}{2} f(x+)$$

$$= \frac{1}{\pi} \int_{-\pi}^{0} [f(x + t) - f(x-)] \frac{\sin((N + 1/2)t)}{2 \sin(t/2)} dt + \frac{1}{2} f(x-)$$

$$+ \frac{1}{\pi} \int_{0}^{\pi} [f(x + t) - f(x+)] \frac{\sin((N + 1/2)t)}{2 \sin(t/2)} dt + \frac{1}{2} f(x+). \qquad (3.13)$$

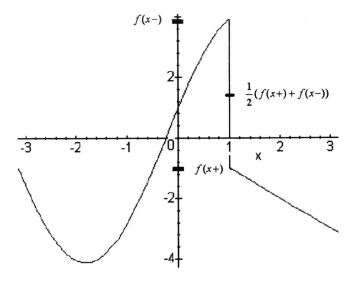

FIGURE 3.2 Convergence of the Fourier series of f to $(1/2)(f(x+) + f(x-))$ at a jump discontinuity.

The proof is complete if we can show that each of the last two integrals in equation 3.13 has limit 0 as $N \to \alpha$. In this event we will have

$$S_N(x) \to \frac{1}{2} (f(x+) + f(x-))$$

as we want to show. To prove that the last integral has limit 0, let

$$g(t) = \frac{f(x + t) - f(x+)}{2 \sin(t/2)} \quad \text{for } 0 < t \le \pi.$$

Notice that

$$\lim_{t \to 0+} g(t) = \lim_{t \to 0+} \frac{f(x + t) - f(x+)}{2 \sin(t/2)} = \lim_{t \to 0+} \frac{f(x + t) - f(x+)}{t} \cdot \frac{t/2}{\sin(t/2)}$$

$$= \lim_{t \to 0+} \frac{f(x + t) - f(x+)}{t} \lim_{t \to 0+} \frac{t/2}{\sin(t/2)} = f'(x+) \cdot 1 = f'(x+).$$

This limit exists because f' is piecewise continuous on $[-\pi, \pi]$. Define

$$g(0) = f'(x+).$$

Because f is piecewise smooth on $[-\pi, \pi]$ and periodic of period 2π, g is piecewise smooth on $[0, \pi]$. By the Riemann-Lebesgue lemma, with $\omega = N + 1/2$,

$$\lim_{\omega \to \infty} \int_0^\pi g(t)\sin(\omega t) \, dt = \lim_{N \to \infty} \int_0^\pi [f(x + t) - f(x+)] \frac{\sin((N + 1/2)(t))}{2 \sin(t/2)} \, dt = 0.$$

This proves that the last integral in equation 3.13 has limit 0 as $N \to \infty$.
    By a similar argument,

$$\lim_{N \to \infty} \int_{-\pi}^0 [f(x + t) - f(x-)] \frac{\sin((N + 1/2)t)}{2 \sin(t/2)} \, dt = 0.$$

Therefore

$$\lim_{N \to \infty} S_N(x) = \frac{1}{2} (f(x+) + f(x-)). \qquad \blacksquare$$

Example 16   Let f be periodic of period 2π and let

$$f(x) = \begin{cases} \dfrac{1}{2} & \text{for } -\pi < x < 0 \\ 1 & \text{for } 0 \le x \le \pi. \end{cases}$$

A graph of f is shown in Figure 3.3(a). The Fourier coefficients are

$$a_0 = \frac{1}{\pi}\left[\int_{-\pi}^{0} \frac{1}{2}\, dx + \int_{0}^{\pi} dx\right] = \frac{3}{2};$$

and for $n = 1, 2, \ldots$

$$a_n = \frac{1}{\pi}\int_{-\pi}^{0} \frac{1}{2}\cos(nx)\, dx + \frac{1}{\pi}\int_{0}^{\pi} \cos(nx)\, dx = 0,$$

and

$$b_n = \frac{1}{\pi}\int_{-\pi}^{0} \frac{1}{2}\sin(nx)\, dx + \frac{1}{\pi}\int_{0}^{\pi} \sin(nx)\, dx = \frac{1}{2n\pi}[1 - \cos(n\pi)].$$

The Fourier series of f on $[-\pi, \pi]$ is

$$\frac{3}{4} + \sum_{n=1}^{\infty} \frac{1}{2n\pi}[1 - (-1)^n]\sin(nx),$$

in which we used the fact that $\cos(n\pi) = (-1)^n$ if n is an integer. Since.

$$1 - (-1)^n = \begin{cases} 2 \text{ if } n \text{ is odd} \\ 0 \text{ if } n \text{ is even} \end{cases}$$

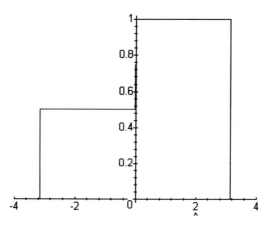

FIGURE 3.3(a)   Graph of the function f of Example 16 for $-\pi < x \leq \pi$.

we need retain only odd values of n in this series, and can write the series as

$$\frac{3}{4} + \sum_{n=1}^{\infty} \frac{1}{(2n - 1)\pi} \sin((2n - 1)x).$$

For $0 < x < \pi$, f is continuous and the series converges to $f(x) = 1$. For $-\pi < x < 0$, f is also continuous and the series converges to $f(x) = 1/2$. At $x = 0$ the series converges to

$$\frac{1}{2}(f(0+) + f(0-)) = \frac{1}{2}\left(1 + \frac{1}{2}\right) = \frac{3}{4}.$$

This is obvious from the series, in which all of the sine terms vanish if $x = 0$. At π and at $-\pi$ the series also converges to 3/4, and this can be verified either by applying the theorem or by again observing that all the sine terms vanish at $\pm\pi$.

On $[-\pi, \pi]$, the Fourier series and the function agree except at 0, $-\pi$, and π.

Figures 3.3(b) through (d) show graphs of the third, seventh, and fifteenth partial sums of the Fourier series, respectively, compared with the graph of f. This provides some pictorial insight into how the Fourier series converges to the function. ∎

In Example 16, the behavior of the partial sums near the point of discontinuity 0 of f is perhaps a little surprising. Near 0, the partial sums have a "jump" which retains approximately the same height no matter how many terms of the

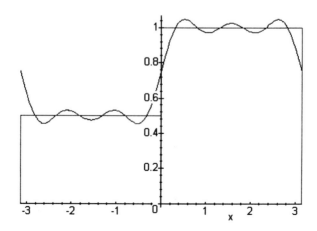

FIGURE 3.3(b)   Third partial sum of the Fourier series of Example 16.

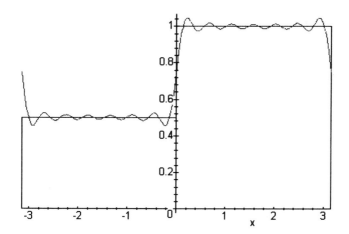

FIGURE 3.3(c) Seventh partial sum of the Fourier series of Example 16.

partial sum are computed. This is called the Gibbs phenomenon, and it is characteristic of the convergence of a Fourier series at a point of discontinuity of the function. The phenomenon is named for Josiah Willard Gibbs, who, about the turn of this century, was the first to completely explain it mathematically, despite the fact that it had been observed by others. The experimental physicist Albert A. Michelson, famed for his experiments on the velocity of light and existence of the aether, had noticed it after he constructed a machine to compute Fourier coefficients. A detailed analysis of the Gibbs phenomenon occurs in [10].

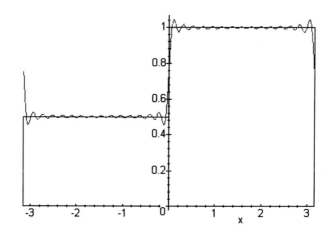

FIGURE 3.3(d) Fifteenth partial sum of the Fourier series of Example 16.

3.3.5   Change of Scale

This discussion of Fourier series can be recast on an interval $[-L, L]$ by a change of variable in the integrals. Suppose f is periodic of period $2L$ (so $f(x + 2L) = f(x)$) and integrable on $[-L, L]$. Since the periodicity condition requires that $f(-L) = f(L)$, we need only define $f(x)$ for $-L < x \le L$.

The *Fourier coefficients of f* on $[-L, L]$ are

$$a_n = \frac{1}{L} \int_{-L}^{L} f(\xi)\cos\left(\frac{n\pi\xi}{L}\right) d\xi \text{ for } n = 0, 1, 2, \ldots$$

and

$$b_n = \frac{1}{L} \int_{-L}^{L} f(\xi)\sin\left(\frac{n\pi\xi}{L}\right) d\xi \text{ for } n = 1, 2, \ldots.$$

The *Fourier series* of f on $[-L, L]$ is

$$\frac{1}{2} a_0 + \sum_{n=1}^{\infty} a_n \cos\left(\frac{n\pi x}{L}\right) + b_n \sin\left(\frac{n\pi x}{L}\right),$$

in which the coefficients are the Fourier coefficients.

The convergence theorem for this series on $[-L, L]$ is an obvious adaptation of the theorem for $[-\pi, \pi]$.

Theorem 4   Let f be piecewise smooth on $[-L, L]$ and periodic of period $2L$. Then at each x the Fourier series of f on $[-L, L]$ converges to

$$\frac{1}{2} (f(x+) + f(x-)). \qquad \blacksquare$$

Example 17   Let f have period 6 and be defined for $-3 < x \le 3$ by

$$f(x) = \begin{cases} 0 \text{ for } -3 < x \le 0 \\ x \text{ for } 0 \le x \le 3. \end{cases}$$

A graph of f is shown in Figure 3.4(a). The Fourier coefficients (with $L = 3$) are computed by routine integrations:

$$a_0 = \frac{1}{3} \int_{-3}^{3} f(x)\, dx = \frac{1}{3} \int_{0}^{3} x\, dx = \frac{3}{2};$$

and for $n = 1, 2, \ldots,$

$$a_n = \frac{1}{3} \int_{0}^{3} x \cos\left(\frac{n\pi x}{3}\right) dx = \frac{3}{n^2\pi^2} [(-1)^n - 1],$$

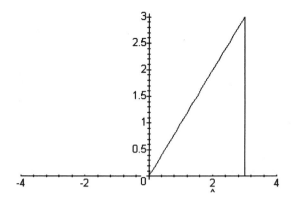

FIGURE 3.4(a)   Graph of the function f of Example 17, for $-3 < x \leq 3$.

and

$$b_n = \frac{1}{3} \int_0^3 x \sin\left(\frac{n\pi x}{3}\right) dx = -\frac{3}{n\pi}(-1)^n.$$

The Fourier series of f on $[-3, 3]$ is

$$\frac{3}{4} + \sum_{n=1}^{\infty} \left[\frac{3}{n^2\pi^2}[(-1)^n - 1]\cos\left(\frac{n\pi x}{3}\right) + \frac{3}{n\pi}(-1)^{n+1}\sin\left(\frac{n\pi x}{3}\right) \right].$$

This series converges to 0 for $-3 < x \leq 0$; to x for $0 \leq x < 3$; and to 3/2 for $x = \pm 3$. The last conclusion is obtained by observing that

$$\frac{1}{2}(f(3+) + f(3-)) = \frac{1}{2}(3 + 0) = \frac{3}{2}$$

and

$$\frac{1}{2}(f(-3+) + f(-3-)) = \frac{1}{2}(0 + 3) = \frac{3}{2}.$$

The Fourier series in this example represents the function exactly on $(-3, 3)$. This is because f is continuous on $(-3, 3)$, in addition to satisfying the other hypotheses of the theorem. Figures 3.4(b) through (d) show graphs of the fifth, tenth, and twentieth partial sums, respectively, of this Fourier series, compared with the graph of f. ∎

EXERCISE 60   In each of problems 1 through 10, find the Fourier series of the function on the interval and determine the sum of this series. In the spirit

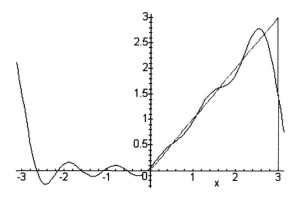

FIGURE 3.4(b)   Fifth partial sum of the Fourier series of Example 17.

of Examples 16 and 17, graph some partial sums of the Fourier series in comparison with a graph of the function.

1. $f(x) = -x, \; -1 < x \le 1$
2. $f(x) = \cosh(\pi x), \; -1 < x \le 1$
3. $f(x) = 1 - |x|, \; -2 < x \le 2$
4. $f(x) = \begin{cases} -4 \text{ for } -\pi < x < 0 \\ 4 \text{ for } 0 \le x \le \pi \end{cases}$
5. $f(x) = \sin(2x)$ for $-\pi < x \le \pi$
6. $f(x) = x^2 - x + 3$ for $-2 < x \le 2$
7. $f(x) = \begin{cases} -x \text{ for } -5 < x < 0 \\ 1 + x^2 \text{ for } 0 \le x \le 5 \end{cases}$

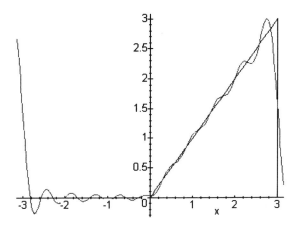

FIGURE 3.4(c)   Tenth partial sum of the Fourier series of Example 17.

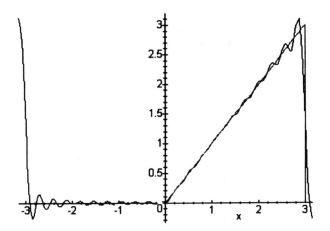

FIGURE 3.4(d)   Twentieth partial sum of the Fourier series of Example 17.

8. $f(x) = \begin{cases} 1 \text{ for } -\pi < x \le 0 \\ 2 \text{ for } 0 < x \le \pi \end{cases}$

9. $f(x) = \cos(x/2) - \sin(x)$ for $-\pi < x \le \pi$

10. $f(x) = \begin{cases} 1 - x \text{ for } -1 < x \le 0 \\ 0 \text{ for } 0 < x \le 1 \end{cases}$

EXERCISE 61   In each of problems 1 through 5, determine the sum of the Fourier series of the function on the interval

1. $f(x) = \begin{cases} 2x \text{ for } -3 < x \le -2 \\ 0 \text{ for } -2 < x < 1 \\ x^2 \text{ for } 1 \le x \le 3 \end{cases}$

2. $f(x) = \begin{cases} \cos(x) \text{ for } -2 < x < 1/2 \\ \sin(x) \text{ for } 1/2 \le x \le 2 \end{cases}$

3. $f(x) = \begin{cases} -2 \text{ for } -4 < x \le 2 \\ 1 + x^2 \text{ for } 2 < x < 3 \\ e^{-x} \text{ for } 3 \le x \le 4 \end{cases}$

4. $f(x) = \begin{cases} 1 \text{ for } -2 < x \le 0 \\ -1 \text{ for } 0 < x < 1/2 \\ x^2 \text{ for } 1/2 \le x \le 2 \end{cases}$

5. $f(x) = \begin{cases} \cos(\pi x) \text{ for } -2 < x < 0 \\ x \text{ for } 0 \le x \le 2 \end{cases}$

EXERCISE 62   Let $f(x) = x^2$ for $-\pi < x \le \pi$. Find the Fourier series of f and sum the series at an appropriately chosen point to determine $\sum_{n=1}^{\infty} 1/n^2$. Use another point to determine $\sum_{n=1}^{\infty} (-1)^n 1/n^2$.

EXERCISE 63 Prove the following theorem on term by term integration of Fourier series. Let f be piecewise continuous on $[-L, L]$ and have Fourier series

$$\frac{1}{2} a_0 + \sum_{n=1}^{\infty} a_n \cos\left(\frac{n\pi x}{L}\right) + b_n \sin\left(\frac{n\pi x}{L}\right).$$

Then for $-L \leq x \leq L$,

$$\int_{-L}^{x} f(s)\, ds = \frac{1}{2} a_0(x + L) + \frac{L}{\pi} \sum_{n=1}^{\infty} \frac{1}{n} \left[a_n \sin\left(\frac{n\pi x}{L}\right) \right.$$

$$\left. - b_n \left\{ \cos\left(\frac{n\pi x}{L}\right) - \cos(n\pi) \right\} \right].$$

This series on the right is that obtained by integrating the Fourier series for f term by term. This equation is valid even at points at which the Fourier series does not converge to $f(x)$. *Hint:* Define

$$F(x) = \int_{-L}^{x} f(s)\, ds - \frac{1}{2} a_0 x.$$

Show that F is continuous and F' is piecewise continuous on $[-L, L]$, with $F(-L) = F(L)$. Write the Fourier series of F and relate its coefficients to those of f by integration by parts.

EXERCISE 64 Assume that the Fourier series of $f(x)$ on $[-L, L]$ converges to $f(x)$ and can be integrated term by term. Multiply

$$f(x) = \frac{1}{2} a_0 + \sum_{n=1}^{\infty} a_n \cos\left(\frac{n\pi x}{L}\right) + b_n \sin\left(\frac{n\pi x}{L}\right)$$

by $f(x)$ and integrate the resulting equation from $-L$ to L to derive Parseval's equation

$$\frac{1}{2} a_0^2 + \sum_{n=1}^{\infty} (a_n^2 + b_n^2) = \frac{1}{L} \int_{-L}^{L} f^2(x)\, dx.$$

EXERCISE 65 Let $f(x) = x^2$ for $[-L, L]$. Write the Fourier series for $f(x)$. Use Parseval's equation to show that

$$\sum_{n=1}^{\infty} \frac{1}{n^4} = \frac{\pi^4}{90}.$$

EXERCISE 66 This exercise and Exercise 67 explore another kind of convergence associated with Fourier series. Instead of convergence of the Fourier series itself, as defined by convergence of its sequence of partial sums, we will consider convergence of the sequence of Cesàro sums of the Fourier series. But first we will write the Fourier series in complex form.

Let f be defined on the real line, periodic of period 2π, and integrable on $[-\pi, \pi]$. By using

$$\cos(x) = \frac{1}{2}(e^{ix} + e^{-ix}), \ \sin(x) = \frac{1}{2i}(e^{ix} - e^{-ix})$$

derive the complex form

$$\sum_{n=-\infty}^{\infty} c_n e^{inx}$$

of the Fourier series for f, where

$$c_n = \frac{1}{2\pi} \int_{-\pi}^{\pi} f(t)e^{-int} \, dt.$$

We will use this idea in Exercise 274, Section 6.10, to give another derivation of Poisson's integral formula for the solution of the Dirichlet problem for a disk.

EXERCISE 67   Continuing from Exercise 66, let S_n be the nth partial sum of the Fourier series for f:

$$S_n(x) = \sum_{j=-n}^{n} c_j e^{ijx}.$$

Define

$$\sigma_n(x) = \frac{1}{n+1} \sum_{j=0}^{n} S_j(x).$$

This is the nth Cesàro sum of the series. It is the average of the first $n + 1$ partial sums.

(a) Prove that

$$\sigma_n(x) = \sum_{j=-n}^{n} \frac{n + 1 - |j|}{n + 1} c_j e^{ijx}.$$

(b) Prove that

$$\sigma_n(x) = \frac{1}{2\pi} \int_{-\pi}^{\pi} f(t) K_n(x - t) \, dt,$$

where

$$K_n(\xi) = \sum_{j=-n}^{n} \frac{n + 1 - |j|}{n + 1} e^{ij\xi}.$$

(c) Prove that

$$\sigma_n(x) = \frac{1}{2\pi} \int_{-\pi}^{\pi} f(x - t) K_n(t) \, dt.$$

(d) Prove that

$$K_n(x) = \begin{cases} \dfrac{1}{n + 1} \left(\dfrac{\sin((n + 1)x/2)}{\sin(x/2)} \right)^2 & \text{for } x \neq 0 \\[2mm] n + 1 \text{ for } x = 0. \end{cases}$$

(e) Prove that, for any number δ with $0 < \delta < \pi$, $K_n(x) \to 0$ uniformly as $n \to \infty$, for x in $[-\pi, \pi]$ but outside $[-\delta, \delta]$. *Hint:* Observe that

$$K_n(x) \leq \frac{1}{n + 1} \frac{1}{\sin^2(x/2)} \leq \frac{1}{n + 1} \frac{1}{\sin^2(\delta/2)}.$$

(f) Prove that

$$\frac{1}{2\pi} \int_{-\pi}^{\pi} K_n(x) \, dx = 1.$$

(g) Prove that, if f is continuous at x, then $\lim_{n \to \infty} \sigma_n(x) = f(x)$. *Hint:* First note that $|f(t)| \leq M$ for $-\pi \leq t \leq \pi$, for some number M. Let $\epsilon > 0$. By continuity of f at x, for some δ, $0 < \delta < \pi$ and $|f(t) - f(x)| < \epsilon/2$ for $|t - x| < \delta$. Further, by part (e), there is some positive N such that $|K_n(t)| < \epsilon/4M$ for t in $[-\pi, \pi]$ but outside $[-\delta, \delta]$, if $n \geq N$. Show that

$$|\sigma_n(x) - f(x)| = \frac{1}{2\pi} \left| \int_{-\pi}^{\pi} (f(x - t) - f(x)) K_n(t) \, dt \right|.$$

Write the last integral as the sum of the integral over $[-\delta, \delta]$ and the integral over the remaining parts of $[-\pi, \pi]$, and use the above bounds to show that this sum is less than ϵ if $n \geq N$.

The conclusion of part (g) shows that the Cesàro sums converge to $f(x)$ wherever f is continuous. A more complete discussion of convergence of Fourier series and Cesàro sums and many other aspects of Fourier series and integrals can be found in Körner [10].

3.4 SINE AND COSINE EXPANSIONS

A function g defined on $[-L, L]$ is *even* if $g(-x) = g(x)$. The graph of an even function is symmetric about the y-axis (Figure 3.5). Examples of even functions are x^n for any even positive integer n, and $\cos(n\pi x/L)$ for any integer n.

We say that g is *odd* if $g(-x) = -g(x)$. The graph of an odd function is symmetric through the origin (Figure 3.6). Examples of odd functions are x^n for any odd positive integer, and $\sin(n\pi x/L)$ for any nonzero integer n.

Even and odd functions behave like even and odd integers under multiplication. A product of two {odd/even} functions is even, and the product of an odd and even function is odd. For example, if g and h are odd functions on $[-L, L]$, then

$$g(-x)h(-x) = [-g(x)][-h(x)] = g(x)h(x)$$

so the product gh is an even function.

If g is even on $[-L, L]$, then

$$\int_{-L}^{L} g(x)\, dx = 2 \int_{0}^{L} g(x)\, dx. \tag{3.14}$$

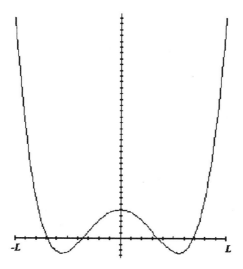

FIGURE 3.5 A typical even function on $[-L, L]$.

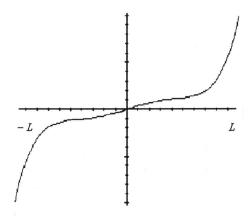

FIGURE 3.6   A typical odd function of $[-L, L]$.

And if h is odd on $[-L, L]$, then

$$\int_{-L}^{L} h(x)\ dx = 0. \tag{3.15}$$

These facts are apparent from Figures 3.5 and 3.6, respectively, and are straight-forward to prove from properties of integrals. We will exploit these properties of even and odd functions to consider Fourier half-range expansions.

3.4.1   Fourier Sine Series

Suppose f is integrable on $[0, L]$. We can write a Fourier sine series for f as follows. Figure 3.7(a) shows a graph of a typical f. First extend f to an odd function h defined on $(-L, L]$ by putting

$$h(x) = \begin{cases} f(x) \text{ for } 0 \le x \le L \\ -f(-x) \text{ for } -L < x < 0. \end{cases}$$

Then h is an odd function (Figure 3.7(b)) which agrees with f on $[0, L]$. Next extend h to an odd periodic function $\tilde{h}$ of period $2L$ (Figure 3.7(c)). Consider the Fourier coefficients of $\tilde{h}$ on $[-L, L]$:

$$a_0 = \frac{1}{L} \int_{-L}^{L} \tilde{h}(\xi)\ d\xi = 0$$

because $\tilde{h}$ is an odd function;

$$a_n = \frac{1}{L} \int_{-L}^{L} \tilde{h}(\xi)\cos\left(\frac{n\pi\xi}{L}\right)\ d\xi = 0$$

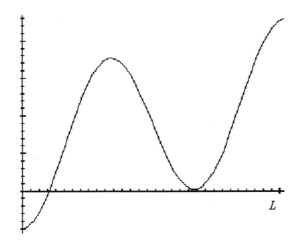

FIGURE 3.7(a) A typical function defined on [0, L].

because $\tilde{h}(x)\cos(n\pi x/L)$ is odd (product of an odd and even function); and

$$b_n = \frac{1}{L} \int_{-L}^{L} \tilde{h}(\xi)\sin\left(\frac{n\pi\xi}{L}\right) d\xi.$$

This number is not necessarily zero. However, because $\tilde{h}(x)\sin(n\pi x/L)$ is an even function (product of two odd functions), and because $\tilde{h}(x) = f(x)$ if $0 \leq x \leq L$, then

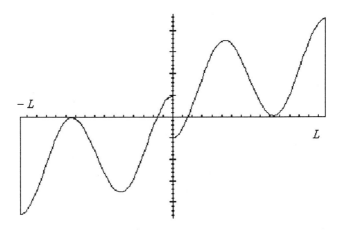

FIGURE 3.7(b) Odd extension of the function of Figure 3.7(a) to $[-L, L]$.

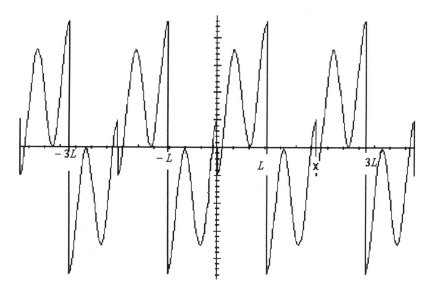

FIGURE 3.7(c)   Periodic extension of the function of Figure 3.7(b).

$$b_n = \frac{2}{L} \int_0^L \tilde{h}(\xi)\sin\left(\frac{n\pi\xi}{L}\right) d\xi = \frac{2}{L} \int_0^L f(\xi)\sin\left(\frac{n\pi\xi}{L}\right) d\xi.$$

The Fourier series of $\tilde{h}$ on $[-L, L]$ therefore contains only sine terms. Since $\tilde{h}(x) = f(x)$ on $[0, L]$, we call this series the Fourier sine series of f on $[0, L]$. To reiterate, the *Fourier sine series* of a function f defined on $[0, L]$ is

$$\sum_{n=1}^{\infty} b_n \sin\left(\frac{n\pi x}{L}\right)$$

where

$$b_n = \frac{2}{L} \int_0^L f(\xi)\sin\left(\frac{n\pi\xi}{L}\right) d\xi.$$

We need not actually define h and $\tilde{h}$ to write this series. These functions were simply a way of motivating the sine series, based on the already defined Fourier series on $[-L, L]$. This device also enables us to immediately write a convergence theorem for sine series by applying the Fourier convergence theorem to the odd periodic extension $\tilde{h}$. If this is done, we conclude that, if f is piecewise smooth on $[0, L]$, the sine series converges to 0 at $x = 0$ and at $x = L$, and to

$$\frac{1}{2}\left(f(x+) + f(x-)\right)$$

for $0 < x < L$.

Example 18   Let $f(x) = e^x$ for $0 \le x \le 2$. Figure 3.8(a) shows a graph of f.
   The Fourier sine series of e^x on $[0, 2]$ converges to 0 at $x = 0$ and $x = 2$, and to e^x for $0 < x < 2$. The coefficients are

$$b_n = \int_0^2 e^\xi \sin\left(\frac{n\pi\xi}{2}\right) d\xi = \frac{2n\pi}{4 + n^2\pi^2}\, [1 - (-1)^n e^2].$$

For $0 < x < 2$, then,

$$e^x = \sum_{n=1}^{\infty} \frac{2n\pi}{4 + n^2\pi^2}\, [1 - (-1)^n e^2]\sin\left(\frac{n\pi x}{2}\right).$$

   Figures 3.8(b) through (f) compare graphs of some partial sums of the sine series, with the graph of f. Note again the appearance of the Gibbs phenomenon. Near 2, the graph experiences a jump whose height remains approximately the same no matter how large n is chosen in the nth partial sum S_n.   ∎

3.4.2   Fourier Cosine Series

We can write a Fourier cosine series of a function f defined on $[0, L]$ by first extending f to an even function g defined on $[-L, L]$:

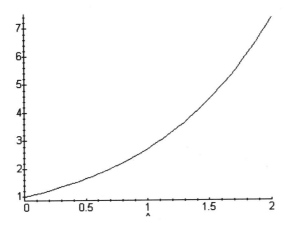

FIGURE 3.8(a)   Graph of f in Example 18.

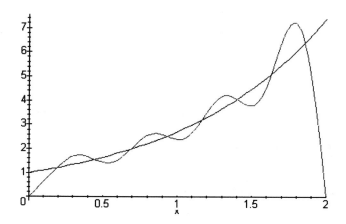

FIGURE 3.8(b)   Eighth partial sum of the sine series of Example 18.

$$g(x) = \begin{cases} f(x) \text{ for } 0 \leq x \leq L \\ f(-x) \text{ for } -L \leq x < 0 \end{cases}$$

and then extending g to a periodic function $\tilde{g}$ of period $2L$. Figure 3.9(a) shows the graph of a typical f, with g in Figure 3.9(b) and $\tilde{g}$ in Figure 3.9(c). The Fourier coefficients of $\tilde{g}$ are

$$a_n = \frac{1}{L} \int_{-L}^{L} \tilde{g}(\xi) \cos\left(\frac{n\pi\xi}{L}\right) d\xi = \frac{2}{L} \int_{0}^{L} f(\xi) \cos\left(\frac{n\pi\xi}{L}\right) d\xi$$

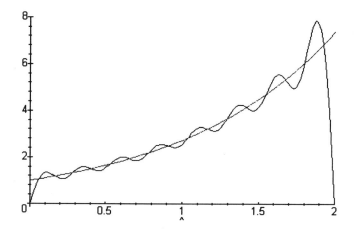

FIGURE 3.8(c)   Fifteenth partial sum of the sine series of Example 18.

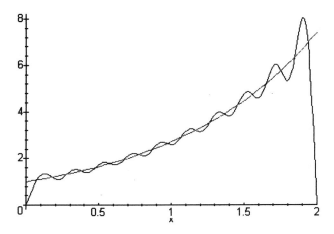

FIGURE 3.8(d) Twentieth partial sum of the sine series of Example 18.

because $\tilde{g}(x)\cos(n\pi x/L)$ is an even function and $\tilde{g}(x) = f(x)$ for $0 \leq x \leq L$; and

$$b_n = \frac{1}{L} \int_{-L}^{L} \tilde{g}(\xi)\sin\left(\frac{n\pi\xi}{L}\right) d\xi = 0$$

because $\tilde{g}(x)\sin(n\pi x/L)$ is an odd function. The Fourier series of $\tilde{g}$ on $[-L, L]$ therefore contains only cosine terms. Since $\tilde{g}(x) = f(x)$ for $0 \leq x \leq L$, this is the Fourier cosine series for f on $[0, L]$.

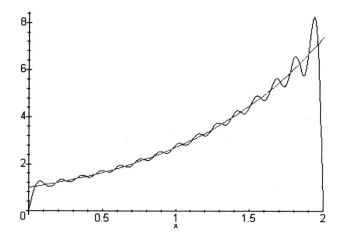

FIGURE 3.8(e) Thirtieth partial sum of the sine series of Example 18.

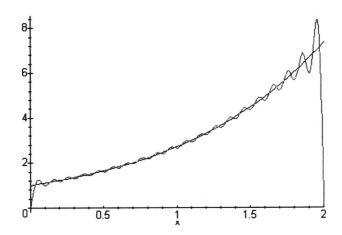

FIGURE 3.8(f)   Fortieth partial sum of the sine series of Example 18.

Thus, the *Fourier cosine series* of *f* on [0, L] is

$$\frac{1}{2} a_0 + \sum_{n=1}^{\infty} a_n \cos \left(\frac{n\pi x}{L} \right),$$

where

$$a_n = \frac{2}{L} \int_0^L f(\xi)\cos \left(\frac{n\pi \xi}{L} \right) d\xi$$

for *n* = 0, 1, 2, ...

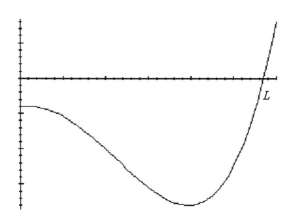

FIGURE 3.9(a)   Typical function defined on [0, L].

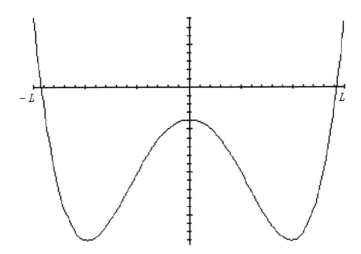

FIGURE 3.9(b)   Even extension of the function of Figure 3.9(a) to $[-L, L]$.

As with the sine series, we do not have to construct $\tilde{g}$ and repeat the motivating argument each time we want a cosine series—simply compute the Fourier cosine coefficients from this integral formula, which is carried out only on $[0, L]$.

If we apply the Fourier convergence theorem to $\tilde{g}$ on $[-L, L]$, we obtain a convergence theorem for Fourier cosine series. If f is piecewise smooth on $[0, L]$, then its Fourier cosine series converges to

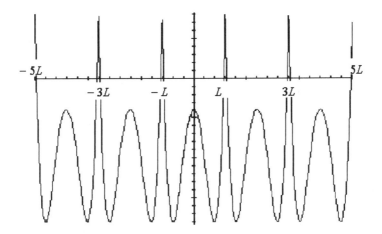

FIGURE 3.9(c)   Periodic extension of the function of Figure 3.9(b).

$$\frac{1}{2}\left(f(x+) + f(x-)\right)$$

for $0 < x < L$. At $x = 0$, the cosine series converges to $f(0+)$, and at $x = L$, to $f(L-)$.

To understand this conclusion about convergence at 0 and L, apply the Fourier convergence theorem to the even periodic extension $\tilde{g}$ of f. At 0 the cosine series of f (which is the Fourier series of $\tilde{g}$ on $[-L, L]$) converges to

$$\frac{1}{2}\left(\tilde{g}(0+) + \tilde{g}(0-)\right).$$

But

$$\tilde{g}(0+) = \lim_{x \to 0+} \tilde{g}(x) = \lim_{x \to 0+} f(x) = f(0+)$$

and

$$\tilde{g}(0-) = \lim_{x \to 0-} \tilde{g}(x) = \lim_{x \to 0+} g(-x) = \lim_{x \to 0+} f(x) = f(0+).$$

Therefore at 0 the cosine series converges to

$$\frac{1}{2}\left(f(0+) + f(0+)\right),$$

which is $f(0+)$.

Similar reasoning proves that the cosine series converges to $f(L-)$ at $x = L$.

Example 19   Let $f(x) = e^x$ for $0 \leq x \leq 2$. The cosine series on $[0, 2]$ converges to e^x for $0 \leq x \leq 2$. To compute this series, evaluate

$$a_0 = \int_0^2 e^x\, dx = e^2 - 1$$

and, for $n = 1, 2, \ldots,$

$$a_n = \int_0^2 e^x \cos\left(\frac{n\pi x}{2}\right) dx = \frac{4}{4 + n^2\pi^2}\left[(-1)^n e^2 - 1\right].$$

Therefore, for $0 \leq x \leq 2$,

$$e^x = \frac{1}{2}(e^2 - 1) + \sum_{n=1}^{\infty} \frac{4}{4 + n^2\pi^2}[(-1)^n e^2 - 1]\cos\left(\frac{n\pi x}{2}\right).$$

Figures 3.10(a) through (c) compare graphs of the second, fourth, and eighth partial sums, respectively, of this cosine series with a graph of f. ∎

EXERCISE 68   In each of problems 1 through 8, write the Fourier sine and the Fourier cosine series of the function. In each case determine the sum of the series and graph some partial sums of the series to compare with the graph of the function.

1. $f(x) = 4$ for $0 \leq x \leq 3$

2. $f(x) = \begin{cases} 1 & \text{for } 0 \leq x \leq 1 \\ -1 & \text{for } 1 < x \leq 2 \end{cases}$

3. $f(x) = \begin{cases} 0 & \text{for } 0 \leq x \leq \pi/2 \\ \sin(x) & \text{for } \pi/2 < x \leq \pi \end{cases}$

4. $f(x) = 2x$ for $0 \leq x \leq 1$

5. $f(x) = x^2$ for $0 \leq x \leq 2$

6. $f(x) = e^{-x}$ for $0 \leq x \leq 1$

7. $f(x) = \sin(3x)$ for $0 \leq x \leq \pi$

8. $f(x) = \begin{cases} x & \text{for } 0 \leq x \leq 1 \\ 2 - x & \text{for } 1 \leq x \leq 2 \end{cases}$

EXERCISE 69   Find the sum of the series $\sum_{n=1}^{\infty} (-1)^n/(4n^2 - 1)$. *Hint:* Expand

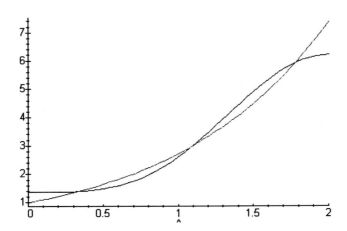

FIGURE 3.10(a)   Second partial sum of the cosine series of Example 19.

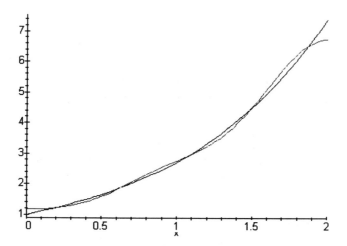

FIGURE 3.10(b)   Fourth partial sum of the cosine series of Example 19.

sin(x) in a cosine series on [0, π] and evaluate this series at an appropriately chosen point.

EXERCISE 70   Here is another way of approaching the coefficients in the sine expansion of f on [0, L]. Suppose

$$f(x) = \sum_{n=1}^{\infty} b_n \sin\left(\frac{n\pi x}{L}\right)$$

and we can interchange an integral with the summation. Multiply this equation

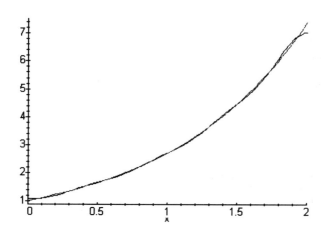

FIGURE 3.10(c)   Eighth partial sum of the cosine series of Example 19.

by $\sin(k\pi x/L)$ and integrate the resulting equation from 0 to L, integrating the series term by term. Solve the resulting equation for b_k to obtain the Fourier sine coefficient of f on $[0, L]$.

EXERCISE 71   In the spirit of Exercise 70, consider a cosine expansion

$$f(x) = \frac{1}{2} a_0 + \sum_{n=1}^{\infty} a_n \cos\left(\frac{n\pi x}{L}\right).$$

Integrate this equation term by term to obtain a formula for a_0. Next, multiply the equation by $\cos(k\pi x/L)$ and integrate from 0 to L, interchanging the sum and the integral. Solve the resulting equation for a_k to obtain the Fourier coefficients in the cosine expansion of f on $[0, L]$.

EXERCISE 72   Derive a version of Parseval's equation (Exercise 64) for Fourier sine series. Derive a version suitable for Fourier cosine series.

EXERCISE 73   Expand $(\pi - x)/2$ in a sine series on $[0, \pi]$ to verify an assertion made in Exercise 58(c).

3.5   BESSEL'S INEQUALITY AND UNIFORM CONVERGENCE

We will derive an important inequality involving the Fourier coefficients of a function. Because the notation is simpler we will develop it first for the Fourier sine coefficients on an interval $[0, \pi]$.

Let f be integrable on $[0, \pi]$. The Fourier sine coefficients of f are

$$b_n = \frac{2}{\pi} \int_0^{\pi} f(\xi)\sin(n\xi) \, d\xi$$

and the Nth partial sum of the sine series is

$$S_N(x) = \sum_{n=1}^{N} b_n \sin(nx).$$

Now

$$0 \le \int_0^\pi [f(x) - S_N(x)]^2 \, dx = \int_0^\pi [f^2(x) + S_N^2(x) - 2f(x)S_N(x)] \, dx$$

$$= \int_0^\pi f^2(x) \, dx + \int_0^\pi \left(\sum_{n=1}^N \sum_{m=1}^N b_n b_m \sin(nx)\sin(mx) \right) dx$$

$$- 2 \int_0^\pi \sum_{n=1}^N f(x) b_n \sin(nx) \, dx$$

$$= \int_0^\pi f^2(x) \, dx + \sum_{n=1}^N \sum_{m=1}^N b_n b_m \int_0^\pi \sin(nx)\sin(mx) \, dx$$

$$- 2 \sum_{n=1}^N b_n \int_0^\pi f(x)\sin(nx) \, dx. \tag{3.16}$$

But

$$\int_0^\pi \sin(nx)\sin(mx) \, dx = \begin{cases} 0 \text{ if } n \ne m \\ \pi/2 \text{ if } n = m. \end{cases}$$

Therefore in inequality 3.16 all terms in the double summation are zero except those for which $m = n$, and for these terms the integrals equal $\pi/2$. Further, in the last summation in inequality 3.16,

$$\int_0^\pi f(x)\sin(nx) \, dx = \frac{\pi}{2} b_n.$$

The inequality therefore becomes

$$0 \le \int_0^\pi f^2(x) \, dx + \sum_{n=1}^N \frac{\pi}{2} b_n^2 - 2 \sum_{n=1}^N \frac{\pi}{2} b_n^2$$

or

$$\sum_{n=1}^N b_n^2 \le \frac{2}{\pi} \int_0^\pi f^2(x).$$

Since this is true for any positive integer N, we conclude that

$$\lim_{N \to \infty} \sum_{n=1}^N b_n^2 \le \frac{2}{\pi} \int_0^\pi f^2(x) \, dx$$

or

$$\sum_{n=1}^{\infty} b_n^2 \leq \frac{2}{\pi} \int_0^{\pi} f^2(x) \, dx. \tag{3.17}$$

This is *Bessel's inequality* for the Fourier sine coefficients, and it implies that the series of squares of the sine coefficients converges.

For the Fourier coefficients of a function f defined on $[-L, L]$, Bessel's inequality takes the form

$$\frac{1}{2} a_0^2 + \sum_{n=1}^{\infty} (a_n^2 + b_n^2) \leq \frac{1}{L} \int_{-L}^{L} f^2(x) \, dx.$$

The proof is like that just done for sine coefficients, but is notationally more complicated because of the presence of both sine and cosine coefficients and the constant term.

Using Bessel's inequality, we can establish a result about uniform convergence of a Fourier series.

Theorem 5 Let f be continuous on $[-L, L]$ and let f' be piecewise continuous. Suppose also that $f(-L) = f(L)$. Then the Fourier series of f on $[-L, L]$ converges absolutely and uniformly to $f(x)$ at each x in $[-L, L]$. ■

Proof: Denote the Fourier coefficients of f using lowercase letters, and those of f' using uppercase letters. We can relate these coefficients using integration by parts. First,

$$A_0 = \frac{1}{L} \int_{-L}^{L} f'(\xi) \, d\xi = \frac{1}{L} (f(L) - f(-L)) = 0.$$

Next,

$$B_n = \frac{1}{L} \int_{-L}^{L} f'(\xi) \sin\left(\frac{n\pi\xi}{L}\right) d\xi = \frac{1}{L} \left[f(\xi) \sin\left(\frac{n\pi\xi}{L}\right) \right]_0^{\pi}$$

$$- \frac{1}{L} \int_0^{\pi} f(\xi) \frac{n\pi}{L} \cos\left(\frac{n\pi\xi}{L}\right) d\xi = - \frac{n\pi}{L} a_n,$$

and by a similar integration,

$$A_n = \frac{n\pi}{L} b_n.$$

Now

$$0 \leq \left(|A_n| - \frac{1}{n} \right)^2 = A_n^2 - \frac{2}{n} |A_n| + \frac{1}{n^2}$$

and

$$0 \leq \left(|B_n| - \frac{1}{n} \right)^2 = B_n^2 - \frac{2}{n} |B_n| + \frac{1}{n^2}.$$

Therefore

$$\frac{1}{n} |A_n| + \frac{1}{n} |B_n| \leq \frac{1}{2} (A_n^2 + B_n^2) + \frac{1}{n^2}.$$

Therefore

$$|a_n| + |b_n| \leq \frac{L}{2\pi} (A_n^2 + B_n^2) + \frac{L}{\pi} \frac{1}{n^2}.$$

Now

$$\sum_{n=1}^{\infty} (A_n^2 + B_n^2)$$

converges by Bessel's inequality applied to the Fourier coefficients of f'. And $\sum_{n=1}^{\infty} 1/n^2$ converges. Therefore

$$\sum_{n=1}^{\infty} (|a_n| + |b_n|)$$

converges. But

$$\left| a_n \cos \left(\frac{n\pi x}{L} \right) + b_n \sin \left(\frac{n\pi x}{L} \right) \right| \leq |a_n| + |b_n|.$$

Therefore the Fourier series of f on $[-L, L]$ converges absolutely and, by a theorem of Weierstrass, uniformly. This completes the proof. ∎

EXERCISE 74   Let $f(x) = \begin{cases} 0 \text{ for } -\pi \leq x \leq 0 \\ \sin(x) \text{ for } 0 \leq x \leq \pi. \end{cases}$ Write the Fourier series of f. Show that this series converges uniformly and absolutely on $[-\pi, \pi]$. Let $F(x) = \int_{-\pi}^{x} f(s) \, ds$. By integrating the Fourier series of f term by term, obtain the Fourier series of F.

EXERCISE 75   Let $f(x) = x$ for $-\pi < x \le \pi$. By writing the Fourier series, show that

$$x = \sum_{n=1}^{\infty} \frac{2}{n} (-1)^{n+1} \sin(nx)$$

for $-\pi < x < \pi$. Now differentiate this Fourier series term by term. Does this differentiated series converge to $f'(x)$? What conclusion does this suggest about term by term differentiation of a Fourier series?

EXERCISE 76   Let $f(x) = |x|$ for $-L \le x \le L$. Show that the Fourier series of f on $[-L, L]$ converges absolutely and uniformly to $f(x)$ on this interval.

EXERCISE 77   Let $f(x) = e^{-x^2}$ for $-L \le x \le L$. Show that the Fourier series of f on $[-L, L]$ converges absolutely and uniformly to $f(x)$ on $[-L, L]$.

EXERCISE 78   Let $f(x) = x^2$ for $-\pi < x \le \pi$. Use the Fourier series of f and Bessel's inequality to show that

$$\sum_{n=1}^{\infty} \frac{1}{n^4} \le \frac{\pi^4}{90}.$$

EXERCISE 79   Let f be piecewise continuous on $[0, \pi]$. Show that

$$\frac{1}{2} a_0^2 + \sum_{n=1}^{\infty} a_n^2 \le \frac{2}{\pi} \int_0^{\pi} f^2(x) \, dx,$$

in which the $a_n's$ are the coefficients in the Fourier cosine expansion of f. This is Bessel's inequality for the coefficients in a cosine expansion.

EXERCISE 80   Use the result of Exercise 79 and the Fourier cosine expansion of x on $[0, \pi]$ to show that

$$\sum_{n=1}^{\infty} \frac{1}{(2n - 1)^4} \le \frac{\pi^4}{96}.$$

3.6   THE FOURIER INTEGRAL

Suppose f is piecewise smooth on every interval $[-L, L]$, and suppose $\int_{-\infty}^{\infty} |f(x)| \, dx$ converges. This condition on the integral of $|f|$ is often described by saying that f is absolutely integrable.

   If f is not periodic then we cannot write $f(x)$ as a Fourier series for all real x. In such a case we may be able to represent f by a Fourier integral, which we will now develop informally.

For any $L > 0$, we can write the Fourier series of f on $[-L, L]$:

$$\frac{1}{2} a_0 + \sum_{n=1}^{\infty} a_n \cos\left(\frac{n\pi x}{L}\right) + b_n \sin\left(\frac{n\pi x}{L}\right),$$

in which the $a_n's$ and $b_n's$ are the Fourier coefficients of f on $[-L, L]$. Insert the Fourier coefficients to write this series as:

$$\frac{1}{2L} \int_{-L}^{L} f(\xi)\, d\xi + \sum_{n=1}^{\infty} \left[\frac{1}{L} \int_{-L}^{L} f(\xi)\cos\left(\frac{n\pi\xi}{L}\right) \cos\left(\frac{n\pi x}{L}\right)\, d\xi\right.$$
$$\left. + \frac{1}{L} \int_{-L}^{L} f(\xi)\sin\left(\frac{n\pi\xi}{L}\right) \sin\left(\frac{n\pi x}{L}\right)\, d\xi\right].$$

We would like to allow $L \to \infty$ to obtain a representation over the entire real line. In order to determine this limit, let

$$\omega_n = \frac{n\pi}{L} \quad \text{and} \quad \Delta\omega = \omega_n - \omega_{n-1} = \frac{\pi}{L}.$$

Upon substituting these quantities, we obtain the expression

$$\frac{1}{2\pi} \left(\int_{-L}^{L} f(\xi)\, d\xi\right) \Delta\omega + \frac{1}{\pi} \sum_{n=1}^{\infty} \left[\left(\int_{-L}^{L} f(\xi)\cos(\omega_n\xi)\, d\xi\right) \cos(\omega_n x)\right.$$
$$\left. + \left(\int_{-L}^{L} f(\xi)\sin(\omega_n\xi)\, d\xi\right) \sin(\omega_n x)\right] \Delta\omega. \quad (3.18)$$

Now let $L \to \infty$. Then $\Delta\omega \to 0$ and

$$\frac{1}{2\pi} \left(\int_{-L}^{L} f(\xi)\, d\xi\right) \Delta\omega \to 0$$

because $\int_{-\infty}^{\infty} |f(x)|\, dx$ converges by assumption. The other terms in the expression 3.18 resemble a Riemann sum for a definite integral. On the basis of this expression, we conjecture that, in the limit as $L \to \infty$, expression 3.18 approaches the limit

$$\frac{1}{\pi} \int_{0}^{\infty} \left[\left(\int_{-\infty}^{\infty} f(\xi)\cos(\omega\xi)\, d\xi\right) \cos(\omega x)\right.$$
$$\left. + \left(\int_{-\infty}^{\infty} f(\xi)\sin(\omega\xi)\, d\xi\right) \sin(\omega x)\right] d\omega. \quad (3.19)$$

This is the Fourier integral representation of f, and it is possible to prove that, under the assumptions made about f, it is equal to

$$\frac{1}{2} (f(x+) + f(x-))$$

for each real x. This conclusion implies that, at any point where f is continuous, the Fourier integral 3.19 converges to $f(x)$.

We may therefore write this integral representation as

$$\frac{1}{2} (f(x+) + f(x-)) = \int_0^\infty [A_\omega \cos(\omega x) + B_\omega \sin(\omega x)] \, d\omega,$$

where

$$A_\omega = \frac{1}{\pi} \int_{-\infty}^\infty f(\xi) \cos(\omega \xi) \, d\xi$$

and

$$B_\omega = \frac{1}{\pi} \int_{-\infty}^\infty f(\xi) \sin(\omega \xi) \, d\xi.$$

These numbers are the *Fourier integral coefficients* of f.

By combining terms in equation 3.19 and using a trigonometric identity (similar to the derivation of equation 3.8), this Fourier integral expansion can be written as

$$\frac{1}{2} (f(x+) + f(x-)) = \frac{1}{\pi} \int_0^\infty \int_{-\infty}^\infty f(\xi) \cos(\omega(\xi - x)) \, d\xi \, d\omega. \qquad (3.20)$$

Example 20   Let $f(x) = \begin{cases} 0 \text{ if } |x| > 1 \\ 1 \text{ if } -1 \le x \le 1. \end{cases}$ Certainly f is piecewise continuous for all x and absolutely integrable.

The Fourier integral coefficients are

$$A_\omega = \frac{1}{\pi} \int_{-\infty}^\infty f(\xi) \cos(\omega \xi) \, d\xi = \frac{1}{\pi} \int_{-1}^1 \cos(\omega \xi) \, d\xi = \frac{2 \sin(\omega)}{\pi \omega}$$

and

$$B_\omega = \frac{1}{\pi} \int_{-\infty}^\infty f(\xi) \sin(\omega \xi) \, d\xi = 0$$

because f is an even function (so $f(\xi)\sin(\omega\xi)$ is odd). For $-\infty < x < \infty$,

$$\frac{1}{2}(f(x+) + f(x-)) = \frac{2}{\pi}\int_{-\infty}^{\infty}\frac{\sin(\omega)}{\omega}\cos(\omega x)\,d\omega.$$

More explicitly,

$$\frac{2}{\pi}\int_{-\infty}^{\infty}\frac{\sin(\omega)}{\omega}\cos(\omega x)\,d\omega = \begin{cases} 1 \text{ for } -1 < x < 1 \\ \dfrac{1}{2} \text{ for } x = \pm 1 \\ 0 \text{ for } |x| > 1. \end{cases}$$ ∎

EXERCISE 81 In each of problems 1 through 6, determine the Fourier integral representation of f, and what this representation converges to

1. $f(x) = \begin{cases} x \text{ for } -\pi \le x \le \pi \\ 0 \text{ for } |x| > \pi \end{cases}$

2. $f(x) = \begin{cases} \cos(x) \text{ for } -\alpha \le x \le \alpha \\ 0 \text{ for } |x| > \alpha, \end{cases}$ in which α is a positive constant

3. $f(x) = e^{-|x|}$

4. $f(x) = xe^{-|x|}$

5. $f(x) = \begin{cases} |x| \text{ for } -\alpha \le x \le \alpha \\ 0 \text{ for } |x| > \alpha, \end{cases}$ with α a positive constant

6. $f(x) = \begin{cases} k \text{ for } 0 < x \le \alpha \\ -k \text{ for } -\alpha \le x < 0 \\ 0 \text{ for } |x| > \alpha, \end{cases}$ with k and α positive constants

EXERCISE 82 Obtain the integral in equation 3.20 from the Fourier integral representation 3.19.

EXERCISE 83 Show that the Fourier integral representation of f can be written

$$\frac{1}{2\pi}\int_{-\infty}^{\infty}\int_{-\infty}^{\infty}e^{i\omega(x-\xi)}f(\xi)\,d\xi\,d\omega.$$

3.6.1 Fourier Cosine and Sine Integrals

Suppose f is piecewise smooth on $[0, \infty)$ and $\int_0^\infty |f(x)|\,dx$ converges. We can expand f in a Fourier sine or cosine integral by reasoning similar to that leading to Fourier sine and cosine series.

For the sine integral, extend f to an odd function g by putting

$$g(x) = \begin{cases} f(x) \text{ for } x \geq 0 \\ -f(-x) \text{ for } x < 0. \end{cases}$$

We can write the Fourier integral for g:

$$\frac{1}{2}(g(x+) + g(x-)) = \int_0^\infty [A_\omega \cos(\omega x) + B_\omega \sin(\omega x)]\, d\omega.$$

Because g is an odd function that agrees with f on $[0, \infty)$,

$$A_\omega = \frac{1}{\pi} \int_{-\infty}^\infty g(\xi)\cos(\omega\xi)\, d\xi = 0$$

and

$$B_\omega = \frac{1}{\pi} \int_{-\infty}^\infty g(\xi)\sin(\omega\xi)\, d\xi = \frac{2}{\pi} \int_0^\infty f(\xi)\sin(\omega\xi)\, d\xi.$$

The integral representation of g therefore contains only sine terms, and for $x \geq 0$ is a sine integral representation of f.

It summary, the *Fourier integral sine representation of f* on $[0, \infty)$ is

$$\int_0^\infty B_\omega \sin(\omega x)\, d\omega$$

where

$$B_\omega = \frac{2}{\pi} \int_0^\infty f(\xi)\sin(\omega\xi)\, d\xi.$$

Under the stated conditions on f, this sine integral converges to 0 at $x = 0$ and to

$$\frac{1}{2}(f(x+) + f(x-))$$

for $x > 0$.

By extending f to an even function on the real line, we obtain the *Fourier integral cosine representation*

$$\int_0^\infty A_\omega \cos(\omega x)\, d\omega$$

in which

$$A_\omega = \frac{2}{\pi} \int_0^\infty f(\xi)\cos(\omega\xi)\ d\xi.$$

This converges to

$$\frac{1}{2}\left(f(x+) + f(x-)\right)$$

for $x > 0$ and to $f(0+)$ at $x = 0$.

Example 21 Let $f(x) = e^{-kx}$ for $x \geq 0$, with k any positive number. Now

$$\int_0^\infty |f(x)|\ dx = \int_0^\infty e^{-kx}\ dx = \frac{1}{k}$$

so f is absolutely integrable on $[0, \infty)$.

For the sine integral expansion, compute

$$B_\omega = \frac{2}{\pi} \int_0^\infty e^{-k\xi} \sin(\omega\xi)\ d\xi = \frac{2}{\pi} \frac{\omega}{k^2 + \omega^2}.$$

For $x > 0$, the sine integral representation is

$$e^{-kx} = \frac{2}{\pi} \int_0^\infty \frac{\omega}{k^2 + \omega^2} \sin(\omega x)\ d\omega. \tag{3.21}$$

For the cosine integral expansion, compute

$$A_\omega = \frac{2}{\pi} \int_0^\infty e^{-k\xi} \cos(\omega\xi)\ d\xi = \frac{2}{\pi} \frac{k}{k^2 + \omega^2}.$$

Since $f(0+) = f(0)$, then for all $x \geq 0$,

$$e^{-kx} = \frac{2k}{\pi} \int_0^\infty \frac{1}{k^2 + \omega^2} \cos(\omega x)\ d\omega. \tag{3.22}$$

The cosine integral converges to e^{-kx} for $x \geq 0$, while the sine integral converges to e^{-kx} for $x > 0$. The integrals 3.21 and 3.22 are called *Laplace's integrals* because the coefficient A_ω in the cosine integral is $2/\pi$ times the Laplace transform of $\cos(\omega x)$, while B_ω in the sine integral is $2/\pi$ times the Laplace transform of $\sin(\omega x)$. ∎

EXERCISE 84   In each of problems 1 through 6, write the Fourier sine integral and the Fourier cosine integral representation of the function. Determine what each integral converges to.

1. $f(x) = \begin{cases} \sinh(x) \text{ for } 0 \leq x \leq k \\ \quad\quad 0 \text{ for } x > k \end{cases}$ with k any positive constant

2. $f(x) = \begin{cases} \cos(\pi x) \text{ for } 0 \leq x \leq 1 \\ \quad\quad 0 \text{ for } x > 1 \end{cases}$

3. $f(x) = e^{-x} \cos(x)$

4. $f(x) = xe^{-3x}$

5. $f(x) = \begin{cases} k \text{ for } 0 \leq x \leq \alpha \\ 0 \text{ for } x > \alpha, \end{cases}$ with k and α positive constants

6. $f(x) = \begin{cases} x^2 \text{ for } 0 \leq x \leq \alpha \\ 0 \text{ for } x > \alpha, \end{cases}$ with α a positive constant

EXERCISE 85   Use the Laplace integrals (Example 21) to compute the Fourier cosine integral of

$$f(x) = \frac{1}{1 + x^2}$$

and the Fourier sine integral of

$$g(x) = \frac{x}{1 + x^2}.$$

EXERCISE 86   Suppose that

$$\lim_{x \to \infty} f(x) = \lim_{x \to \infty} f'(x) = 0$$

and that $f''(x)$ exists for $x \geq 0$ and $f'(0) = 0$. Show that the integrand in the Fourier cosine integral representation of $f''(x)$ is equal to $-\omega^2$ times the integrand in the Fourier cosine integral representation of $f(x)$.

EXERCISE 87   Assume that

$$f(x) = \frac{1}{\pi} \int_0^\infty A(\omega)\cos(\omega x) \, d\omega$$

in which

$$A(\omega) = \int_{-\infty}^\infty f(t)\cos(\omega t) \, dt.$$

(a)   Show that

$$x^2 f(x) = \frac{1}{\pi} \int_0^\infty A^*(\omega)\cos(\omega x) \, d\omega,$$

in which $A^*(\omega) = -A''(\omega)$.

(b)   Show that

$$x f(x) = \frac{1}{\pi} \int_0^\infty B^*(\omega)\sin(\omega x) \, d\omega,$$

where $B^*(\omega) = -A'(\omega)$.

3.7   THE COMPLEX FOURIER INTEGRAL AND THE FOURIER TRANSFORM

In addition to Fourier series and integrals, we will use the Fourier transform. This will require that we first develop a complex version of the Fourier integral.

The Fourier integral can be written in complex form by using Euler's formula

$$e^{it} = \cos(t) + i \sin(t).$$

Replacing t with $-t$ we have

$$e^{-it} = \cos(t) - i \sin(t).$$

Solve these equations for $\cos(t)$ to obtain

$$\cos(t) = \frac{1}{2} (e^{it} + e^{-it}).$$

Now suppose f is absolutely integrable and piecewise smooth on the real line. From equation 3.20,

$$\frac{1}{2} (f(x+) + f(x-)) = \frac{1}{\pi} \int_0^\infty \int_{-\infty}^\infty f(\xi)\cos(\omega(\xi - x)) \, d\xi \, d\omega$$

for all x. Insert

$$\cos(\omega(\xi - x)) = \frac{1}{2} (e^{i\omega(\xi - x)} + e^{-i\omega(\xi - x)})$$

into this equation to obtain

$$\frac{1}{2}(f(x+) + f(x-)) = \frac{1}{\pi} \int_0^\infty \int_{-\infty}^\infty f(\xi) \frac{1}{2} (e^{i\omega(\xi-x)} + e^{-i\omega(\xi-x)}) \, d\xi \, d\omega$$

$$= \frac{1}{2\pi} \int_0^\infty \int_{-\infty}^\infty f(\xi) e^{i\omega(\xi-x)} \, d\xi \, d\omega$$

$$+ \frac{1}{2\pi} \int_0^\infty \int_{-\infty}^\infty f(\xi) e^{-i\omega(\xi-x)} \, d\xi \, d\omega. \qquad (3.23)$$

In the first integral on the right side of equation 3.23, put $\omega = -w$:

$$\int_0^\infty \int_{-\infty}^\infty f(\xi) e^{i\omega(\xi-x)} \, d\xi \, d\omega = \int_0^{-\infty} \int_{-\infty}^\infty f(\xi) e^{-iw(\xi-x)} \, d\xi (-1) \, dw$$

$$= \int_{-\infty}^0 \int_{-\infty}^\infty f(\xi) e^{-iw(\xi-x)} \, d\xi \, dw.$$

Substitute this into equation 3.23, but replace the variable of integration w with ω to have the same variable of integration in both integrals. We obtain

$$\frac{1}{2}(f(x+) + f(x-)) = \frac{1}{2\pi} \int_{-\infty}^0 \int_{-\infty}^\infty f(\xi) e^{-i\omega(\xi-x)} \, d\xi \, d\omega + \frac{1}{2\pi} \int_0^\infty \int_{-\infty}^\infty f(\xi) e^{-i\omega(\xi-x)} \, d\xi \, d\omega.$$

Finally, combine the integrals with respect to ω to write

$$\frac{1}{2}(f(x+) + f(x-)) = \frac{1}{2\pi} \int_{-\infty}^\infty \int_{-\infty}^\infty f(\xi) e^{-i\omega(\xi-x)} \, d\xi \, d\omega. \qquad (3.24)$$

This is the complex Fourier integral representation of f.

Thus far we have simply written the Fourier integral in complex exponential form. The integral on the right side of equation 3.24 converges to the quantity on the left, under conditions stated for the convergence of the Fourier integral.

Now write equation 3.24 as

$$\frac{1}{2}(f(x+) + f(x-)) = \frac{1}{2\pi} \int_{-\infty}^\infty \left(\int_{-\infty}^\infty f(\xi) e^{-i\omega\xi} \, d\xi \right) e^{i\omega x} \, d\omega. \qquad (3.25)$$

Define

$$\mathfrak{F}[f](\omega) = \int_{-\infty}^\infty f(\xi) e^{-i\omega\xi} \, d\xi. \qquad (3.26)$$

$\mathfrak{F}[f]$ is the *Fourier transform* of f. This is a function of ω, and is defined for all ω such that this integral converges.

As we might expect with a transform defined by an integral, the Fourier transform is linear. This means that, if f and g have Fourier transforms, and α and β are real numbers, then

$$\mathfrak{F}[\alpha f + \beta g] = \alpha \mathfrak{F}[f] + \beta \mathfrak{F}[g].$$

It is often convenient to use the alternative notation of denoting the Fourier transform of a function by a carat over the function:

$$\mathfrak{F}[f] = \hat{f}.$$

In this notation,

$$\mathfrak{F}[f](\omega) = \hat{f}(\omega),$$

a simplification that is useful in carrying out calculations.

Substituting this notation into equation 3.25 yields

$$\frac{1}{2}\,(f(x+) + f(x-)) = \frac{1}{2\pi} \int_{-\infty}^{\infty} \hat{f}(\omega)e^{i\omega x}\,d\omega. \qquad (3.27)$$

This is the *inverse Fourier Transform* of f. Equations 3.26 and 3.27 define a *Fourier transform pair*. Given f satisfying certain conditions, we can in theory compute its Fourier transform $\mathfrak{F}[f]$ by integration (equation 3.26). Equation 3.27 enables us to recover a function from its Fourier transform.

This recovery process has some ambiguity, since there are many functions having the same Fourier transform. For example, if we change the value of $f(x)$ at finitely many points to obtain a new function, both functions will have the same Fourier transform. However, if we begin with a function f, compute its Fourier transform $\hat{f}$, and then apply equation 3.27 to $\hat{f}$, we will recover $f(x)$ at each x where f is continuous. More precisely, we will state the following result without proof.

Theorem 6   Let f and g be real or complex-valued functions which are continuous on the real line, and suppose that f and g are absolutely integrable. Suppose $\hat{f} = \hat{g}$. Then $f = g$.   ∎

In practice an inverse Fourier transform is often computed using a table and properties of the transform, or a computer routine.

Example 22   Let $f(x) = \begin{cases} 0 & \text{for } x < 0 \\ e^{-ax} & \text{for } x \geq 0, \end{cases}$ in which a is a positive number. The Fourier transform of f is

$$\hat{f}(\omega) = \int_{-\infty}^{\infty} f(\xi) e^{-i\omega\xi} \, d\xi = \int_{0}^{\infty} e^{-a\xi} e^{-i\omega\xi} \, d\xi = \int_{0}^{\infty} e^{-(a+i\omega)\xi} \, d\xi$$

$$= \lim_{r \to \infty} \int_{0}^{r} e^{-(a+i\omega)\xi} \, d\xi = \lim_{r \to \infty} \frac{1}{a + i\omega} [1 - e^{-(a+i\omega)r}] = \frac{1}{a + i\omega}.$$

In this limit,

$$e^{-(a+i\omega)r} = e^{-ar}[\cos(\omega r) - i \sin(\omega r)] \to 0$$

as $r \to \infty$ because $a > 0$. $\qquad\blacksquare$

The Fourier transform of a function is often a complex-valued function, as occurs in this example. Using the transform notation $\mathfrak{F}$, we can write the conclusion of the preceding example as

$$\mathfrak{F}[f](\omega) = \frac{1}{a + i\omega}.$$

If we denote the inverse Fourier transform by $\mathfrak{F}^{-1}$, then

$$\mathfrak{F}^{-1}\left[\frac{1}{a + i\omega}\right](x) = f(x),$$

where f is the function defined in Example 22.

When the Fourier transform is used to solve differential equations, we often exploit the following operational property. It states that under certain conditions, the Fourier transform of the nth derivative of f is the nth power of $i\omega$, multiplied by the Fourier transform of f.

Theorem 7 (Operational Rule) Let n be a positive integer. Suppose the nth derivative $f^{(n)}$ is piecewise continuous on the real line, and $\int_{-\infty}^{\infty} |f^{(n-1)}(x)| \, dx$ converges. Assume also that

$$\lim_{x \to \infty} f^{(k)}(x) = \lim_{x \to -\alpha} f^{(k)}(x) = 0$$

for $k = 0, 1, \ldots, n - 1$. Then

$$\mathfrak{F}[f^{(n)}](\omega) = (i\omega)^n \hat{f}(\omega). \quad\blacksquare \qquad\qquad (3.28)$$

In the notation of the theorem, $f^{(0)}$ (the "zeroeth" derivative) is just f itself. This notation makes the statement of theorems such as this one more efficient.

EXERCISE 88   Prove Theorem 7. *Hint*: Use integration by parts.

We conclude this section with a technical lemma which we will use later.

Lemma 4

1. Let $g(x) = e^{-x^2/2}$. Then

$$\hat{g}(\omega) = \sqrt{2\pi}\, e^{-\omega^2/2}.$$

2. Let f be continuous on the real line and suppose $\int_{-\infty}^{\infty} |f(x)|\, dx$ converges. Let k be a positive number. For real x, denote $f_k(x) = kf(kx)$. Then

$$\hat{f_k}(\omega) = \hat{f}(\omega/k).$$

3. Let f be continuous on the real line and suppose $\int_{-\infty}^{\infty} |f(x|\, dx$ converges. For a given real number t, define $g(x) = f(x - t)$. Then

$$\hat{g}(\omega) = e^{-i\omega t}\hat{f}(\omega). \qquad \blacksquare$$

Proof of (1): With x instead of ξ as integration variable, begin by writing

$$\hat{g}(\omega) = \int_{-\infty}^{\infty} e^{-x^2/2}e^{-i\omega x}\, dx = \int_{-\infty}^{\infty} e^{-(x+i\omega)^2/2-\omega^2/2}\, dx = e^{-\omega^2/2} \int_{-\infty}^{\infty} e^{-(x+i\omega)^2/2}\, dx.$$

We will evaluate the last integral by Cauchy's Theorem from complex analysis. A reader who is unfamiliar with complex integration can skip this derivation. We will assume for specificity that $\omega > 0$, but the argument is easily adapted to the case that $\omega < 0$.

Define $h(z) = e^{-z^2/2}$. This function is analytic on the entire complex plane. Let C be the simple closed curve shown in Figure 3.11, with r and R positive numbers. By Cauchy's Theorem,

$$\oint_C h(z)\, dz = 0.$$

Thus

$$\sum_{j=1}^{4} \int_{C_j} e^{-z^2/2}\, dz = 0. \qquad (3.29)$$

Consider these integrals individually. First,

$$\int_{C_1} e^{-z^2/2}\, dz = \int_{-r}^{R} e^{-x^2/2}\, dx.$$

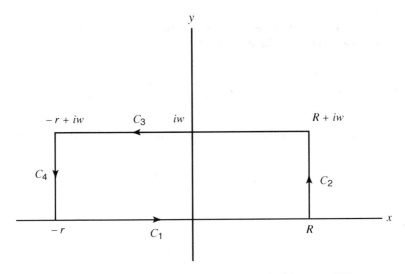

FIGURE 3.11 Contour used in the proof of Lemma 4(1).

As $r \to \infty$ and $R \to \infty$,

$$\int_{C_1} e^{-z^2/2} \, dz \to \int_{-\infty}^{\infty} e^{-x^2/2} \, dx = \sqrt{2\pi},$$

assuming familiarity with this standard result.

On C_2, $z = R + it$, with t varying from 0 to ω. Therefore

$$\left| \int_{C_2} e^{-z^2/2} \, dz \right| \leq (\text{length of } C_2) \max_{z \text{ on } C_2} |g(z)| = \omega \max_{0 \leq t \leq \omega} \left| e^{-(R+it)^2/2} \right|$$

$$= \omega \max_{0 \leq t \leq \omega} \left| e^{(-R^2+t^2)/2} e^{-iRt} \right| = \omega e^{(-R^2+\omega^2)/2} \to 0 \text{ as } R \to \infty.$$

On C_4, $z = -r + it$ with t varying from 0 to ω. A similar analysis to that just done for C_2 gives

$$\left| \int_{C_4} e^{-z^2/2} \, dz \right| \to 0 \text{ as } r \to \infty.$$

Finally, on C_3, $z = x + i\omega$, with x varying from R to $-r$ to preserve the counterclockwise orientation on C. Thus,

$$\int_{C_4} e^{-z^2/2} \, dz = \int_R^{-r} e^{-(x+i\omega)^2/2} \, dx \rightarrow \int_\infty^{-\infty} e^{-(x+i\omega)^2/2} \, dx = -\int_{-\infty}^\infty e^{-(x+i\omega)^2/2} \, dx$$

as $r \rightarrow \infty$ and $R \rightarrow \infty$. In the limit as $r \rightarrow \infty$ and $R \rightarrow \infty$, equation 3.29 yields

$$-\int_{-\infty}^\infty e^{-(x+i\omega)^2/2} \, dx + \sqrt{2\pi} = 0.$$

Therefore

$$\hat{g}(\omega) = \sqrt{2\pi} e^{-\omega^2/2},$$

completing the proof of (1).

Proof of (2): This conclusion is proved by a simple change of variable. First,

$$\hat{f_k}(\omega) = \int_{-\infty}^\infty f_k(\xi) e^{-i\omega\xi} \, d\xi = \int_{-\infty}^\infty kf(k\xi) e^{-i\omega\xi} \, d\xi.$$

Upon letting $z = k\xi$ we obtain

$$\hat{f_k}(\omega) = \int_{-\infty}^\infty kf(z) e^{-i\omega z/k} \frac{1}{k} \, dz = \int_{-\infty}^\infty f(z) e^{-i\omega z/k} \, dz = \hat{f}(\omega/k).$$

Proof of (3): Like (2), this is also a simple change of variable. Compute

$$\hat{g}(\omega) = \int_{-\infty}^\infty g(\xi) e^{-i\omega\xi} \, d\xi = \int_{-\infty}^\infty f(\xi - t) e^{-i\omega\xi} \, d\xi.$$

Let $z = \xi - t$ to obtain

$$\hat{g}(\omega) = \int_{-\infty}^\infty f(z) e^{-i\omega(z+t)} \, dz = e^{-i\omega t} \int_{-\infty}^\infty f(z) e^{-i\omega z} \, dz = e^{-i\omega t} \hat{f}(\omega). \qquad \blacksquare$$

Conclusion (3) is called the time shifting property of the Fourier transform.

EXERCISE 89   In problems 1 through 6, find the complex Fourier integral representation of the function.

1. $f(x) = \begin{cases} 1 \text{ for } -\pi \le x \le \pi \\ \quad 0 \text{ for } |x| > \pi \end{cases}$

2. $f(x) = \begin{cases} \cos(x) \text{ for } -\alpha \le x \le \alpha \\ \quad\quad 0 \text{ for } |x| > \alpha \end{cases}$

3. $f(x) = e^{-|x|}$

4. $f(x) = xe^{-|x|}$

5. $f(x) = \begin{cases} |x| \text{ for } -\alpha \le x \le \alpha \\ \quad 0 \text{ for } |x| > \alpha \end{cases}$

6. $f(x) = \begin{cases} \quad k \text{ for } 0 < x \le \alpha \\ -k \text{ for } -\alpha \le x < 0 \\ \quad 0 \text{ for } |x| > \alpha \end{cases}$

EXERCISE 90   In each of problems 1 through 6, find the Fourier transform of f.

1. $f(x) = \begin{cases} k \text{ for } -\alpha \le x \le \alpha \\ 0 \text{ for } |x| > \alpha \end{cases}$

2. $f(x) = \begin{cases} \quad k \text{ for } a \le x \le b \\ 0 \text{ for } x < a \text{ and for } x > b \end{cases}$

3. $f(x) = \begin{cases} \sin(x) \text{ for } -\alpha \le x \le \alpha \\ \quad 0 \text{ for } |x| > \alpha \end{cases}$

4. $f(x) = e^{-|x|}$

5. $f(x) = e^{-|x|} \cos(x)$

6. $f(x) = \begin{cases} e^{-x} \sin(x) \text{ for } x \ge 0 \\ \quad\quad 0 \text{ for } x < 0 \end{cases}$

EXERCISE 91   The amplitude spectrum of f is a graph of the magnitude of its Fourier transform. Generate the amplitude spectrum of each of the functions in problems 1 through 6 of Exercise 90.

In each of the following exercises, assume that f has a Fourier transform.

EXERCISE 92   Prove that

$$\mathfrak{F}[e^{ikx} f(x)](\omega) = \hat{f}(\omega - k)$$

for any real number k. This result is called frequency shifting.

EXERCISE 93   Let k be any nonzero real number. Prove that

$$\mathfrak{F}[f(kx)](\omega) = \frac{1}{|k|} \hat{f}\left(\frac{\omega}{k}\right).$$

This result is called scaling.

EXERCISE 94   Prove that

$$\mathfrak{F}[f(-x)](\omega) = \hat{f}(-\omega).$$

This is called time reversal.

EXERCISE 95   Prove that

$$\mathfrak{F}[\hat{f}(x)](\omega) = 2\pi f(-\omega).$$

This is called symmetry of the Fourier transform.

EXERCISE 96   Let k be any real number. Prove that

$$\mathfrak{F}[f(x)\cos(kx)](\omega) = \frac{1}{2} [\hat{f}(\omega + k) + \hat{f}(\omega - k)].$$

This result is called modulation.

EXERCISE 97   Write $\hat{f}(\omega) = A(\omega) + iB(\omega)$. Show that

$$A(\omega) = \int_{-\infty}^{\infty} f(\xi)\cos(\omega\xi) \, d\xi$$

and

$$B(\omega) = -\int_{-\infty}^{\infty} f(\xi)\sin(\omega\xi) \, d\xi.$$

Relate these conclusions to the Fourier integral coefficients of f.

EXERCISE 98   Let a be a positive number. Show that

$$\mathfrak{F}[e^{-at^2}](\omega) = \sqrt{\frac{\pi}{a}} \, e^{-\omega^2/4a}.$$

Hint: Write the integral defining $\hat{f}(\omega)$, with $f(t) = e^{-at^2}$. Compute $d/d\omega \, \hat{f}(\omega)$ by differentiating under the integral sign and integrate by parts to show that

$$\frac{d}{d\omega}\hat{f}(\omega) = -\frac{\omega}{2a}\hat{f}(\omega).$$

Solve this ordinary differential equation for $\hat{f}(\omega)$ and evaluate the constant by using the standard integral

$$\int_{-\infty}^{\infty} e^{-x^2}\, dx = \sqrt{\pi}.$$

EXERCISE 99   Let n be a positive integer. Let f be piecewise continuous on $[-L, L]$ for every positive L and suppose that $\int_{-\infty}^{\infty} |t^n f(t)|\, dt$ converges. Prove that

$$\mathfrak{F}[t^n f(t)](\omega) = i^n \frac{d^n}{d\omega^n}\hat{f}(\omega).$$

EXERCISE 100   Use the results of Exercises 98 and 99 to derive the formula

$$\mathfrak{F}[te^{-at^2}](\omega) = -\frac{i\omega}{2a}\sqrt{\frac{\pi}{a}}\, e^{-\omega^2/4a}.$$

EXERCISE 101   Prove that $\hat{f}(-\omega) = \overline{\hat{f}(\omega)}$, where the bar indicates complex conjugation.

EXERCISE 102   Let f be piecewise continuous on every interval $[-L, L]$. Suppose $\int_{-\infty}^{\infty} |f(t)|\, dt$ converges. Suppose also that $\hat{f}(0) = 0$. Prove that

$$\mathfrak{F}\left[\int_{-\infty}^{t} f(\tau)\, d\tau\right](\omega) = \frac{1}{i\omega}\hat{f}(\omega).$$

EXERCISE 103   Prove Parseval's Theorem. Let f be real-valued, continuous and piecewise smooth on the real line. Assume that f and f^2 are absolutely integrable. Then

$$\int_{-\infty}^{\infty} f^2(x)\, dx = \frac{1}{2\pi}\int_{-\infty}^{\infty} |\hat{f}(\omega)|^2\, d\omega.$$

Hint: Use equation 3.27 to write $f(x) = (1/2\pi)\int_{-\infty}^{\infty} \hat{f}(\omega)e^{i\omega x}\, d\omega$. Multiply both sides of this equation by $f(x)$ and integrate both sides of the resulting equation with respect to x from $-\infty$ to ∞, interchanging the order of integration in the double integral on the right.

EXERCISE 104 Prove Plancherel's Theorem. Let f and g be real-valued, continuous, and piecewise smooth on the real line, and suppose f, f^2, g, and g^2 are absolutely integrable. Then

$$\int_{-\infty}^{\infty} f(x)g(x)\, dx = \frac{1}{2\pi} \int_{-\infty}^{\infty} \hat{f}(\omega)\overline{\hat{g}(\omega)}\, d\omega.$$

3.8 CONVOLUTION

The convolution operation plays an important role in applications of the Fourier transform to partial differential equations. Let f and g be real or complex-valued functions defined on the real line. We will say that f *has a convolution with* g if

1. $\int_a^b f(x)\, dx$ and $\int_a^b g(x)\, dx$ exist for every interval $[a, b]$, and
2. for every real number t,

$$\int_{-\infty}^{\infty} |f(t - x)g(x)|\, dx$$

 converges.

In this event, the *convolution $f * g$ of f with g* is the function defined by

$$f * g(t) = \int_{-\infty}^{\infty} f(t - x)g(x)\, dx$$

for every real t. Here $f * g(t)$ denotes the value of the function $f * g$ at t. Some authors emphasize this by writing $f * g(t)$ as $(f * g)(t)$.

We will show that, if f has a convolution with g, then g has a convolution with f, and $f * g = g * f$. Thus the convolution operation $*$ is commutative. Further, conclusion (2) of Theorem 8 states that convolution is a linear operation.

Theorem 8 Suppose f has a convolution with g. Then

1. g has a convolution with f, and

$$f * g = g * f.$$

2. If f and g both have convolutions with h, and α and β are numbers, then $\alpha f + \beta g$ also has a convolution with h, and

$$(\alpha f + \beta g) * h = \alpha(f * h) + \beta(g * h). \qquad \blacksquare$$

Proof: Let $z = t - x$ to write, for any positive r and R, and any real t,

$$\int_{-r}^{R} f(t - x)g(x) \, dx = \int_{t+r}^{t-R} f(z)g(t - z)(-1) \, dz$$

$$= \int_{t-R}^{t+r} g(t - z)f(z) \, dz \rightarrow \int_{-\infty}^{\infty} g(t - z)f(z) \, dz$$

as $r \rightarrow \infty$ and $R \rightarrow \infty$. But in this limit,

$$\int_{-r}^{R} f(t - x)g(x) \, dx \rightarrow f * g(t).$$

Thus

$$f * g(t) = \int_{-\infty}^{\infty} f(t - x)g(x) \, dx = \int_{-\infty}^{\infty} g(t - z)f(z) \, dz = g * f(t).$$

The linearity, conclusion (2), follows from elementary properties of integrals, given that the improper integrals involved converge. $\qquad \blacksquare$

The next two theorems develop some additional properties of $f * g$.

Theorem 9 Suppose f and g are integrable on each interval $[a, b]$ and that $\int_{-\infty}^{\infty} |g(x)| \, dx$ converges.

1. If f is bounded on the real line, then $f * g$ exists and is a bounded function on the real line.
2. If f is bounded and continuous on the real line, then $f * g$ is bounded and continuous.
3. If f is bounded and uniformly continuous on the real line, then $f * g$ is uniformly continuous. $\qquad \blacksquare$

Proof: For (1), suppose that $|f(x)| \leq M$ for all real x. Then, for all real x and t,

$$|f(t - x)g(x)| \leq M|g(x)|.$$

Since $\int_{-\infty}^{\infty} |g(x)|\, dx$ converges, so does $\int_{-\infty}^{\infty} |f(t-x)g(x)|\, dx$. Then f has a convolution with g. Further, for any real t,

$$|f * g(t)| = \left| \int_{-\infty}^{\infty} f(t-x)g(x)\, dx \right| \leq \int_{-\infty}^{\infty} |f(t-x)g(x)|\, dx \leq M \int_{-\infty}^{\infty} |g(x)|\, dx,$$

hence $f * g$ is bounded on the entire real line.

To prove (2), suppose that f is bounded and continuous. As before, suppose $|f(x)| \leq M$ for all real x. By (1), $f * g$ is bounded. To prove continuity, we first need an estimate. Let t and s be real numbers. Then

$$|f * g(t) - f * g(s)| = \left| \int_{-\infty}^{\infty} f(t-x)g(x)\, dx - \int_{-\infty}^{\infty} f(s-x)g(x)\, dx \right|$$

$$= \left| \int_{-\infty}^{\infty} (f(t-x) - f(s-x))g(x)\, dx \right|$$

$$\leq \int_{-\infty}^{\infty} |f(t-x) - f(s-x)||g(x)|\, dx$$

$$\leq \int_{-R}^{R} |f(t-x) - f(s-x)||g(x)|\, dx$$

$$+ \int_{R}^{\infty} |f(t-x) - f(s-x)||g(x)|\, dx$$

$$+ \int_{-\infty}^{-R} |f(t-x) - f(s-x)||g(x)|\, dx.$$

$$\leq \sup_{-R \leq x \leq R} |f(t-x) - f(s-x)| \int_{-R}^{R} |g(x)|\, dx$$

$$+ 2M \int_{R}^{\infty} |g(x)|\, dx + 2M \int_{-\infty}^{-R} |g(x)|\, dx$$

$$\leq \sup_{-R \leq x \leq R} |f(t-x) - f(s-x)| \int_{-\infty}^{\infty} |g(x)|\, dx$$

$$+ 2M \int_{R}^{\infty} |g(x)|\, dx + 2M \int_{-\infty}^{-R} |g(x)|\, dx.$$

Denote $\int_{-\infty}^{\infty} |g(x)| \, dx = I$. In summary, we now have, for any $R > 0$, and any real numbers t and s,

$$|f * g(t) - f * g(s)| \leq I \left(\sup_{-R \leq x \leq R} |f(t - x) - f(s - x)| \right)$$

$$+ 2M \left(\int_{-\infty}^{-R} |g(x)| \, dx + \int_{R}^{\infty} |g(x)| \, dx \right). \quad (3.30)$$

We will now show that $f * g$ is continuous at t, using estimate 3.30 to show that we can make $f * g(s)$ as close as we like to $f * g(t)$ by choosing s sufficiently close to t. To this end, let $\epsilon > 0$. Since $\int_{-\infty}^{\infty} |g(x)| \, dx$ converges, we can produce a positive number R such that

$$\int_{-\infty}^{-R} |g(x)| \, dx + \int_{R}^{\infty} |g(x)| \, dx < \frac{\epsilon}{4M + 1}.$$

Next, f is continuous on $[t - 1 - R, t + 1 + R]$ and hence is uniformly continuous on this interval. Thus we can find a positive δ such that, for α and β in this interval and $|\alpha - \beta| < \delta$,

$$|f(\alpha) - f(\beta)| < \frac{\epsilon}{2I + 1}.$$

The way we defined this interval, if now $-R \leq x \leq R$ and $|t - s| < \delta$, then

$$|f(t - x) - f(s - x)| < \frac{\epsilon}{2I + 1}.$$

Thus, for $|t - s| < \delta$, inequality 3.30 yields

$$|f * g(t) - f(g(s))| < I \left(\frac{\epsilon}{2I + 1} \right) + 2M \left(\frac{\epsilon}{4M + 1} \right) < \frac{\epsilon}{2} + \frac{\epsilon}{2} = \epsilon$$

and hence $f * g$ is continuous at t.

If f is bounded and uniformly continuous, a modification of this continuity argument shows that $f * g$ is uniformly continuous on the real line. ∎

EXERCISE 105   Complete the proof of Theorem 9 by showing that, under the conditions of the theorem, $f * g$ is uniformly continuous if f is bounded and uniformly continuous on the real line.

Theorem 10   Suppose f and g are bounded and continuous on the real line, and that both $\int_{-\infty}^{\infty} |f(x)| \, dx$ and $\int_{-\infty}^{\infty} |g(x)| \, dx$ converge. Then

1.

$$\int_{-\infty}^{\infty} f * g(t) \, dt = \int_{-\infty}^{\infty} f(t) \, dt \int_{-\infty}^{\infty} g(t) \, dt.$$

2. For any real ω,

$$\widehat{f * g}(\omega) = \hat{f}(\omega)\hat{g}(\omega). \qquad \blacksquare$$

Conclusion (2) states that the Fourier transform of the convolution of f and g is equal to the product of the transforms of f and g. This equation is often known as the *convolution theorem*, and it is a fundamental result which we will use in solving partial differential equations.

We will not provide a complete proof of (1), but here is an outline of the argument. Write

$$\int_{-\infty}^{\infty} f * g(t) \, dt = \int_{-\infty}^{\infty} \left(\int_{-\infty}^{\infty} f(t - x)g(x) \, dx \right) dt = \int_{-\infty}^{\infty} \left(\int_{-\infty}^{\infty} f(t - x)g(x) \, dt \right) dx$$

$$= \int_{-\infty}^{\infty} \left(\int_{-\infty}^{\infty} f(t - x) \, dt \right) g(x) \, dx.$$

Now, for any real x,

$$\int_{-\infty}^{\infty} f(t - x) \, dt = \int_{-\infty}^{\infty} f(t) \, dt.$$

Thus

$$\int_{-\infty}^{\infty} f * g(t) \, dt = \int_{-\infty}^{\infty} \left(\int_{-\infty}^{\infty} f(t) \, dt \right) g(x) \, dx = \int_{-\infty}^{\infty} f(t) \, dt \int_{-\infty}^{\infty} g(x) \, dx$$

$$= \int_{-\infty}^{\infty} f(t) \, dt \int_{-\infty}^{\infty} g(t) \, dt.$$

This argument may seem plausible enough, but a careful reader should have the feeling that something tricky happened somewhere in the middle. This intuition is correct. The crux of the argument, a change in the order of integration, is not valid without justification. We will not pursue the proof of a result on change of order of integration. However, the hypotheses of the theorem are sufficient to justify this interchange.

Using conclusion (1) of the theorem, we can prove the convolution theorem.

Proof of (2): Let

$$F(t) = e^{-i\omega t}f(t), \ G(t) = e^{-i\omega t}g(t)$$

for real t and ω. Then

$$\widehat{f * g}(\omega) = \int_{-\infty}^{\infty} f * g(t)e^{-i\omega t} \, dt = \int_{-\infty}^{\infty} \left(\int_{-\infty}^{\infty} f(t - x)g(x) \, dx \right) e^{-i\omega t} \, dt$$

$$= \int_{-\infty}^{\infty} \left(\int_{-\infty}^{\infty} e^{-i\omega t}f(t - x)g(x) \, dx \right) dt$$

$$= \int_{-\infty}^{\infty} \left(\int_{-\infty}^{\infty} e^{-i\omega(t-x)}f(t - x)e^{-i\omega x}g(x) \, dx \right) dt.$$

Upon recognizing that the integral within the large parentheses is the convolution of F with G, we now have, by application of part (1) of this theorem,

$$\widehat{f * g}(\omega) = \int_{-\infty}^{\infty} F * G(t) \, dt = \int_{-\infty}^{\infty} F(t) \, dt \int_{-\infty}^{\infty} G(t) \, dt$$

$$= \int_{-\infty}^{\infty} f(t)e^{-i\omega t} \, dt \int_{-\infty}^{\infty} g(t)e^{-i\omega t} \, dt = \hat{f}(\omega)\hat{g}(\omega). \qquad \blacksquare$$

EXERCISE 106   Let $f(t) = e^{-t}$ and

$$g(t) = \begin{cases} t \text{ for } |t| \leq 1 \\ 0 \text{ for } |t| > 1. \end{cases}$$

Compute $f * g$.

EXERCISE 107   Let k be a positive number. Define $f(t) = e^{-|t|}$ and $g(t) = e^{-k|t|}$. Determine $f * g$.

EXERCISE 108   Let k be a positive number. Define

$$f(t) = \begin{cases} t \text{ for } |t| \leq k \\ 0 \text{ for } |t| > k. \end{cases}$$

Determine $f * f$.

EXERCISE 109 Suppose f and g are bounded and continuous on the real line and that $\int_{-\infty}^{\infty} |f(t)|\, dt$ and $\int_{-\infty}^{\infty} |g(t)|\, dt$ converge. Prove that

$$\int_{-\infty}^{\infty} |f * g(t)|\, dt \le \int_{-\infty}^{\infty} |f(t)|\, dt \int_{-\infty}^{\infty} |g(t)|\, dt.$$

EXERCISE 110 Let f have a convolution with g. Define the support of a function to be the closure of the set of points at which the function does not vanish:

$$S(F) = \overline{\{x \mid F(x) \ne 0\}}.$$

Prove that

$$S(f * g) \subset \{x + y \mid x \in S(f) \text{ and } y \in S(g)\}.$$

EXERCISE 111 Suppose $f * g, f' * g$ and $f * g'$ all exist. Prove that

$$(f * g)' = f' * g + f' * g'.$$

EXERCISE 112 (Convolution and the Dirac Delta Function). The Heaviside function H is defined by

$$H(t) = \begin{cases} t \text{ for } t \ge 0 \\ 0 \text{ for } t < 0. \end{cases}$$

In terms of H, the Dirac delta function is defined by

$$\delta(t) = \lim_{\epsilon \to 0+} \frac{1}{2\epsilon} [H(t + \epsilon) - H(t - \epsilon)].$$

Strictly speaking, this is not a function, but is an object called a distribution. It is instructive to attempt a graph of $\delta(t)$, noting that for $\epsilon > 0$,

$$H(t + \epsilon) - H(t - \epsilon) = \begin{cases} 1 \text{ for } -\epsilon \le t < \epsilon \\ 0 \text{ for } t \ge \epsilon \text{ and for } t < -\epsilon. \end{cases}$$

Assuming that the limit can be interchanged with the operation of taking the Fourier transform, define

$$\hat{\delta}(\omega) = \lim_{\epsilon \to 0} \frac{1}{2\epsilon} [\hat{H}(t + \epsilon) - \hat{H}(t - \epsilon)].$$

(a)   Show that

$$\mathfrak{F}[H(t + \epsilon) - H(t - \epsilon)](\omega) = 2 \, \frac{\sin(\epsilon\omega)}{\omega}.$$

(b)   Conclude from (a) and the definition of $\hat{\delta}(\omega)$ that

$$\hat{\delta}(\omega) \equiv 1.$$

(c)   Show that, if f has a convolution with δ, then

$$f * \delta = \delta * f = f.$$

This means that the Dirac delta function behaves like a group identity element under the convolution operation.

3.9   FOURIER SINE AND COSINE TRANSFORMS

We have seen how the Fourier integral suggests the Fourier transform. In a similar way, Fourier sine and cosine integrals suggest the Fourier sine and cosine transforms.

Suppose f is piecewise smooth and absolutely integrable on $[0, \infty)$. The *Fourier sine transform* of f is denoted $\mathfrak{F}_S[f]$ and is defined by

$$\mathfrak{F}_S[f](\omega) = \int_0^\infty f(x)\sin(\omega x) \, dx. \qquad (3.31)$$

This is a function of ω.

The Fourier sine transform of f is often denoted $\hat{f}_S$.

To see how the definition arises from the Fourier sine integral, recall that the sine integral representation of f is

$$\frac{1}{2} \left(f(x+) + f(x-) \right) = \int_0^\infty B_\omega \sin(\omega x) \, d\omega$$

where

$$B_\omega = \frac{2}{\pi} \int_0^\infty f(\xi)\sin(\omega\xi) \, d\xi.$$

Now recognize that

$$B_\omega = \frac{2}{\pi} \hat{f}_S(\omega)$$

so

$$\frac{1}{2}(f(x+) + f(x-)) = \frac{2}{\pi} \int_0^\infty \hat{f}_s(\omega)\sin(\omega x) \, d\omega. \qquad (3.32)$$

Equation 3.32 defines the *inverse Fourier sine transform*, and it recovers f from the sine transform of f, in the sense of returning $f(x)$ where f is continuous, and the average of left and right limits where f has a jump discontinuity.

Equations 3.31 and 3.32 form a *Fourier sine transform pair*. The first equation gives the Fourier sine transform of f, and the second enables us to recover f from its sine transform (in the sense of recovering a function that agrees with f wherever f is continuous).

In a similar way, we can define the *Fourier cosine transform* by

$$\mathfrak{F}_C[f](\omega) = \int_0^\infty f(x)\cos(\omega x) \, dx. \qquad (3.33)$$

The cosine transform of $f(x)$ is also denoted $\hat{f}_C(\omega)$.

This definition arises from the Fourier cosine integral. The Fourier cosine integral representation of f is

$$\frac{1}{2}(f(x+) + f(x-)) = \int_0^\infty A_\omega \cos(\omega x) \, d\omega$$

in which

$$A_\omega = \frac{2}{\pi} \int_0^\infty f(x)\cos(\omega x) \, dx.$$

Since

$$A_\omega = \frac{2}{\pi} \hat{f}_C(\omega)$$

then

$$\frac{1}{2}(f(x+) + f(x-)) = \frac{2}{\pi} \int_0^\infty \hat{f}_C(\omega)\cos(\omega x) \, d\omega. \qquad (3.34)$$

The right side of equation 3.34 is the *inverse Fourier cosine transform* of f. It enables us to recover f from its transform (again, recovering $f(x)$ at each x at which f is continuous).

Equations 3.33 and 3.34 form a *transform pair for the Fourier cosine transform*.

The operational rules used in applying these transforms to partial differential equations are given by the following theorem.

Theorem 11   Let f and f' be continuous on $[0, \infty)$. Assume that $f(x) \to 0$ and $f'(x) \to 0$ as $x \to \infty$. Suppose f'' is piecewise continuous on $[0, \infty)$. Then

1.

$$\mathfrak{F}_S[f''](\omega) = -\omega^2 \hat{f}_S(\omega) + \omega f(0).$$

2.

$$\mathfrak{F}_C[f''](\omega) = -\omega^2 \hat{f}_C(\omega) - f'(0). \qquad \blacksquare$$

Both conclusions are proved by two integrations by parts.

EXERCISE 113   In each of problems 1 through 5, calculate the Fourier sine transform and the Fourier cosine transform of the function.

1. $f(x) = e^{-x}$
2. $f(x) = xe^{-ax}$, with a any positive number
3. $f(x) = \begin{cases} \cos(x) \text{ for } 0 \le x \le \alpha \\ \qquad 0 \text{ for } x > \alpha \end{cases}$
4. $f(x) = \begin{cases} \quad k \text{ for } 0 \le x \le \alpha \\ -k \text{ for } \alpha < x \le 2\alpha \\ \quad 0 \text{ for } x > \alpha \end{cases}$
5. $f(x) = e^{-x}\cos(x)$

EXERCISE 114   Prove Theorem 11(1).

EXERCISE 115   Prove Theorem 11(2).

EXERCISE 116   Show that, under appropriate conditions on f,

$$\mathfrak{F}_S[f^{(4)}(x)](\omega) = \omega^4 \hat{f}_S(\omega) - \omega^3 f(0) + \omega f''(0),$$

in which $f^{(4)}$ is the fourth derivative of f. The phrase "appropriate conditions" means conditions sufficient to apply the operational rule for the Fourier sine transform to $(f'')''$.

EXERCISE 117   Prove that, under appropriate conditions on f,

$$\mathfrak{F}_C[f^{(4)}(x)](\omega) = \omega^4 \hat{f}_C(\omega) + \omega^2 f'(0) - f^{(3)}(0).$$

4

THE WAVE EQUATION

This chapter is devoted to solutions and properties of solutions of the constant coefficient, hyperbolic, second order linear partial differential equation. We will consider this equation in its canonical form.

In addition to its mathematical interest and importance, we have seen that this hyperbolic equation has physical significance in modeling wave motion, and for this reason is referred to as the wave equation. It is both interesting and insightful to view conclusions drawn from the equation in this context, a practice we will indulge as we solve the equation under a variety of conditions. We will begin with solutions of the Cauchy problem for the wave equation on the entire real line.

4.1 THE CAUCHY PROBLEM AND d'ALEMBERT'S SOLUTION

Begin with the wave equation 2.18 from Section 2.5.1:

$$u_{tt} = c^2 u_{xx}$$

with independent variables x and t and with c a positive constant. The connection with wave motion suggests that we think of t as time and x as a space dimension in coupling physical intuition and with the mathematics of this equation.

If we write

$$c^2 u_{xx} - u_{tt} = 0$$

and compare this equation with equation 2.1 of Section 2.1 (with t in place of

158

y) we have $A = c^2$, $B = 0$, and $C = -1$. Then $B^2 - AC = c^2 > 0$ and the wave equation is hyperbolic. The characteristic equations 2.6 and 2.7 are

$$\frac{dt}{dx} = \frac{1}{c} \quad \text{and} \quad \frac{dt}{dx} = -\frac{1}{c}$$

with general solutions defined implicitly by $x - ct = k_1$ and $x + ct = k_2$, respectively. The straight lines defined by these equations are the characteristics of the wave equation. Let

$$\xi = x - ct, \, \eta = x + ct$$

and

$$U(\xi, \eta) = u(x, t).$$

By the chain rule,

$$u_x = U_\xi \xi_x + U_\eta \eta_x = U_\xi + U_\eta,$$

$$u_{xx} = (U_{\xi\xi} + U_{\xi\eta}) + (U_{\eta\xi} + U_{\eta\eta}) = U_{\xi\xi} + 2U_{\xi\eta} + U_{\eta\eta},$$

$$u_t = U_\xi(-c) + U_\eta(c)$$

and

$$u_{tt} = -c(U_{\xi\xi}(-c) + U_{\xi\eta}(c)) + c(U_{\eta\xi}(-c) + U_{\eta\eta}(c))$$
$$= c^2 U_{\xi\xi} - 2c^2 U_{\xi\eta} + c^2 U_{\eta\eta}.$$

Then

$$c^2 u_{xx} - u_{tt} = 0 = -4c^2 U_{\xi\eta}$$

and we have derived the canonical form of the wave equation 2.18:

$$U_{\xi\eta} = 0. \tag{4.1}$$

To solve this equation, and hence also equation 2.18, we will follow a line of reasoning developed by the French mathematician Jean le Rond d'Alembert in the 1740s. Begin by writing the canonical form 4.1 as

$$(U_\eta)_\xi = 0.$$

This implies that U_η is independent of ξ and hence is a function of η only, say

$$U_\eta = w(\eta).$$

Integration of this equation with respect to η yields

$$U(\xi, \eta) = \int w(\eta) \, d\eta + F(\xi),$$

in which the "constant" of the integration with respect to η may be a function of ξ. Since $\int w(\eta) \, d\eta$ is just another function of η, say $G(\eta)$, U must have the form

$$U(\xi, \eta) = F(\xi) + G(\eta),$$

in which F and G can be any functions of a single variable having two continuous derivatives.

We conclude that any function of the form

$$u(x, t) = F(x - ct) + G(x + ct) \tag{4.2}$$

with F and G twice continuously differentiable, is a solution of the wave equation 2.18, and, conversely, any solution of equation 2.18 has this form. This conclusion is important in its own right. Equation 4.2 is *d'Alembert's solution* to the wave equation.

We will use d'Alembert's solution to solve the Cauchy problem for the wave equation:

$$u_{tt} = c^2 u_{xx} \text{ for } -\infty < x < \infty, \, t > 0$$

$$u(x, 0) = \varphi(x), \, u_t(x, 0) = \psi(x) \text{ for } -\infty < x < \infty.$$

Interpreting t as time, these conditions are the *initial conditions*, giving the displacement and velocity of the string at time zero. Intuitively this information and the wave equation governing the motion should be enough to determine the shape of the string at any later time. We will show that indeed this Cauchy problem has a unique solution that depends continuously on the initial data φ and ψ.

We know that any solution of the wave equation 2.18 has the form

$$u(x, t) = F(x - ct) + G(x + ct).$$

The idea is to choose F and G to produce a solution satisfying the initial conditions. Now

$$u(x, 0) = F(x) + G(x) = \varphi(x) \tag{4.3}$$

and

$$u_t(x, 0) = -cF'(x) + cG'(x) = \psi(x). \tag{4.4}$$

Integrate equation 4.4 and rearrange terms to obtain

$$-F(x) + G(x) = \frac{1}{c} \int_0^x \psi(s)\, ds - F(0) + G(0).$$

Add this equation to equation 4.3 to obtain

$$2G(x) = \varphi(x) + \frac{1}{c} \int_0^x \psi(s)\, ds - F(0) + G(0).$$

Therefore

$$G(x) = \frac{1}{2} \varphi(x) + \frac{1}{2c} \int_0^x \psi(s)\, ds - \frac{1}{2} F(0) + \frac{1}{2} G(0). \tag{4.5}$$

But then, from equation 4.3,

$$F(x) = \varphi(x) - G(x) = \frac{1}{2} \varphi(x) - \frac{1}{2c} \int_0^x \psi(s)\, ds + \frac{1}{2} F(0) - \frac{1}{2} G(0). \tag{4.6}$$

Finally, use equations 4.5 and 4.6 to write the solution 4.2 as

$$u(x, t) = F(x - ct) + G(x + ct)$$

$$= \frac{1}{2} \varphi(x - ct) - \frac{1}{2c} \int_0^{x-ct} \psi(s)\, ds + \frac{1}{2} (F(0) - G(0))$$

$$+ \frac{1}{2} \varphi(x + ct) + \frac{1}{2c} \int_0^{x+ct} \psi(s)\, ds - \frac{1}{2} (F(0) - G(0))$$

or, more succinctly,

$$u(x, t) = \frac{1}{2} (\varphi(x - ct) + \varphi(x + ct)) + \frac{1}{2c} \int_{x-ct}^{x+ct} \psi(s)\, ds. \tag{4.7}$$

This is *d'Alembert's formula* for the solution of the Cauchy problem for the wave equation.

Example 23   Solve the Cauchy problem

$$u_{tt} = 9u_{xx} \text{ for } -\infty < x < \infty, t > 0$$

$$u(x, 0) = \cos(x),\ u_t(x, 0) = \sin(2x) \text{ for } -\infty < x < \infty.$$

$$c = 3 \qquad u(x,0) = \cos x \qquad -\infty < x < \infty$$
$$u_t(x,0) = \sin 2x$$

With $\varphi(x) = \cos(x)$, $\psi(x) = \sin(2x)$, and $c = 3$, the solution is

$$u(x, t) = \frac{1}{2}(\cos(x - 3t) + \cos(x + 3t)) + \frac{1}{6}\int_{x-3t}^{x+3t} \sin(2s)\, ds$$

$$= \frac{1}{2}(\cos(x - 3t) + \cos(x + 3t)) - \frac{1}{12}(\cos(2(x + 3t))$$

$$- \cos(2(x - 3t))).$$

This solution can be written

$$u(x, t) = \cos(x)\cos(3t) + \frac{1}{6}\sin(2x)\sin(6t). \qquad \blacksquare$$

The d'Alembert formula shows that a solution of the Cauchy problem for the wave equation exists and is unique. We will use this formula to show that the solution depends continuously on the initial data. This means that small changes in the initial data incur correspondingly small changes in the solution, as we might expect from the vibrating string interpretation of the Cauchy problem.

Theorem 12 Let u_1 be the solution of $u_{tt} = c^2 u_{xx}$ on $-\infty < x < \infty$, $t > 0$, satisfying

$$u_1(x, 0) = \varphi_1(x), \frac{\partial u_1}{\partial t}(x, 0) = \psi_1(x).$$

Let u_2 be the solution satisfying

$$u_2(x, 0) = \varphi_2(x), \frac{\partial u_2}{\partial t}(x, 0) = \psi_2(x).$$

Let $\epsilon > 0$ and $T > 0$. Then there exists a positive number δ such that, if

$$|\varphi_1(x) - \varphi_2(x)| < \delta \text{ and } |\psi_1(x) - \psi_2(x)| < \delta \qquad (4.8)$$

for $-\infty < x < \infty$ and $0 \leq t \leq T$, then

$$|u_1(x, t) - u_2(x, t)| < \epsilon$$

for $-\infty < x < \infty$ and $0 \leq t \leq T$. $\qquad \blacksquare$

Proof: We know that

$$u_1(x, t) = \frac{1}{2}(\varphi_1(x - ct) + \varphi_1(x + ct)) + \frac{1}{2c}\int_{x-ct}^{x+ct}\psi_1(s)\,ds$$

and

$$u_2(x, t) = \frac{1}{2}(\varphi_2(x - ct) + \varphi_2(x + ct)) + \frac{1}{2c}\int_{x-ct}^{x+ct}\psi_2(s)\,ds.$$

Suppose now that we are given positive numbers ϵ and T. Then

$$|u_1(x, t) - u_2(x, t)| \le \frac{1}{2}|\varphi_1(x - ct) - \varphi_2(x - ct)| + \frac{1}{2}|\varphi_1(x + ct)$$

$$- \varphi_2(x + ct)| + \frac{1}{2c}\int_{x-ct}^{x+ct}|\psi_1(s) - |\psi_2(s)|\;ds < \frac{1}{2}\delta + \frac{1}{2}\delta$$

$$+ \frac{1}{2c}(2cT)\delta = (1 + T)\delta$$

if $0 \le t \le T$ and inequalities 4.8 are satisfied. We will therefore have

$$|u_1(x, t) - u_2(x, t)| < \epsilon$$

for all x and $0 \le t \le T$ if we choose δ as any positive number such that

$$(1 + T)\delta < \epsilon.$$

Thus we may choose δ as any number satisfying $0 < \delta < \epsilon/(1 + T)$.   ■

We say that the Cauchy problem for the wave equation on the real line is
well posed because (1) a solution exists, (2) the solution is uniquely determined
by the Cauchy data (initial conditions), and (3) the solution depends continuously on the initial conditions. A problem is *ill posed* if one of these criteria
fails to be met. We will see in Chapter 5 that the Cauchy problem for the heat
equation is ill posed.

EXERCISE 118   Let w be a solution of the problem

$$u_{tt} = c^2 u_{xx} \text{ for } -\infty < x < \infty, t > 0$$

$$u(x, 0) = \varphi(x), u_t(x, 0) = 0 \text{ for } -\infty < x < \infty.$$

Let v be a solution of

$$u_{tt} = c^2 u_{xx} \text{ for } -\infty < x < \infty, \, t > 0$$

$$u(x, 0) = 0, \, u_t(x, 0) = \psi(x) \text{ for } -\infty < x < \infty.$$

Prove that $w + v$ is a solution of

$$u_{tt} = c^2 u_{xx} \text{ for } -\infty < x < \infty, \, t > 0$$

$$u(x, 0) = \varphi(x), \, u_t(x, 0) = \psi(x) \text{ for } -\infty < x < \infty.$$

EXERCISE 119   In each of problems 1 through 9, solve

$$u_{tt} = c^2 u_{xx} \text{ for } -\infty < x < \infty, \, t > 0$$

$$u(x, 0) = \varphi(x), \, u_t(x, 0) = \psi(x) \text{ for } -\infty < x < \infty$$

for the given φ, ψ and c.

1. $\varphi(x) = \cos(3x)$, $\psi(x) = x$, $c = 7$
2. $\varphi(x) = x^2$, $\psi(x) = \sin(2x)$, $c = 4$
3. $\varphi(x) = e^{-|x|}$, $\psi(x) = \sin^2(x)$, $c = 3$
4. $\varphi(x) = \cosh(x)$, $\psi(x) = 2x$, $c = 2$
5. $\varphi(x) = \cos(x) - \sin(x)$, $\psi(x) = \sin(x)$, $c = 2$
6. $\varphi(x) = 2 + x$, $\psi(x) = e^x$, $c = 1$
7. $\varphi(x) = \cos(x)$, $\psi(x) = xe^{-x}$, $c = 4$
8. $\varphi(x) = \sin(3x)$, $\psi(x) = \cos(3x)$, $c = 1$
9. $\varphi(x) = x^3$, $\psi(x) = x \cos(x)$, $c = 3$

EXERCISE 120   Write the solution of the Cauchy problem for the wave equation for $c = 1$, $\varphi(x) = \sin(x)$, and $\psi(x) = 0$. Then write the solution with $\varphi(x) = \sin(x) + \epsilon$, $\psi(x) = 0$, with ϵ any positive number. Show that these solutions differ in magnitude by ϵ for all x and $t \geq 0$.

EXERCISE 121   Write the solution of the Cauchy problem for $c = 1$, $\varphi(x) = \cos(x)$, and $\psi(x) = x$. Then write the solution with $\varphi(x) = \cos(x) + \epsilon$ and $\psi(x) = x + \epsilon$, with ϵ any positive number. Show that these solutions differ in magnitude by no more than ϵT for all x, if $0 \leq t \leq T$.

EXERCISE 122   Let $u(x, t)$ be a solution of the problem

$$u_{xx} = u_{tt} + f(x, t) \text{ for } -\infty < x < \infty, \, t > 0$$

$$u(x, 0) = u_t(x, 0) = 0.$$

Let $w(x, t, T)$ be a solution of the problem

$$w_{xx} - w_{tt} = 0 \text{ for } -\infty < x < \infty, t > T$$

$$w(x, T, T) = 0, w_t(x, T, T) = -f(x, T).$$

Prove that

$$u(x, t) = \int_0^t w(x, t, T) \, dT.$$

This conclusion is known as Duhamel's Principle for the wave equation.

EXERCISE 123   Consider the problem

$$u_{tt} = c^2 u_{xx} + f(x, t) \text{ for } -\infty < x < \infty, t > 0$$

$$u_t(x, 0) = 0 \text{ for } -\infty < x < \infty$$

$$u(0, t) = 0 \text{ for } t > 0.$$

Derive the solution

$$u(x, t) = \frac{1}{2c} \int_0^t \int_{x-c(t-\tau)}^{x+c(t-\tau)} f(\xi, \tau) \, d\xi \, d\tau.$$

4.1.1   An Alternate Derivation of d'Alembert's Formula

We will use the Fourier transform to give an alternate derivation of d'Alembert's formula. We want to solve:

$$u_{tt} = c^2 u_{xx}$$

$$u(x, 0) = \varphi(x), u_t(x, 0) = \psi(x)$$

for $-\infty < x < \infty$, $t > 0$. Begin by applying the Fourier transform to the wave equation, thinking of x as the variable in which the transform is carried out and carrying t along as a parameter. Formally,

$$\mathfrak{F}[u_{tt}] = c^2 \mathfrak{F}[u_{xx}].$$

Now,

$$\mathfrak{F}[u_{tt}](\omega) = \int_{-\infty}^{\infty} \frac{\partial^2 u}{\partial t^2}(x, t)e^{-i\omega x} \, dx = \frac{\partial^2}{\partial t^2} \int_{-\infty}^{\infty} u(x, t)e^{-i\omega x} \, dx = \frac{\partial^2 \hat{u}}{\partial t^2}(\omega, t).$$

The operations $\partial^2/\partial t^2$ and $\int_{-\infty}^{\infty} \cdots dx$ commute (that is, the partial differentiation with respect to t can be taken outside of the integration with respect to x) because x and t are independent and u is assumed continuous with continuous first partial derivatives.

Computing $\mathfrak{F}[u_{xx}](\omega)$ is another matter, since now the differentiation is with respect to x, the variable with respect to which the transform is taken. Now we must apply the operational formula for the Fourier transform (Theorem 7, Section 3.7). With $n = 2$ in that formula (for the second derivative), we obtain

$$\mathfrak{F}[u_{xx}](\omega) = (i\omega)^2\hat{u}(\omega, t) = -\omega^2\hat{u}(\omega, t).$$

In order to apply the operational formula, we seek solutions satisfying the hypotheses of Theorem 7. Specifically, we assume that

$$\int_{-\infty}^{\infty} |u_x(x, t)| \, dx$$

converges for $t \geq 0$, and that $u(x, t) \to 0$ and $u_x(x, t) \to 0$ as $x \to \pm\infty$, for $t \geq 0$.

We now have

$$\hat{u}_{tt}(\omega, t) = -c^2\omega^2\hat{u}(\omega, t),$$

or

$$\hat{u}_{tt} + c^2\omega^2\hat{u} = 0.$$

Think of this as an ordinary differential equation in t for $\hat{u}(\omega, t)$, with ω carried along as a parameter. The general solution for $\hat{u}(\omega, t)$ is

$$\hat{u}(\omega, t) = a(\omega)\cos(\omega ct) + b(\omega)\sin(\omega ct).$$

The coefficients in this solution may depend on ω. To solve for these coefficients, first apply the Fourier transform to the initial condition $u(x, 0) = \varphi(x)$ to get

$$\hat{u}(\omega, 0) = \hat{\varphi}(\omega).$$

But $\hat{u}(\omega, 0) = a(\omega)$, so

$$a(\omega) = \hat{\varphi}(\omega),$$

the Fourier transform of the initial position function.

Next apply the transform to the initial condition $u_t(x, 0) = \psi(x)$:

$$\hat{u}_t(\omega, 0) = \hat{\psi}(\omega).$$

Here we used the fact that $\partial/\partial t$ passes through the transform. But $\hat{u}_t(\omega, 0) = c\omega b(\omega)$, so

$$b(\omega) = \frac{1}{c\omega} \hat{\psi}(\omega).$$

Therefore

$$\hat{u}(\omega, t) = \hat{\varphi}(\omega)\cos(c\omega t) + \frac{1}{c\omega} \hat{\psi}(\omega)\sin(c\omega t).$$

This gives the transform of the solution in terms of quantities that are known, or computable from given information. From this expression we must extract $u(x, t)$. We will exploit the facts that

$$\cos(T) = \frac{1}{2} (e^{iT} + e^{-iT}) \text{ and } \sin(T) = \frac{1}{2i} (e^{iT} - e^{-iT}).$$

We will also find it convenient to define

$$\Psi(x) = \int_{-\infty}^{x} \psi(s) \, ds.$$

In this way $\Psi'(x) = \psi(x)$ and we can use the operational formula for the Fourier transform to write

$$\hat{\psi}(\omega) = \hat{\Psi}'(\omega) = i\omega\hat{\Psi}(\omega).$$

This eliminates the $1/\omega$ factor in one term of $\hat{u}(\omega, t)$ and enables us to write

$$\hat{u}(\omega, t) = \hat{\varphi}(\omega)\cos(c\omega t) + \frac{1}{c} i\hat{\Psi}(\omega)\sin(c\omega t)$$

$$= \hat{\varphi}(\omega)\frac{1}{2}(e^{ic\omega t} + e^{-ic\omega t}) + \frac{1}{2c}\hat{\Psi}(\omega)(e^{ic\omega t} - e^{-ic\omega t})$$

$$= \frac{1}{2}(\hat{\varphi}(\omega)e^{ic\omega t} + \hat{\varphi}(\omega)e^{-ic\omega t}) + \frac{1}{2c}(\hat{\Psi}(\omega)e^{ic\omega t} - \hat{\Psi}(\omega)e^{-ic\omega t}).$$

Now recognize that $\hat{\varphi}(\omega)e^{ic\omega t}$ is the Fourier transform of the translated function $\varphi(x + ct)$, by Lemma 4(3) of Section 3.7. With similar recognition of the other terms in the last line of the last equation, we have

$$\hat{u}(\omega, t) = \mathfrak{F}\left[\frac{1}{2}(\varphi(x + ct) + \varphi(x - ct)) + \frac{1}{2c}(\Psi(x + ct) - \Psi(x - ct))\right](\omega, t).$$

By the uniqueness of the inverse Fourier transform of a continuous function,

$$u(x, t) = \frac{1}{2}(\varphi(x + ct) + \varphi(x - ct)) + \frac{1}{2c}(\Psi(x + ct) - \Psi(x - ct)).$$

Finally,

$$\Psi(x + ct) - \Psi(x - ct) = \int_{-\infty}^{x+ct} \psi(s)\,ds - \int_{-\infty}^{x-ct} \psi(s)\,dx = \int_{x-ct}^{x+ct} \psi(s)\,ds$$

and we have d'Alembert's solution.

This derivation by Fourier transform has been informal and we have not verified that all the hypotheses of theorems are in effect before applying the theorem. For example, we put $\Psi(x) = \int_{-\infty}^{x} \psi(s)\,ds$, and Ψ need not enjoy the same properties assumed for ψ. We can, however, regard the calculation just done as entirely formal. It has led us to a function $u(x, t)$ defined in terms of φ and ψ. Once we obtain u, it is routine to verify that u is the solution of the initial value problem, with certain assumptions on the differentiability of φ and ψ.

4.2   D'ALEMBERT'S SOLUTION AS A SUM OF FORWARD AND BACKWARD WAVES

D'Alembert's formula for the solution of the Cauchy problem for the wave equation can be written

$$u(x, t) = \frac{1}{2} \left(\varphi(x - ct) - \frac{1}{c} \int_0^{x-ct} \psi(s) \, ds \right)$$

$$+ \frac{1}{2} \left(\varphi(x + ct) + \frac{1}{c} \int_0^{x+ct} \psi(s) \, ds \right)$$

$$= F(x - ct) + B(x + ct), \tag{4.9}$$

in which

$$F(x) = \frac{1}{2} \varphi(x) - \frac{1}{2c} \int_0^x \psi(s) \, ds \tag{4.10}$$

and

$$B(x) = \frac{1}{2} \varphi(x) + \frac{1}{2c} \int_0^x \psi(s) \, ds. \tag{4.11}$$

The graph of $F(x - ct)$ is the graph of F translated ct units to the right. We may therefore think of $F(x - ct)$ as a *forward wave* (F for forward), propagating the graph of F to the right with velocity c. The graph of $B(x + ct)$ is the graph of B translated ct units to the left, so $B(x + ct)$ is a *backward wave* (B for backward), propagating the graph of B to the left with velocity c. The resulting motion is a superposition of these forward and backward waves. The fact that these waves are moving with constant velocity means that, for solutions of the wave equation, disturbances are propagated with finite speed. This is a characteristic property of solutions of the hyperbolic wave equation. We will see in Chapter 5 that solutions of the parabolic heat equation exhibit different behavior.

Each of Figures 4.1, 4.2, and 4.3 displays these combinations of forward and backward wave motion for the case $c = 1$ and $\psi(x) = 0$, for a given initial displacement $\varphi(x)$. In each case, envision the string as being released from rest from the given initial displacement position.

In Figure 4.1 the initial displacement function is

$$\varphi(x) = \begin{cases} \sin(x) & \text{for } -\pi \le x \le \pi \\ 0 & \text{for } |x| > \pi. \end{cases}$$

Initially the string takes the shape of the sine function between $-\pi$ and π, and is not displaced elsewhere. Figure 4.1(a) shows this shape of the string before it is released. Figure 4.1(b) shows the position of the string at time $t = 3/4$

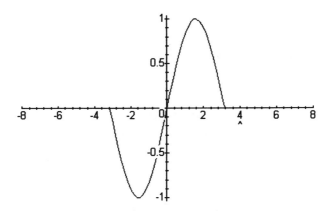

FIGURE 4.1(a)   Initial position of the string.

second. This is a sum of the forward wave at that time (move the initial wave of Figure 4.1(a) to the right ct units, or 3/4 units, and then move the initial wave to the left 3/4 units, and add these forward and backward waves). Figure 4.1(c) is the string at $t = 1$. This is a superposition of the original position moved right 1 unit with the original position moved left 1 unit. Figures 4.1(d) through 4.1(h) show the position of the string at $t = $ 1.5, 2, 2.5, 3.5, and 5 seconds, respectively. Notice that in (g) enough time has passed for the resulting wave to separate into two copies of the original wave, one moving to the right, the other moving to the left. As time increases (h) these waves move further apart, one continuing to the right, the other to the left. These graphs are what an observer would see at the given times if the string were placed in the initial position $y = \varphi(x)$ and released from the rest.

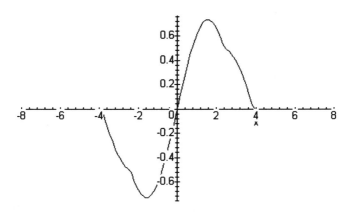

FIGURE 4.1(b)   Position at $t = 3/4$.

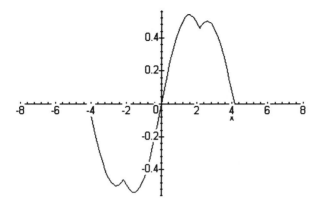

FIGURE 4.1(c) $t = 1$.

Figures 4.2 repeats this visualization of the superposition of forward and backward waves for the initial position function

$$\varphi(x) = \begin{cases} x^2 + 3x \text{ for } -3 \leq x \leq 0 \\ 4 \sin(x) \text{ for } 0 \leq x \leq 2\pi \\ 0 \text{ for } x < -3 \text{ and } x > 2\pi. \end{cases}$$

Figure 4.2(a) is this original position function (shape of the string at $t = 0$). Figures 4.2(b) through (h) show a sum of this graph displaced left and right, at times 1/2, 1, 3/2, 2, 4, 5, and 6 seconds, respectively. Because the original position is nonzero only on a finite interval, eventually the forward and backward waves separate and move independently in opposite directions. This can be seen at times $t = 5$ and $t = 6$ in this example.

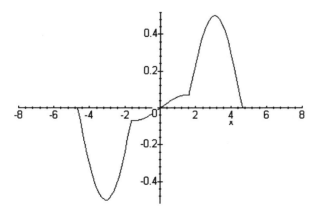

FIGURE 4.1(d) $t = 1.5$.

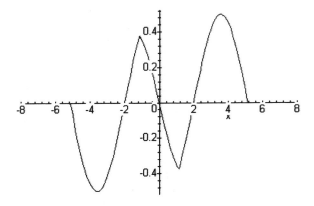

FIGURE 4.1(e) $t = 2$.

Figure 4.3(a) is the initial position function

$$\varphi(x) = \begin{cases} \cos(x) \text{ for } -\pi/2 \le x \le \pi/2 \\ \quad\quad 0 \text{ for } |x| \ge \pi/2 \end{cases}$$

and in graphs (b) through (e) the wave is shown at times $t = 1/2$, 1, 3/2, and 2 seconds, respectively. By the two-second mark the forward and backward waves have separated. Again, this occurs here because the initial position function is zero outside of a bounded interval.

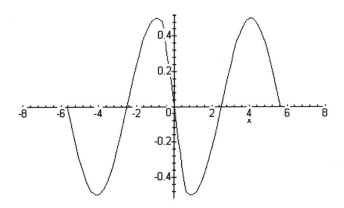

FIGURE 4.1(f) $t = 2.5$.

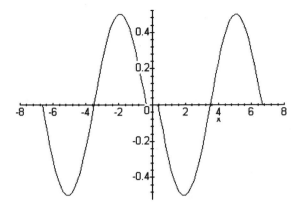

FIGURE 4.1(g) $t = 3.5$.

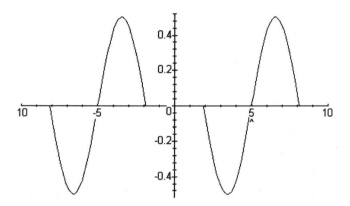

FIGURE 4.1(h) $t = 5$.

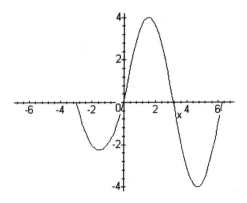

FIGURE 4.2(a)   Initial position.

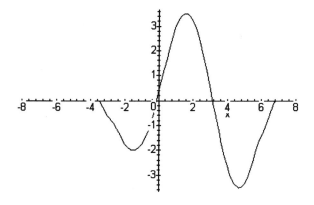

FIGURE 4.2(b) $t = 1/2.$

EXERCISE 124   In problems 1 through 6, write the solution of the Cauchy problem

$$u_{tt} = u_{xx} \text{ for } -\infty < x < \infty, t > 0$$

$$u(x, 0) = \varphi(x), u_t(x, 0) = 0 \text{ for } -\infty < x < \infty$$

as the sum of a forward wave and a backward wave. In the spirit of Figures 4.1–4.3, graph the initial position and then graph the solution at selected times,

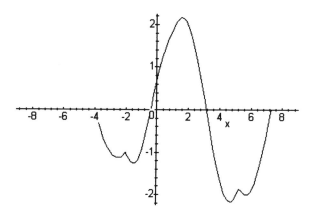

FIGURE 4.2(c) $t = 1.$

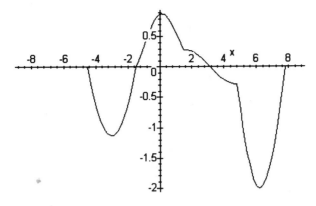

FIGURE 4.2(d) $t = 3/2$.

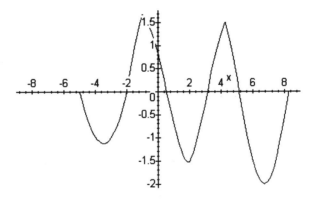

FIGURE 4.2(e) $t = 2$.

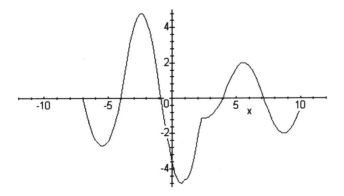

FIGURE 4.2(f) $t = 4$.

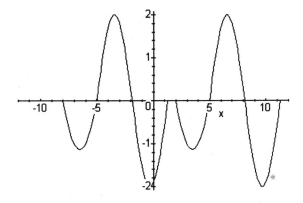

FIGURE 4.2(g) $t = 5$.

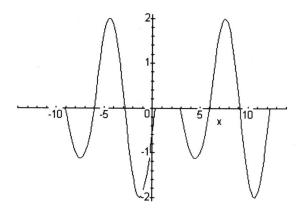

FIGURE 4.2(h) $t = 6$.

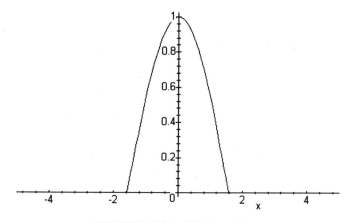

FIGURE 4.3(a)   Initial position.

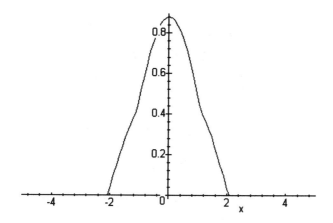

FIGURE 4.3(b) $t = 1/2$.

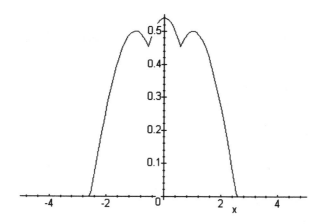

FIGURE 4.3(c) $t = 1$.

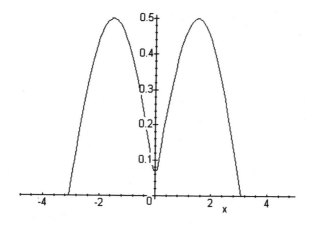

FIGURE 4.3(d) $t = 3/2$.

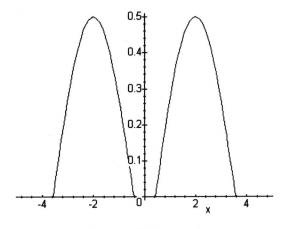

FIGURE 4.3(e) $t = 2$.

showing the solution as a superposition of forward and backward waves along the real line.

1. $\varphi(x) = \begin{cases} \sin(2x) \text{ for } -\pi \le x \le \pi \\ \quad\quad 0 \text{ for } |x| > \pi, \end{cases}$

2. $\varphi(x) = \begin{cases} 1 - |x| \text{ for } -1 \le x \le 1 \\ \quad\quad 0 \text{ for } |x| > 1, \end{cases}$

3. $\varphi(x) = \begin{cases} \cos(x) \text{ for } -\pi/2 \le x \le \pi/2 \\ \quad\quad 0 \text{ for } |x| > \pi/2 \end{cases}$

4. $\varphi(x) = \begin{cases} 1 - x^2 \text{ for } |x| \le 1 \\ \quad 0 \text{ for } |x| > 1 \end{cases}$

5. $\varphi(x) = \begin{cases} x^2 - x - 2 \text{ for } -1 \le x \le 2 \\ \quad\quad 0 \text{ for } x < -1 \text{ or } x > 2 \end{cases}$

6. $\varphi(x) = \begin{cases} x^3 - x^2 - 4x + 4 \text{ for } -2 \le x \le 2 \\ \quad\quad\quad 0 \text{ for } |x| > 2 \end{cases}$

4.3   DOMAIN OF DEPENDENCE AND THE CHARACTERISTIC TRIANGLE

Consider again the Cauchy problem for the wave equation on the real line. We would like to consider what information is needed to determine the value $u(x_0, t_0)$ of the solution at a point x_0 and time t_0.

A point (x, t) lies in the x, t-plane, with the first coordinate specifying a location on the line, the second the time at which we are looking at that location. For example, if we move up the line $x = x_0$ along points (x_0, t), we remain at the location $x = x_0$ on the real line, but as the height above the horizontal axis increases we are looking at this point of the string at later times. A graph in this x, t-plane is not to be confused with a graph of $y = u(x, t)$ at a particular time. This graph is drawn in the x, y-plane and gives a picture of the string at that time.

At (x_0, t_0), the Cauchy problem for the wave equation has solution

$$u(x_0, t_0) = \frac{1}{2} (\varphi(x_0 - ct_0) + \varphi(x_0 + ct_0)) + \frac{1}{2c} \int_{x_0 - ct_0}^{x_0 + ct_0} \psi(s) \, ds,$$

where φ gives the initial position and ψ gives the initial velocity. The terms $\varphi(x_0 - ct_0)$ and $\varphi(x_0 + ct_0)$ are the initial position function evaluated at two numbers, $x_0 - ct_0$ and $x_0 + ct_0$, and depend only on these two numbers. The integral term in d'Alembert's formula depends on values of $\psi(s)$ for $x_0 - ct_0 \leq s \leq x_0 + ct_0$, and hence is influenced by the behavior of ψ on the entire interval $[x_0 - ct_0, x_0 + ct_0]$.

The lines $x - ct =$ constant and $x + ct =$ constant are the characteristics of the wave equation. These are straight lines having, respectively, slope $1/c$ and $-1/c$ in the x, t-plane. The two unique characteristics passing through (x_0, t_0) are the lines

$$x - ct = x_0 - ct_0 \text{ and } x + ct = x_0 + ct_0$$

shown in Figure 4.4, and they intersect the x-axis at $(x_0 - ct_0, 0)$ and $(x_0 + ct_0, 0)$, respectively. These two points, together with (x_0, t_0), form the vertices of a solid triangle called the *characteristic triangle*, and $u(x_0, t_0)$ depends on the initial position function φ evaluated at the base vertices of this triangle, and on the integral of the initial velocity over the base side of this triangle. Because of this, the interval $[x_0 - ct_0, x_0 + ct_0]$ is called the *domain of dependence* of the point (x_0, t_0).

Now consider again equation 4.9, in which we wrote the d'Alembert solution as the sum of a forward wave F and a backward wave B:

$$u(x, t) = F(x - ct) + B(x + ct)$$

with F and B defined by equations 4.10 and 4.11. In particular,

$$u(x_0, t_0) = F(x_0 - ct_0) + B(x_0 + ct_0).$$

Along the characteristic $x - ct = x_0 - ct_0$, $F(x - ct)$ has the constant value $F(x_0 - ct_0)$. This means that a disturbance initially at $(x_0 - ct_0, 0)$ is carried forward with constant velocity c along this characteristic (Figure 4.5). Along the characteristic $x + ct = x_0 + ct_0$ $B(x + ct)$ is the constant $B(x_0 + ct_0)$. A

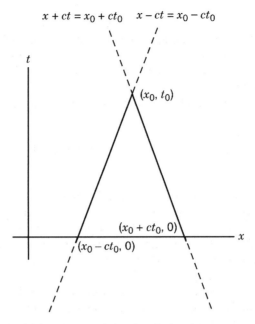

FIGURE 4.4   Characteristic triangle for the wave equation.

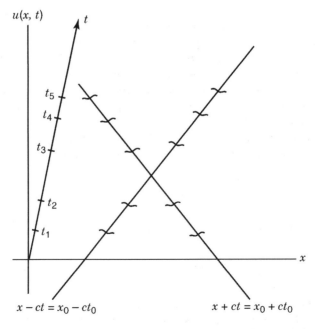

FIGURE 4.5   Initial disturbances carried along characteristics.

disturbance initially at $(x_0 + ct_0, 0)$ is carried backward with velocity c unchanged along this characteristic (also Figure 4.5). *A solution of the wave equation propagates disturbances with constant velocity* c *along its characteristics.*

Influence of an Interval in the Space Variable on the Resulting Wave Motion We can also consider the influence on the ensuing wave motion of an arbitrary interval on the horizontal axis.

Consider points $(a, 0)$ and $(b, 0)$ as in Figure 4.6. There are two characteristics through each point. These divide the half plane $t > 0$ into six regions, labeled *I* through *VI*. Region *I* is the characteristic triangle having base vertices $(a, 0)$ and $(b, 0)$. Consider in turn a typical point in each region, and how the wave (solution) at this point is influenced by what is happening on $[a, b]$.

If (x, t) is a point in region *II*, its domain of dependence (Figure 4.7) does not intersect the characteristic triangle *I*. The same is true for a point in region *III*. A point x on the string at a time t, with (x, t) in one of these regions, is not influenced in any way by a disturbance at time $t = 0$ in $[a, b]$.

As shown in Figure 4.8, a backward wave originating at a point in $[a, b]$ can reach a point in region *IV*, and a forward wave can reach a point in region *V*. A point x on the string, at a time t such that (x, t) is in region *IV*, is reached by a backward wave originating at time zero at a point in $[a, b]$. If (x, t) is in region *V*, this point on the string is reached by a forward wave originating at time zero at a point in $[a, b]$.

If (x, t) is in region *VI* (of Figure 4.6), enough time has passed (t is sufficiently large) that the forward and backward waves in the d'Alembert solution have already passed, and x is not further influenced by a displacement in $[a, b]$ at time zero, but remains at rest after a displacement of $(1/2c) \int_a^b \psi(s) \, ds$.

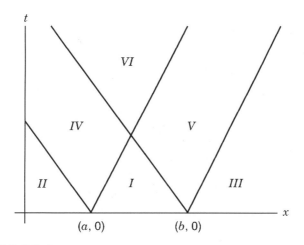

FIGURE 4.6 Characteristics partitioning the half-plane $t > 0$.

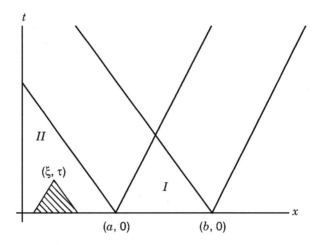

FIGURE 4.7   Domain of dependence of a point in region *II*.

Finally, region *I* is the characteristic triangle and the solution at a point in this region is influenced by both forward and backward waves originating at a point on [*a*, *b*].

EXERCISE 125   Suppose the initial displacement and velocity functions vanish outside an interval of finite length. Specifically, let *a* be a positive number and suppose $\varphi(x) = 0$ and $\psi(x) = 0$ for $|x| > a$. Prove that the solution of the Cauchy problem

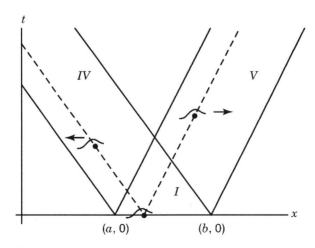

FIGURE 4.8   Forward and backward waves originating in [*a*, *b*].

$$u_{tt} = c^2 u_{xx} \text{ for } -\infty < x < \infty, t > 0$$

$$u(x, 0) = \varphi(x), u_t(x, 0) = \psi(x) \text{ for } -\infty < x < \infty$$

vanishes outside of $[-a - ct, a + ct]$.

EXERCISE 126   Suppose there is some interval $[a, b]$ on the real line outside of which φ and ψ are identically zero. Let u be a solution of the Cauchy problem

$$u_t = u_{xx} \text{ for } -\infty < x < \infty, t > 0$$

$$u(x, 0) = \varphi(x), u_t(x, 0) = \psi(x),$$

and suppose u is continuous with continuous first and second partial derivatives. Show that

$$\int_{-\infty}^{\infty} [u_x^2 + u_t^2]_{t=T} \, dx = \int_{-\infty}^{\infty} [u_x^2 + u_t^2]_{t=0} \, dx$$

for every $T > 0$. *Hint:* First establish the identity

$$2u_t(u_{xx} - u_{tt}) = (2u_t u_x)_x - (u_x^2 + u_t^2)_t.$$

Form the double integral of this identity over the rectangular region $-R - 2T \leq x \leq R + 2T, 0 \leq t \leq T$ and apply Green's Theorem.

The integral

$$\int_{\alpha}^{\beta} [u_x^2 + u_t^2]_{t=T} \, dx$$

is called the energy of $u(x, t)$ in the interval $[\alpha, \beta]$ at time T. The above conclusion states that the energy of the solution over the entire line is constant.

4.4   THE WAVE EQUATION ON A HALF-LINE

Consider the problem

$$u_{tt} = c^2 u_{xx} \text{ for } x > 0, t > 0$$

$$u(x, 0) = \varphi(x), u_t(x, 0) = \psi(x) \text{ for } x \geq 0$$

$$u(0, t) = 0 \text{ for } t \geq 0.$$

This problem models vibrations along a string stretched from 0 to infinity. Because $x \geq 0$ this string has a left end, and the condition $u(0, t) = 0$ for all time means that this end is fixed (the string is fastened there). The condition $u(0, t) = 0$ is called a *boundary condition* and for this reason this problem is called an *initial-boundary value problem*.

We will show how to use the d'Alembert formula for the initial value problem on the entire line, to write a solution of this problem on the half-line. Define odd extensions of φ and ψ to the entire line:

$$\Phi(x) = \begin{cases} \varphi(x) \text{ for } x \geq 0 \\ -\varphi(-x) \text{ for } x < 0 \end{cases}$$

and

$$\Psi(x) = \begin{cases} \psi(x) \text{ for } x \geq 0 \\ -\psi(-x) \text{ for } x < 0. \end{cases}$$

On the half-line $x \geq 0$, Φ agrees with φ, and Ψ with ψ. Further, for all $x \neq 0$,

$$\Phi(-x) = -\Phi(x)$$

and

$$\Psi(-x) = -\Psi(x).$$

Now consider the Cauchy problem for the entire line:

$$u_{tt} = c^2 u_{xx} \text{ for } -\infty < x < \infty$$

$$u(x, 0) = \Phi(x), \ u_t(x, 0) = \Psi(x) \text{ for } -\infty < x < \infty.$$

The solution of this problem is

$$u(x, t) = \frac{1}{2} (\Phi(x - ct) + \Phi(x + ct)) + \frac{1}{2c} \int_{x-ct}^{x+ct} \Psi(s) \, ds. \qquad (4.12)$$

u satisfies the wave equation for $-\infty < x < \infty$ and $t > 0$ and hence also for $x > 0$ and $t > 0$. Further, if $x \geq 0$ then

$$u(x, 0) = \frac{1}{2} (\Phi(x) + \Phi(x)) = \Phi(x) = \varphi(x)$$

and

$$u_t(x, 0) = \frac{1}{2} (-c\Phi'(x) + c\Phi'(x)) + \frac{1}{2c} (c\Psi(x) - (-c)\Psi(x)) = \Psi(x) = \psi(x).$$

Therefore u satisfies the initial conditions of the problem on the half-line. Finally, we claim that $u(0, t) = 0$. Compute

$$u(0, t) = \frac{1}{2} (\Phi(-ct) + \Phi(ct)) + \frac{1}{2c} \int_{-ct}^{ct} \Psi(s) \, ds.$$

But

$$\Phi(-ct) + \Phi(ct) = -\varphi(ct) + \varphi(ct) = 0$$

because we defined Φ as the odd extension of φ; and

$$\int_{-ct}^{ct} \Psi(s) \, ds = 0$$

because we also defined Ψ to be an odd function.

We conclude that solution 4.12 of the extended initial value problem on the entire line is also the solution of the initial-boundary value problem on the half-line.

This method of solution is sometimes called the *method of images*. In effect we cast an image of the problem across the y-axis to create a problem on the whole real line, which we have solved previously. If the image is cast correctly (in this case, using an odd function), the solution on the line gives the solution on the half-line. A similar strategy, called the method of electrostatic images because of its physical motivation, will be exploited in Section 6.14.2 to determine Green's functions and solve Dirichlet problems.

Example 24 Solve

$$u_{tt} = u_{xx} \text{ for } x > 0, t > 0$$

$$u(x, 0) = 1 - e^{-x}, u_t(x, 0) = \cos(x) \text{ for } x \geq 0$$

$$u(0, t) = 0 \text{ for } t \geq 0.$$

Here $\varphi(x) = 1 - e^{-x}$ and $\psi(x) = \cos(x)$. Form the odd extensions of these functions to the entire line:

$$\Phi(x) = \begin{cases} 1 - e^{-x} & \text{for } x \geq 0 \\ -1 + e^{x} & \text{for } x < 0 \end{cases}$$

and

$$\Psi(x) = \begin{cases} \cos(x) \text{ for } x \geq 0 \\ -\cos(x) \text{ for } x < 0. \end{cases}$$

Then $\Phi(-x) = -\Phi(x)$ and $\Psi(-x) = -\Psi(x)$. Graphs of both extensions are shown in Figure 4.9.

The solution to the initial-boundary value problem on the half-line is

$$u(x, t) = \frac{1}{2} (\Phi(x - t) + \Phi(x + t)) + \frac{1}{2} \int_{x-t}^{x+t} \Psi(s) \, ds$$

for $x \geq 0$, $t \geq 0$.

It is natural to write this solution in terms of the original initial position and velocity functions φ and ψ, and we can do this because Φ and Ψ are defined in terms of these functions.

If $x - t \geq 0$ then

$$u(x, t) = \frac{1}{2} (\Phi(x - t) + \Phi(x + t)) + \frac{1}{2} \int_{x-t}^{x+t} \Psi(s) \, ds$$

$$= \frac{1}{2} (1 - e^{-(x-t)} + 1 - e^{-(x+t)}) + \frac{1}{2} \int_{x-t}^{x+t} \cos(s) \, ds.$$

This yields

$$u(x, t) = 1 - e^{-x} \cosh(t) + \cos(x)\sin(t) \text{ for } x - t \geq 0.$$

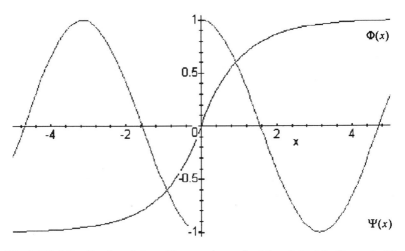

FIGURE 4.9 Graphs of the odd extensions of $\varphi(x)$ and $\psi(x)$ in Example 24.

Now suppose $x - t < 0$. Keep in mind that $x + t > 0$ if $x > 0$ and $t > 0$. Now

$$u(x,\ t) = \frac{1}{2}\ (\Phi(x\ -\ t)\ +\ \Phi(x\ +\ t))\ +\ \frac{1}{2} \int_{x-t}^{x+t} \Psi(s)\ ds$$

$$= \frac{1}{2}\ (-1\ +\ e^{x-t}\ +\ 1\ -\ e^{-(x+t)})\ +\ \frac{1}{2} \int_{x-t}^{0} -\ \cos(s)\ ds\ +\ \frac{1}{2} \int_{0}^{x+t} \cos(s)\ ds$$

$$= \frac{1}{2}\ (e^x e^{-t}\ -\ e^{-x} e^{-t})\ -\ \left[\frac{1}{2}\ \sin(s)\right]_{x-t}^{0}\ +\ \left[\frac{1}{2}\ \sin(s)\right]_{0}^{x+t}$$

$$= e^{-t}\ \sinh(x)\ +\ \sin(x)\cos(t).$$

In summary, the solution is

$$u(x,\ t) = \begin{cases} 1\ -\ e^{-x}\ \cosh(t)\ +\ \cos(x)\sin(t) \text{ for } x \geq t \geq 0 \\ e^{-t}\ \sinh(x)\ +\ \sin(x)\cos(t) \text{ for } 0 < x < t. \end{cases} \qquad \blacksquare$$

EXERCISE 127   In each of problems 1 through 10, write the solution of the problem.

$$u_{tt} = c^2 u_{xx} \text{ for } x > 0,\ t > 0$$

$$u(x,\ 0) = \varphi(x),\ u_t(x,\ 0) = \psi(x) \text{ for } x \geq 0$$

$$u(0,\ t) = 0 \text{ for } t \geq 0.$$

1. $\varphi(x) = x^2$, $\psi(x) = \sin(x)$, $c = 1$
2. $\varphi(x) = x^2$, $\psi(x) = x$, $c = 4$
3. $\varphi(x) = 1 - e^x$, $\psi(x) = x^2$, $c = 2$
4. $\varphi(x) = 1 - \cos(x)$, $\psi(x) = e^{-x}$, $c = 4$
5. $\varphi(x) = x \sin(x)$, $\psi(x) = x^2$, $c = 2$
6. $\varphi(x) = \sinh^2(x)$, $\psi(x) = x$, $c = 7$
7. $\varphi(x) = x^3$, $\psi(x) = e^{-x}$, $c = 3$
8. $\varphi(x) = x^2 - x$, $\psi(x) = x - 1$, $c = 3$
9. $\varphi(x) = \cosh(x) - 1$, $\psi(x) = \sin(x)$, $c = 5$
10. $\varphi(x) = x^3 + x$, $\psi(x) = \cos(2x)$, $c = 6$

4.5 A NONHOMOGENEOUS INITIAL-BOUNDARY VALUE PROBLEM ON A HALF-LINE

Consider the problem

$$u_{tt} = c^2 u_{xx} \text{ for } x > 0, t > 0$$

$$u(x, 0) = \varphi(x), u_t(x, 0) = \psi(x) \text{ for } x \geq 0$$

$$u(0, t) = f(t) \text{ for } t \geq 0.$$

This is an initial-boundary value problem that differs from that of the preceding section if $f(t)$ is not identically zero. We say that this problem has a nonhomogeneous boundary condition. This problem models vibrations in a string stretched from 0 to infinity, with given initial displacement and velocity, and with its behavior at the left end governed by the function f. Imagine a person standing at the left end of the string and moving it according to the function f.

We can solve this problem using d'Alembert's formula for the initial value problem on the entire line, but now we must take into account the behavior of the left end of the string.

Suppose first that $x_0 \geq ct_0$. Now $[x_0 - ct_0, x_0 + ct_0]$ lies entirely on the nonnegative part of the x-axis, and not enough time has passed for the initial displacement $f(t)$ at time zero to reach any point in this interval. In this case $u(x_0, t_0)$ is not influenced by f and the d'Alembert formula holds. By equation 4.9,

$$u(x_0, t_0) = F(x_0 - ct_0) + B(x_0 + ct_0) \text{ for } x_0 \geq ct_0. \qquad (4.13)$$

F and B are the forward and backward waves defined by equations 4.10 and 4.11

Now suppose $x_0 < ct_0$. As shown in Figure 4.10, the characteristic triangle through (x_0, t_0) intersects the t-axis and has part of its base on the negative x-axis. Now d'Alembert's formula does not apply because $\varphi(x)$ and $\psi(x)$ are not defined for $x_0 - ct_0 < x < 0$; hence $F(x_0 - ct_0)$ is not defined. However, putting $x = 0$ into equation 4.13 formally yields

$$u(0, t_0) = f(t_0) = F(-ct_0) + B(ct_0).$$

This suggests that we can extend F to this negative value by defining

$$F(-ct_0) = f(t_0) - B(ct_0).$$

Both function values on the right are well defined. Further, since t_0 can be any

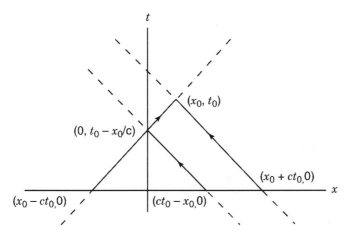

FIGURE 4.10 Reflection of a backward wave by the t-axis.

positive number, we can think of ct_0 as any positive number and use this equation as a model to define

$$F(-x) = f\left(\frac{x}{c}\right) - B(x) \qquad (4.14)$$

for any positive number x. This extends F to negative values. (For simplicity we are using the same symbol F for the extended function.) Now put, for $x_0 - ct_0 < 0$,

$$F(x_0 - ct_0) = F(-(ct_0 - x_0)) = f\left(\frac{ct_0 - x_0}{c}\right) - B(ct_0 - x_0)$$

or

$$F(x_0 - ct_0) = f\left(t_0 - \frac{x_0}{c}\right) - B(ct_0 - x_0).$$

Substituting this into d'Alembert's solution 4.9 we have

$$u(x_0, t_0) = f\left(t_0 - \frac{x_0}{c}\right) - B(ct_0 - x_0) + B(x_0 + ct_0) \text{ for } x_0 - ct_0 < 0.$$

In view of the definition of the backward wave B, this equation can be written

$$u(x_0, t_0) = f\left(t_0 - \frac{x_0}{c}\right) + \frac{1}{2}\left(\varphi(x_0 + ct_0) - \varphi(ct_0 - x_0)\right)$$

$$+ \frac{1}{2c}\int_{ct_0 - x_0}^{x_0 + ct_0} \psi(s)\, ds \text{ for } x_0 + ct_0. \quad (4.15)$$

Using equations 4.13 and 4.15 we can now compute $u(x_0, t_0)$ for $x_0 - ct_0 \geq 0$ and also for $x_0 - ct_0 < 0$, in terms of the initial and boundary data of the problem. The value of the solution in the case $x_0 - ct_0 < 0$ depends on the initial position at $x_0 + ct_0$ and at $ct_0 - x_0$ and on the initial velocity on $[ct_0 - x_0, x_0 + ct_0]$. This is the domain of dependence of the solution when $x_0 < ct_0$.

Now refer again to Figure 4.10. The characteristic through (x_0, t_0) and $(x_0 - ct_0, 0)$ intersects the t-axis at $(0, t_0 - x_0/c)$ when $x_0 < ct_0$. The reflection of $(x_0 - ct_0, 0)$ across the t-axis is the point $(ct_0 - x_0, 0)$, and the line through this point and $(0, t_0 - x_0/c)$ is again a characteristic. We may think of the solution 4.15 for $x_0 < ct_0$ as the result of a backward wave from $(x_0 + ct_0, 0)$, together with a backward wave from $(ct_0 - x_0, 0)$, reflected by the t-axis at $(0, t_0 - x_0/c)$.

We have used the zero subscript in this discussion to discuss the solution at a particular point and maintain x and t as variables. However, we now drop the subscript and write the solution at any (x, t) with $x \geq 0, t \geq 0$:

$$u(x, t) = \frac{1}{2}\left(\varphi(x - ct) + \varphi(x + ct)\right) + \frac{1}{2c}\int_{x - ct}^{x + ct} \psi(s)\, ds \text{ for } x \geq ct \quad (4.16)$$

and

$$u(x, t) = f\left(t - \frac{x}{c}\right) + \frac{1}{2}\left(\varphi(x + ct) - \varphi(ct - x)\right)$$

$$+ \frac{1}{2c}\int_{ct - x}^{x + ct} \psi(s)\, ds \text{ for } x < ct. \quad (4.17)$$

From equation 4.17, $u(0, t) = f(t)$ for $t \geq 0$, so this solution satisfies the boundary condition.

Example 25   Solve

$$u_{tt} = 4u_{xx} \text{ for } x > 0, t > 0$$

$$u(x, 0) = \sin(3x), u_t(x, 0) = x \text{ for } x \geq 0$$

$$u(0, t) = 1 - e^{-t} \text{ for } t \geq 0.$$

In the context of the preceding discussion, $\varphi(x) = \sin(3x)$, $\psi(x) = x$, $f(t) = 1 - e^{-t}$, and $c = 2$. We can write the solution using equations 4.16 and 4.17:

$$u(x, t) = \frac{1}{2}(\sin(3(x - 2t)) + \sin(3(x + 2t))) + \frac{1}{4}\int_{x-2t}^{x+2t} s\, ds \text{ if } x - 2t \geq 0$$

$$u(x, t) = \left(1 - e^{-t+x/2}\right) + \frac{1}{2}(\sin(3(x + 2t)) - \sin(3(2t - x)))$$

$$+ \frac{1}{4}\int_{2t-x}^{x+2t} s\, ds \text{ for } x - 2t < 0.$$

After some manipulation we can write this solution as

$$u(x, t) = \begin{cases} \sin(3x)\cos(6t) + xt & \text{for } x - 2t \geq 0 \\ \sin(3x)\cos(6t) + xt + 1 - e^{-t}e^{x/2} & \text{for } x - 2t < 0. \end{cases}$$

It is routine to show that this function satisfies the initial and boundary conditions. ∎

EXERCISE 128   In each of problems 1 through 10, write the solution of $u_{tt} = c^2 u_{xx}$ for $x > 0$, $t > 0$ with the given initial and boundary information.

1. $\varphi(x) = x$, $\psi(x) = e^{-x}$, $f(t) = t^2$, $c = 1$
2. $\varphi(x) = x^2$, $\psi(x) = 2x$, $f(t) = t$, $c = 3$
3. $\varphi(x) = \sin(x)$, $\psi(x) = x$, $f(t) = 1 - e^t$, $c = 7$
4. $\varphi(x) = 1 - \cos(x)$, $\psi(x) = \cos(2x)$, $f(t) = \sin(t)$, $c = 5$
5. $\varphi(x) = \cos(x)$, $\psi(x) = x^2$, $f(t) = t + \cos(t)$, $c = 3$
6. $\varphi(x) = \sin^2(x)$, $\psi(x) = xe^x$, $f(t) = t^2$, $c = 1$
7. $\varphi(x) = e^{-x}$, $\psi(x) = \sin(x)$, $f(t) = 1 - t$, $c = 3$
8. $\varphi(x) = x^3$, $\psi(x) = x$, $f(t) = \sin(3t)$, $c = 4$
9. $\varphi(x) = x + x^2$, $\psi(x) = 1$, $f(t) = 2t$, $c = 2$
10. $\varphi(x) = 1 - x$, $\psi(x) = \cos(2x)$, $f(t) = e^{-t}$, $c = 5$

4.6   A NONHOMOGENEOUS WAVE EQUATION

Consider the problem

$$u_{tt} = c^2 u_{xx} + P(x, t) \text{ for } -\infty < x < \infty, t > 0$$

$$u(x, 0) = \varphi(x), u_t(x, 0) = \psi(x) \text{ for } -\infty < x < \infty. \tag{4.18}$$

In considering the wave equation as modeling vibrations along a string, $P(x, t)$ can be thought of as an external driving or damping force. We will exploit the characteristic triangle to write the solution of this initial value problem.

Suppose we want the solution at (x_0, t_0). Let Δ denote the characteristic triangle having vertices (x_0, t_0), $(x_0 - ct_0, 0)$, and $(x_0 + ct_0, 0)$, as in Figure 4.11. Δ includes the sides L, M, and I of the triangle.

Calculate the integral of $-P$ over this region:

$$-\iint_\Delta P(x, t)\, dA = \iint_\Delta (c^2 u_{xx} - u_{tt})\, dA = \iint_\Delta \left(\frac{\partial}{\partial x}(c^2 u_x) - \frac{\partial}{\partial t}(u_t) \right) dA.$$

Apply Greene's theorem to the last integral, with x and t as the variables instead of x and y. This converts the double integral to a line integral around the boundary C of Δ. This piecewise smooth curve, which consists of three line segments, is oriented counterclockwise. We obtain

$$-\iint_\Delta P(x, t)\, dA = \oint_C u_t\, dx + c^2 u_x\, dt.$$

Now evaluate the line integral over each of the line segments comprising C. On I, $t = 0$ (so $dt = 0$) and x varies from $x_0 - ct_0$ to $x_0 + ct_0$. Then

$$\int_I u_t\, dx + c^2 u_x\, dt = \int_{x_0 - ct_0}^{x_0 + ct_0} u_t(x, 0)\, dx = \int_{x_0 - ct_0}^{x_0 + ct_0} \psi(s)\, ds.$$

On M, $x - ct = x_0 - ct_0$ so $dx = c\, dt$ and

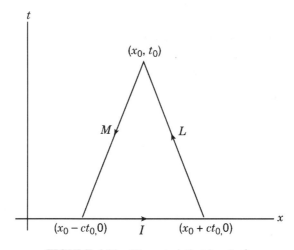

$$t$$

$$(x_0, t_0)$$

$$M \qquad\qquad L$$

$$x$$

$$(x_0 - ct_0, 0) \qquad I \qquad (x_0 + ct_0, 0)$$

FIGURE 4.11   Characteristic triangle Δ.

$$\int_L u_t \, dx + c^2 u_x \, dt = \int_L u_t c \, dt + c^2 u_x \frac{1}{c} \, dx = c \int_L u_t \, dt + u_x \, dx$$

$$= c \int_L du = c(u(x_0 - ct_0, 0) - u(x_0, t_0))$$

because in the counterclockwise orientation M extends from (x_0, t_0) to $(x_0 - ct_0, 0)$.

On L, $x + ct = x_0 + ct_0$, so $dx = -c \, dt$ and

$$\int_M u_t \, dx + c^2 u_x \, dt = \int_M u_t(-c) \, dt + c^2 u_x \left(-\frac{1}{c} \right) dx = -c \int_M du$$

$$= -c(u(x_0, t_0) - u(x_0 + ct_0, 0)).$$

Upon summing these line integrals we obtain

$$-\iint_\Delta P(x, t) \, dA = \int_{x_0 - ct_0}^{x_0 + ct_0} \psi(s) \, ds + cu(x_0 - ct_0, 0)$$

$$- 2cu(x_0, t_0) + cu(x_0 + ct_0, 0).$$

Now $u(x_0 - ct_0, 0) = \varphi(x_0 - ct_0)$ and $u(x_0 + ct_0, 0) = \varphi(x_0 + ct_0)$. We can therefore solve for $u(x_0, t_0)$ in the last equation to obtain

$$u(x_0, t_0) = \frac{1}{2} (\varphi(x_0 + ct_0) + \varphi(x_0 - ct_0)) + \frac{1}{2c} \int_{x_0 - ct_0}^{x_0 + ct_0} \psi(s) \, ds$$

$$+ \frac{1}{2c} \iint_\Delta P(x, t) \, dA.$$

We have used x_0 and t_0 in this derivation to be able to use (x, t) for a point of Δ. However, once we have this formula it is convenient to drop the subscript notation and write the solution as a function of x and t:

$$u(x, t) = \frac{1}{2} (\varphi(x + ct) + \varphi(x - ct)) + \frac{1}{2c} \int_{x - ct}^{x + ct} \psi(s) \, ds$$

$$+ \frac{1}{2c} \iint_\Delta P(\xi, \eta) \, d\xi \, d\eta. \quad (4.19)$$

The solution to this nonhomogeneous problem is the d'Alembert formula for the homogeneous Cauchy problem ($P(x, t)$ identically zero), plus a constant times the integral of the forcing term over the characteristic triangle.

Example 26   Solve

$$u_{tt} = 4u_{xx} + x\cos(t) \text{ for } -\infty < x < \infty, \, t > 0$$

$$u(x, 0) = e^{-x}, \, u_t(x, 0) = \sin(x) \text{ for } -\infty < x < \infty.$$

Figure 4.12 shows the characteristic triangle Δ having (x, t) as a vertex. Evaluate

$$\int\int_{\Delta} \xi \cos(\eta) \, d\xi \, d\eta = \int_0^t \left(\int_{x-2t+2\eta}^{x+2t-2\eta} \xi \, d\xi \right) \cos(\eta) \, d\eta = \int_0^t 4x(t - \eta)\cos(\eta) \, d\eta$$

$$= 4x(1 - \cos(t)).$$

The solution is

$$u(x, t) = \frac{1}{2} (e^{-x-2t} + e^{-x+2t}) + \frac{1}{4} \int_{x-2t}^{x+2t} \sin(s) \, ds + x(1 - \cos(t))$$

$$= e^{-x} \cosh(t) - \frac{1}{4} (\cos(x + 2t) - \cos(x - 2t)) + x(1 - \cos(t)).$$

Some simplification yields

$$u(x, t) = e^{-x} \cosh(t) + \frac{1}{2} \sin(x)\sin(2t) + x(1 - \cos(t)). \qquad \blacksquare$$

EXERCISE 129   In each of problems 1 through 10, solve problem 4.18 for the given initial data and $P(x, t)$.

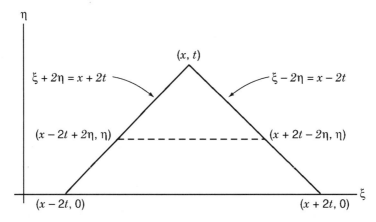

FIGURE 4.12   Determining the limits on the double integral in Example 26.

1. $\varphi(x) = x$, $\psi(x) = e^{-x}$, $P(x, t) = x + t$, $c = 4$
2. $\varphi(x) = \sin(x)$, $\psi(x) = 2x$, $P(x, t) = 2xt$, $c = 2$
3. $\varphi(x) = x^2 - x$, $\psi(x) = \cos(2x)$, $P(x, t) = t \cos(x)$, $c = 8$
4. $\varphi(x) = x^2$, $\psi(x) = xe^{-x}$, $P(x, t) = x \sin(t)$, $c = 4$
5. $\varphi(x) = \cosh(x)$, $\psi(x) = 1$, $P(x, t) = 3xt^2$, $c = 3$
6. $\varphi(x) = 1 + x$, $\psi(x) = \sin(x)$, $P(x, t) = x - \cos(t)$, $c = 7$
7. $\varphi(x) = \cos(2x)$, $\psi(x) = 1 - \cos(x)$, $P(x, t) = t^2$, $c = 2$
8. $\varphi(x) = x^3$, $\psi(x) = \sin^2(x)$, $P(x, t) = e^{-x} \cos(t)$, $c = 1$
9. $\varphi(x) = x \sin(x)$, $\psi(x) = e^{-x}$, $P(x, t) = xt$, $c = 2$
10. $\varphi(x) = 1 - x^2$, $\psi(x) = x \sin(x)$, $P(x, t) = t \sin(x)$, $c = 6$

4.7 THE WAVE EQUATION ON A BOUNDED INTERVAL

We will now consider a problem involving the wave equation on a bounded interval $[0, L]$:

$$u_{tt} = c^2 u_{xx} \text{ for } 0 < x < L, \, t > 0$$

$$u(x, 0) = \varphi(x), \, u_t(x, 0) = \psi(x) \text{ for } 0 \le x \le L \qquad (4.20)$$

$$u(0, t) = u(L, t) = 0 \text{ for } t \ge 0.$$

This models the motion of a string given initial position φ and initial velocity ψ, but stretched between 0 and L with these ends held motionless. This is an initial-boundary value problem with homogeneous boundary conditions.

One way to proceed is to attempt to adapt d'Alembert's formula, as we did for the problem on a half-line. First we must appropriately extend φ and ψ to the entire line, since we need to be able to evaluate these functions (or, more properly, their extensions) at points $x - ct$ and $x + ct$, and these can be any real numbers. Extend φ and ψ to odd functions defined on $[-L, L]$:

$$\varphi_o(x) = \begin{cases} \varphi(x) \text{ for } 0 \le x \le L \\ -\varphi(-x) \text{ for } -L \le x < 0 \end{cases}$$

and

$$\psi_o(x) = \begin{cases} \psi(x) \text{ for } 0 \le x \le L \\ -\psi(-x) \text{ for } -L \le x < 0. \end{cases}$$

The idea, which we have seen before (Figures 3.7(a) and (b)) is to reflect the graph of the function from 0 to L through the origin to produce an odd function defined on $[-L, L]$. Notice that $\varphi_o(x) = \varphi(x)$ and $\psi_o(x) = \psi(x)$ for $0 \le x \le L$.

Extend each of these functions periodically over the entire real line, with period $2L$. We saw this before in Figure 3.7(c), in which the idea is to duplicate the graph from $[-L, L]$ over the intervals $[L, 3L]$, $[3L, 5L]$, ..., $[-3L, -L]$, $[-5L, -3L]$, ... Let φ_p be the periodic extension of φ_o and ψ_p that of ψ_o. Now φ_p and ψ_p are defined over the real line. Further,

$$\varphi_p(x) = \varphi(x),\ \psi_p(x) = \psi(x)\ \text{for}\ 0 \leq x \leq L$$

and, for all x, and any integer n,

$$\varphi_p(x) = \varphi_p(x + 2nL)\ \text{and}\ \psi_p(x) = \psi_p(x + 2nL).$$

Now consider the familiar initial value problem

$$u_{tt} = c^2 u_{xx}\ \text{for}\ -\infty < x < \infty,\ t > 0$$

$$u(x, 0) = \varphi_p(x),\ u_t(x, 0) = \psi_p(x)\ \text{for}\ -\infty < x < \infty,$$

defined over the entire real line. We know that this problem has the unique solution

$$u(x, t) = \frac{1}{2}\left(\varphi_p(x - ct) + \varphi_p(x + ct)\right) + \frac{1}{2c}\int_{x-ct}^{x+ct} \psi_p(s)\, ds. \qquad (4.21)$$

We claim that this is also the solution of the initial-boundary value problem on $[0, L]$.

EXERCISE 130   Carry out the details of showing that equation 4.21 defines the solution of the initial-boundary value problem 4.20.

Unless φ and ψ have special properties, it is generally not possible to write $u(x, t)$, as given by equation 4.21, as a "simple" function of x and t. The solution 4.21 can, however, be used to evaluate $u(x, t)$ at any given position and time.

Example 27   Consider the initial-boundary value problem

$$u_{tt} = u_{xx}\ \text{for}\ 0 < x < 2,\ t > 0$$

$$u(x, 0) = x(2 - x),\ u_t(x, 0) = 0\ \text{for}\ 0 \leq x \leq 2$$

$$u(0, t) = u(2, t) = 0\ \text{for}\ t \geq 0.$$

Figure 4.13(a) shows a graph of $\varphi(x) = x(2 - x)$ on $[0, 2]$, Figure 4.13(b)

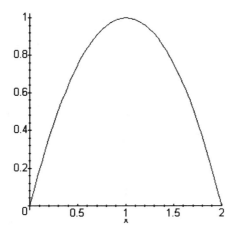

FIGURE 4.13(a)   Initial position function in Example 27.

its odd extension to $[-2, 2]$, and Figure 4.13(c) the extension of this function to the real line, with period 4. This periodic function is denoted φ_p.

Suppose we want $u(1, 6)$, the solution at $x = 1$ at time 6. The solution is

$$u(1, 6) = \frac{1}{2} \left(\varphi_p(-5) + \varphi_p(7) \right) = \frac{1}{2} \left(-1 + (-1) \right) = -1,$$

since

$$\varphi_p(-5) = \varphi_p(-5 + 4) = \varphi_p(-1) = -\varphi(1) = -1$$

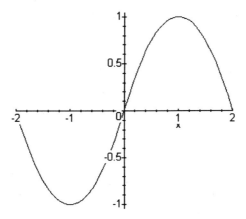

FIGURE 4.13(b)   Odd extension to $[-2, 2]$ of the initial position in Example 27.

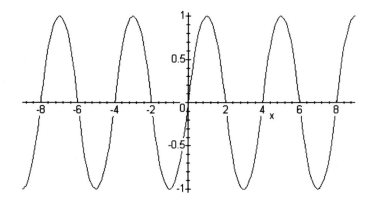

FIGURE 4.13(c)   Periodic extension of the function of Figure 4.14(b).

and

$$\varphi_p(7) = \varphi_p(-1 + 8) = \varphi_p(-1) = -1.$$

Similarly,

$$u(1/2, 3/4) = \frac{1}{2} (\varphi_p(-1/4) + \varphi_p(5/4)) = \frac{1}{2} \left(-\frac{7}{16} + \frac{15}{16} \right) = \frac{1}{4}. \qquad \blacksquare$$

EXERCISE 131   In each of problems 1 through 8, determine the requested values of the solution of the initial-boundary value problem 4.20, with the given information and $c = 1$.

1. $\varphi(x) = \sin(\pi x)$, $\psi(x) = 0$, $L = 1$; $u(1/3, 4)$, $u(1/2, 2)$, $u(3/4, 3)$
2. $\varphi(x) = x(3 - x)$, $\psi(x) = 0$, $L = 3$; $u(1/4, 1)$, $u(1, 2)$, $u(3/2, 2)$, $u(2, 3)$
3. $\varphi(x) = x(1 - x^2)$, $\psi(x) = 0$, $L = 1$; $u(1/5, 1)$, $u(3/4, 2)$, $u(3/5, 3)$
4. $\varphi(x) = x^2(4 - x)$, $\psi(x) = 0$, $L = 4$; $u(1/2, 2)$, $u(1, 1)$, $u(1, 4)$, $u(2, 2)$, $u(3, 2)$
5. $\varphi(x) = x(4 - x^2)$, $\psi(x) = 0$, $L = 2$; $u(1, 1/2)$, $u(1/2, 3)$, $u(1/3, 1)$, $u(1, 4)$
6. $\varphi(x) = 0$, $\psi(x) = e^{-x}$, $L = \pi$; $u(1/4, 1)$, $u(1/3, 1)$, $u(2, 5)$, $u(3, 2)$
7. $\varphi(x) = 0$, $\psi(x) = x$, $L = 2\pi$; $u(1, 5)$, $u(2, 1)$, $u(4, 4)$, $u(3, 5)$
8. $\varphi(x) = 0$, $\psi(x) = 3x^2$, $L = 2\pi$; $u(1, 4)$, $u(2, 1/2)$, $u(3, 1)$, $u(6, 3)$

EXERCISE 132   Consider the general problem

$$u_{tt} = c^2 u_{xx} + f(x, t) \text{ for } 0 < x < L, t > 0$$

$$u(0, t) = A(t), u(L, t) = B(t) \text{ for } t \geq 0$$

$$u(x, 0) = \varphi(x), u_t(x, 0) = \psi(x) \text{ for } 0 \leq x \leq L$$

assuming f, A, B, φ, and ψ are continuous and that the conditions are compatible at the end points. Assuming that this problem has a solution, show that this solution is unique. *Hint:* Suppose there are two solutions u and v and let $w = u - v$. Determine the initial boundary value problem satisfied by w and let

$$E(t) = \frac{1}{2} \int_0^L (w_t^2 + c^2 w_x^2) \, dx$$

for $t \geq 0$. Calculate $E'(t)$, using an integration by parts on the term involving w_{xt}, and show that $E'(t) = 0$ and hence $E(t)$ is constant. Finally, show that $E(t)$ is identically zero and conclude that $w_t^2 + c^2 w_x^2$ is identically zero. Show that $w(x, t) \equiv 0$.

EXERCISE 133   The partial differential equation

$$u_{tt} = u_{xx} - au_t - bu \text{ for } 0 < x < L, t > 0$$

is called the telegraph equation. Here a and b are positive constants. Suppose

$$u(0, t) = u(L, t) = 0 \text{ for } t \geq 0$$

$$u(x, 0) = \varphi(x), u_t(x, 0) = \psi(x) \text{ for } 0 \leq x \leq L.$$

(a) Prove that, for any $T > 0$,

$$\int_0^L (u_x^2 + u_t^2 + bu^2)_{t=T} \, dx \leq \int_0^L (u_x^2 + u_t^2 + bu^2)_{t=0} \, dx.$$

Hint: First show that

$$(2u_t u_x)_x = (u_x^2 + u_t^2 + bu^2)_t + 2au_t^2.$$

(b) Use the integral inequality from (a) to show that the initial boundary value problem for the telegraph equation can have only one solution.

4.8   A GENERAL PROBLEM ON A BOUNDED INTERVAL

In this section we will show how characteristics can be exploited to solve a general initial-boundary value problem on a bounded interval:

$$u_{tt} = c^2 u_{xx} \text{ for } 0 < x < L, \, t > 0$$

$$u(x, 0) = \varphi(x), \, u_t(x, 0) = \psi(x) \text{ for } 0 < x < L$$

$$u(0, t) = a(t), \, u(L, t) = b(t) \text{ for } t > 0.$$

This problem models the motion of a string with given initial position φ and initial velocity ψ, and having free ends whose motion is dictated by control functions a at the left end and b at the right end.

The characteristics of the wave equation are the lines

$$x - ct = \text{constant}, \, x + ct = \text{constant}.$$

The solution is based on segments of characteristics and the way they partition the strip $0 \leq x \leq L, \, t \geq 0$ into triangles and quadrilaterals, which are labeled *I, II, III,* ... in Figure 4.14.

Begin with the characteristic $x - ct = 0$ of slope $1/c$ through the origin. Think of a beam of light emanating from the origin and shining along this characteristic and think of the t-axis and the line $x = L$ as reflective walls. The beam will strike the wall $x = L$ at $(L, L/c)$ and reflect back upward to the left along the characteristic $x + ct = 2L$, where it will impact the t-axis at $(0, 2L/c)$. From here the beam will reflect upward to the right along the characteristic $x - ct = -2L$ and strike the wall $x = L$ at $(L, 3L/c)$. Here the beam will reflect upward to the left along the characteristic $x + ct = 4L$ to strike the t-axis at $(0, 4L/c)$, and the reflection process continues indefinitely.

Similarly, a beam originating at $(L, 0)$ and directed along the characteristic $x + ct = L$ will strike the t-axis at $(0, L/c)$, then reflect upward to the right along the characteristic $x - ct = -L$ to strike the line $x = L$ at $(L, 2L/c)$. From here it will reflect back along the characteristic $x + ct = 3L$ to the t-axis, and continue on.

We know $u(x, t)$ on the sides of the strip from the initial and boundary conditions. On the bottom side,

$$u(x, 0) = \varphi(x).$$

On the left side,

$$u(0, t) = a(t)$$

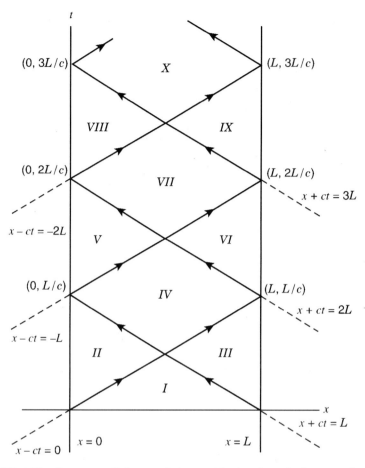

FIGURE 4.14   Segments of characteristics partitioning the strip $0 \le x \le L$, $t \ge 0$.

and on the right side,

$$u(L, t) = b(t).$$

Now we want to find $u(x, t)$ at interior points of the strip.

The key lies in the following observation. Consider the characteristic quadrilateral shown in Figure 4.15, formed from segments of four characteristics. P_1 and P_2 are opposite vertices, as are Q_1 and Q_2. We claim that, if u is any solution of the wave equation $u_{tt} = c^2 u_{xx}$ (regardless of initial and boundary conditions), then

$$u(P_1) + u(P_2) = u(Q_1) + u(Q_2). \tag{4.22}$$

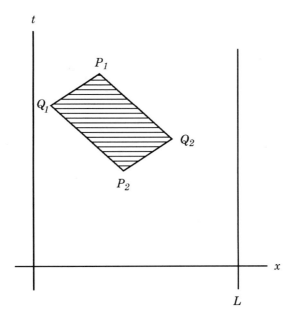

FIGURE 4.15 Typical characteristic quadrilateral.

EXERCISE 134   Derive equation 4.22. *Hint:* Recall equation 4.2.

The ramification of equation 4.22 is that, if we know $u(x, t)$ at three vertices of a characteristic quadrilateral, then we can determine its value at the fourth vertex.

Now begin working through the regions shown in Figure 4.14.

If (x, t) is in region *I*, then $u(x, t)$ is given by d'Alembert's formula, equation 4.7.

Next suppose $P:(x, t)$ is in region *II*. Form a characteristic quadrilateral as in Figure 4.16, having one vertex on the line $x = 0$ and two vertices on the piece of the characteristic from the origin bounding region *I*. From equation 4.22,

$$u(P) = u(A) + u(B) - u(C).$$

But, $u(A)$ is known because A is on the left boundary of the strip. And $u(B)$ and $u(C)$ are known because these are on the boundary of region *I*, where we know $u(x, t)$. We can therefore solve for $u(x, t)$ at any point in region *II*.

If $P:(x, t)$ is in region *III*, the construction shown in Figure 4.17 enables us to use equation 4.22 to again evaluate $u(x, t)$ in terms of known values of u at three vertices of a characteristic quadrilateral.

Now suppose $P:(x, t)$ is in region *IV*. Construct a characteristic quadrilateral as shown in Figure 4.18 and again use equation 4.22 to solve for $u(x, t)$ in

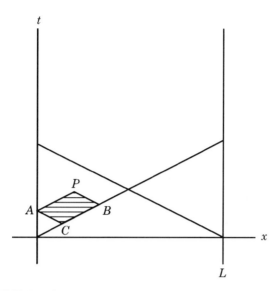

FIGURE 4.16 Characteristic quadrilateral in region *II*.

terms of previously determined values of u at the other three vertices of the quadrilateral.

Once we know $u(x, t)$ in region *IV*, we can continue these constructions to solve for $u(x, t)$ at points of *V* and *VI*; then move on to region *VII*; then *VIII* and *IX*, and so on.

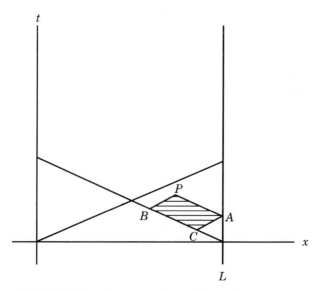

FIGURE 4.17 Characteristic quadrilateral in region *III*.

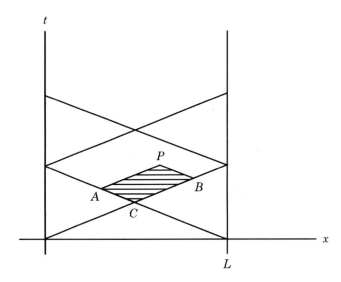

FIGURE 4.18   Characteristic quadrilateral in region *IV.*

Example 28   Consider the problem

$$u_{tt} = 9u_{xx} \text{ for } 0 < x < 4, \, t > 0$$

$$u(x, 0) = \varphi(x) = x \sin(\pi x) \text{ for } 0 < x < 4$$

$$u_t(x, 0) = \psi(x) = x(x - 4) \text{ for } 0 < x < 4$$

$$u(0, t) = u(4, t) = 9\pi t^2 \text{ for } t > 0.$$

We will construct $u(x, t)$ at various points (x, t). Since (x, t) is a point at which we will find the solution, we will use ξ and η as coordinates in the plane in computing equations of lines and points.

Characteristics through the origin and $(4, 0)$ divide the strip $0 \leq \xi \leq 4$, $\eta \geq 0$ into four regions, labeled, *I, II, III,* and *IV,* as we did previously in Figure 4.14.

First let $P : (x, t)$ be in region *I.* Here d'Alembert's formula applies and

$$u(x, t) = \frac{1}{2} (\varphi(x + 3t) + \varphi(x - 3t)) + \frac{1}{6} \int_{x-3t}^{x+3t} \psi(s) \, ds$$

$$= \frac{1}{2} (x + 3t)\sin(\pi(x + 3t)) + \frac{1}{2} (x - 3t)\sin(\pi(x - 3t))$$

$$+ x^2 t + 3t^3 - 4xt. \tag{4.23}$$

Next suppose $P : (x, t)$ is in region *II.* Figure 4.19 shows the characteristic

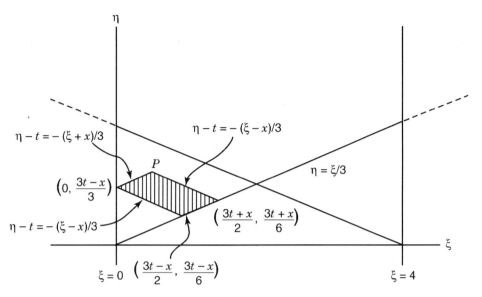

FIGURE 4.19 Characteristic quadrilateral in region *II* in Example 28.

quadrilateral having *P* as one vertex, together with the relevant lines and points, which are found by routine analytic geometry. From equation 4.22,

$$u(x, t) = u\left(0, t - \frac{x}{3}\right) + u\left(\frac{3t + x}{2}, \frac{3t + x}{6}\right) - u\left(\frac{3t - x}{2}, \frac{3t - x}{6}\right).$$

We can compute $u(0, t - x/3)$ from the boundary condition at $x = 0$, and we can compute

$$u\left(\frac{3t + x}{2}, \frac{3t + x}{6}\right) \text{ and } u\left(\frac{3t - x}{2}, \frac{3t - x}{6}\right)$$

from equation 4.23, which gives the solution in region *I*. This is a lengthy but routine calculation, and we obtain

$$u(x, t) = 9\pi \left(\frac{x - 3t}{3}\right)^2 + \frac{1}{2}(x + 3t)\sin(\pi(x + 3t))$$

$$- \frac{1}{2}(x - 3t)\sin(\pi(x - 3t)) + 3xt^2 + \frac{1}{9}x^3 - 4xt.$$

This gives $u(x, t)$ for (x, t) in region *II*.

Now suppose $P:(x, t)$ is in region *III*. Figure 4.20 shows the characteristic quadrilateral and the relevant points and lines for this case. Again applying equation 4.22,

$$u(x, t) = u\left(\frac{4 - 3t + x}{2}, \frac{4 + 3t - x}{6}\right) + u\left(4, \frac{-4 + 3t + x}{3}\right)$$
$$- u\left(\frac{12 - x - 3t}{2}, \frac{x + 3t - 4}{6}\right).$$

We know $u(4, (-4 + 3t + x)/3)$ from the given boundary condition at $x = 4$, and we know

$$u\left(\frac{4 + x - 3t}{2}, \frac{4 + 3t - x}{6}\right) \text{ and } u\left(\frac{12 - x - 3t}{2}, \frac{x + 3t - 4}{6}\right)$$

from the solution in region *I*. As in region *II*, there is a lengthy but routine calculation, after which we find

$$u(x, t) = \frac{1}{2}(x - 3t)\sin(\pi(x - 3t)) - \frac{1}{2}(8 - x - 3t)\sin(\pi(8 - x - 3t))$$
$$+ 9\pi\left(\frac{-4 + x + 3t}{3}\right)^2 + \frac{64}{9} + 12t^2 + 4xt + \frac{4}{3}x^2 - 3xt^2$$
$$- \frac{1}{9}x^3 - 16t - \frac{16}{3}x.$$

This gives the solution at points in region *III*.

We can now work our way up the strip, using equation 4.22 to calculate $u(x, t)$ in region *IV* of Figure 4.14, then in regions *V, VI,* and *VII,* and so on. ∎

The calculations involved in obtaining solution values, as just done in Example 28, can be quite involved and should, if possible, be done using computer software.

In the discussion thus far we have glossed over a significant point—compatibility of the initial and boundary conditions at $(0, 0)$ and at $(L, 0)$. To illustrate, consider $u(0, 0)$. On the other hand, this is $u(x, 0)$ at $x = 0$; hence $u(0, 0) = \varphi(0)$. But $u(0, 0)$ is also $u(0, t)$ at $t = 0$, so $u(0, 0) = a(0)$. This makes sense only if

$$\varphi(0) = a(0).$$

In general, to have a solution u that is continuous with continuous first and

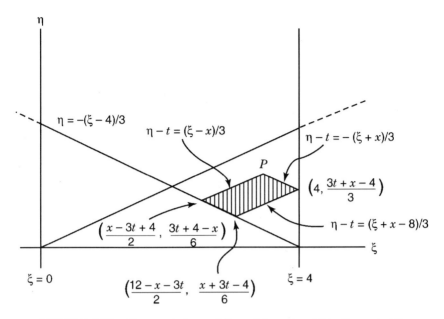

FIGURE 4.20   Characteristic quadrilateral in region *III* in Example 28.

second partial derivatives on $0 \le x \le L$, $t \ge 0$, the following *compatibility conditions* must be satisfied by the initial and boundary data:

$$a(0) = \varphi(0), \ a'(0) = \psi(0), \ a''(0) = c^2\varphi''(0) \tag{4.24}$$

at the left end, and

$$b(0) = \varphi(L), \ b'(0) = \psi(L), \ b''(0) = c^2\varphi''(L) \tag{4.25}$$

at the right end of the interval. It is easy to check that these conditions are satisfied in Example 28.

EXERCISE 135   Derive the compatibility conditions 4.24 and 4.25.

EXERCISE 136   Obtain $u(x, t)$ for (x, t) in regions *I*, *II*, and *III* for each of the following problems. Verify that the initial and boundary conditions are compatible in each problem.

1.
$$u_{tt} = u_{xx} \text{ for } 0 < x < 2, \ t > 0$$

$$u(x, 0) = x(2 - x), \ u_t(x, 0) = 0 \text{ for } 0 < x < 2$$

$$u(0, t) = -t^2 = u(2, t) \text{ for } t > 0.$$

2. $u_{tt} = 4u_{xx}$ for $0 < x < 3$, $t > 0$

$u(x, 0) = \sin(\pi x)$, $u_t(x, 0) = 0$ for $0 < x < 3$

$u(0, t) = u(3, t) = t^3$ for $t > 0$.

3. $u_{tt} = 9u_{xx}$ for $0 < x < 2$, $t > 0$

$u(x, 0) = x^2(2 - x)$, $u_t(x, 0) = 0$ for $0 < x < 2$

$u(0, t) = 18t^2$, $u(2, t) = -36t^2$ for $t > 0$.

4. $u_{tt} = 16u_{xx}$ for $0 < x < 1$, $t > 0$

$u(x, 0) = \sin(\pi x)$, $u_t(x, 0) = 4x$ for $0 < x < 1$

$u(0, t) = 16\pi t^2$, $u(1, t) = -16\pi t^2 + 4t$ for $t > 0$.

5. $u_{tt} = u_{xx}$ for $0 < x < 2$, $t > 0$

$u(x, 0) = x(2 - x)^2$, $u_t(x, 0) = x^2$ for $0 < x < 2$

$u(0, t) = -4t^2$, $u(2, t) = 2t^2 + 4t$ for $t > 0$.

EXERCISE 137   Obtain $u(x, t)$ for (x, t) in regions *I*, *II*, *III*, and *IV* for the problem:

$$u_{tt} = 4u_{xx} \text{ for } 0 < x < 1, t > 0$$

$$u(x, 0) = x \cos(\pi x) + x, \; u_t(x, 0) = x^2 \text{ for } 0 < x < 1$$

$$u(0, t) = 0, \; u(1, t) = 2\pi t^2 + t \text{ for } t > 0.$$

EXERCISE 138   In Example 28, obtain $u(x, t)$ for (x, t) in region *IV*.

4.9   FOURIER SERIES SOLUTION ON A BOUNDED INTERVAL

Consider again problem 4.20:

$$u_{tt} = c^2 u_{xx} \text{ for } 0 < x < L, t > 0$$

$$u(x, 0) = \varphi(x), \; u_t(x, 0) = \psi(x) \text{ for } 0 \le x \le L$$

$$u(0, t) = u(L, t) = 0 \text{ for } t \ge 0.$$

This initial-boundary value problem models the motion of an elastic string stretched from 0 to L, with initial position given by φ and initial velocity by ψ.

In the preceding section we solved a general version of this problem by exploiting the characteristics of the wave equation. In this section we will use separation of variables, or the Fourier method, which we first saw in Section 3.1.

Attempt a solution of the form $u(x, t) = X(x)T(t)$, the product of a function of x and a function of t. Substitute this into the wave equation to obtain

$$XT'' = c^2 X''T$$

or

$$\frac{T''}{c^2 T} = \frac{X''}{X}.$$

In this equation the left side is a function just of t, and the right side just of x. We have separated the variables. Because x and t are independent, we can put any positive value of t we like into the left side, and the right side must equal the resulting number for all x in $(0, L)$. Therefore X''/X is constant for $0 < x < L$, and so T''/c^2T equals the same constant for $t > 0$. This means that, for some constant λ,

$$\frac{T''}{c^2 T} = \frac{X''}{X} = -\lambda.$$

λ is called the *separation constant*. The negative sign on λ is a convention, and we would reach the same conclusion if we used λ in place of $-\lambda$.

We now have

$$X'' + \lambda X = 0 \text{ and } T'' + \lambda c^2 T = 0.$$

Since $u(0, t) = X(0)T(t) = 0$ for all t, we infer that $X(0) = 0$. Similarly, $u(L, t) = X(L)T(t) = 0$ implies that $X(L) = 0$. (Actually, we could have $T(t) = 0$ for all t if $u(x, t) = 0$, which is the correct solution if the initial displacement and velocity are identically zero. A string lying horizontally, initially at rest, given no displacement, and not acted upon by any driving force, does not move.)

The problem for X is therefore

$$X'' + \lambda X = 0; \quad X(0) = X(L) = 0. \tag{4.26}$$

We want values of λ (called *eigenvalues*) for which there are nontrivial solutions for X (called *eigenfunctions*) satisfying this ordinary differential equation and the boundary conditions.

We solved this problem for λ and X in Section 3.1, in the case $L = \pi$. In the interest of completeness, we will carry out the details of solution for the present case. Consider possibilities for λ.

If $\lambda = 0$, then $X'' = 0$ and $X(x) = ax + b$ for some constants a and b. But $X(0) = b = 0$, and $X(L) = aL = 0$ implies that $a = 0$ also. There is no nontrivial solution for X in problem 4.26 in the case $\lambda = 0$, so 0 is not an eigenvalue of this problem.

If $\lambda < 0$, say $\lambda = -k^2$, then $X'' - k^2X = 0$, so

$$X = ae^{kx} + be^{-kx}.$$

Now $X(0) = a + b = 0$ implies that $a = -b$, so

$$X = 2a \sinh(kx).$$

But then $X(L) = 2a \sinh(kL) = 0$ can be satisfied only if $a = 0$. This case leads only to trivial solutions for X; hence problem 4.26 has no negative eigenvalue.

If $\lambda > 0$, say $\lambda = k^2$, then $X'' + k^2X = 0$ and

$$X = a \cos(kx) + b \sin(kx).$$

Now $X(0) = a = 0$, so $X = b \sin(kx)$. Finally, we must come to grips with the condition

$$X(L) = b \sin(kL) = 0.$$

This equation is satisfied if $b = 0$, but then $X \equiv 0$ and we will have obtained no values of λ for which there are nontrivial solutions of problem 4.26. There is another possibility—we can require that kL be an integer multiple of π,

$$kL = n\pi \text{ for } n = 1, 2, 3, \ldots .$$

Then

$$\lambda_n = k^2 = n^2\pi^2/L^2 \text{ for } n = 1, 2, \ldots$$

is an eigenvalue for each positive integer n, and corresponding eigenfunctions are constant multiples of

$$X_n(x) = \sin(n\pi x/L).$$

Now we know the admissible values of λ and corresponding solutions for X satisfying problem 4.26. With these values of λ, the problem for T has the form

$$T'' + \frac{n^2\pi^2c^2}{L^2}\,T = 0$$

with general solution

$$T_n(t) = a_n \cos\left(\frac{n\pi ct}{L}\right) + b_n \sin\left(\frac{n\pi ct}{L}\right)$$

for $n = 1, 2, \ldots$.

For each positive integer n, we now have infinitely many functions

$$u_n(x, t) = X_n(x)T_n(t) = \left[a_n \cos\left(\frac{n\pi ct}{L}\right) + b_n \sin\left(\frac{n\pi ct}{L}\right)\right] \sin\left(\frac{n\pi x}{L}\right) \quad (4.27)$$

which satisfy the wave equation and the boundary conditions $u(0, t) = u(L, t) = 0$. The remaining issue is to find a solution satisfying the initial conditions.

Probably no one of the functions 4.27 satisfies the initial conditions. For example,

$$u_m(x, 0) = \varphi(x) = a_m \sin(m\pi x/L)$$

can hold for $0 < x < L$ and some constant a_m only if the initial position is a constant multiple of this eigenfunction, which need not be the case. Even if this were the case, the other initial condition

$$\left.\frac{\partial u_m}{\partial t}\right]_{t=0} = \psi(x) = \frac{m\pi c}{L}\,b_m \sin(m\pi x/L)$$

will not hold for $0 < x < L$ unless the initial velocity is also a constant multiple of this eigenfunction. Thus one of the functions 4.27 is a solution of the initial-boundary value problem only if both the initial position and velocity happen to be constant multiples of the same eigenvalue.

We could try a finite sum of the functions 4.27, setting

$$u(x, t) = \sum_{n=1}^{N} u_n(x, t).$$

This will also satisfy the wave equation and the boundary conditions. However, this will require that:

$$u(x, 0) = \varphi(x) = \sum_{n=1}^{N} a_n \sin(n\pi x/L)$$

and

$$\left. \frac{\partial u}{\partial t} \right]_{t=0} = \psi(x) = \sum_{n=1}^{N} \frac{n\pi c}{L} b_n \sin(n\pi x/L)$$

for $0 < x < L$. Except for very special choices of φ and ψ, we cannot choose the $a_n's$ and $b_n's$ for these equations to hold for $0 < x < L$. For example, if the initial position is given by $\varphi(x) = x(L - x)$, then we would have to have

$$u(x, 0) = x(L - x) = \sum_{n=1}^{N} a_n \sin(n\pi x/L) \text{ for } 0 < x < L,$$

and no finite sum of constant multiples of these sine functions can equal this polynomial over an entire interval.

EXERCISE 139   Prove the assertion just made that $x(L - x)$ cannot be written as a finite sum of these eigenfunctions for $0 < x < L$.

We therefore take a bold step and attempt a superposition of the functions 4.27 over all positive integers n:

$$u(x, t) = \sum_{n=1}^{\infty} \left[a_n \cos\left(\frac{n\pi ct}{L}\right) + b_n \sin\left(\frac{n\pi ct}{L}\right) \right] \sin\left(\frac{n\pi x}{L}\right). \quad (4.28)$$

We must choose the coefficients so that u satisfies the initial conditions. First, we need

$$u(x, 0) = \sum_{n=1}^{\infty} a_n \sin\left(\frac{n\pi x}{L}\right) = \varphi(x).$$

This is possible, if we choose a_n as the coefficient in the Fourier sine expansion of φ on $[0, L]$:

$$a_n = \frac{2}{L} \int_0^L \varphi(\xi)\sin\left(\frac{n\pi\xi}{L}\right) d\xi.$$

To solve for the $b_n's$, formally differentiate the series term by term to obtain

$$u_t(x, t) = \sum_{n=1}^{\infty} \frac{n\pi c}{L} \left[-a_n \sin\left(\frac{n\pi ct}{L}\right) + b_n \cos\left(\frac{n\pi ct}{L}\right) \right] \sin\left(\frac{n\pi x}{L}\right).$$

We need

$$u_t(x, 0) = \sum_{n=1}^{\infty} \frac{n\pi c}{L} b_n \sin\left(\frac{n\pi x}{L}\right) = \psi(x). \quad (4.29)$$

This is the Fourier sine expansion of ψ on $[0, L]$ if we choose

$$\frac{n\pi c}{L} b_n = \frac{2}{L} \int_0^L \psi(\xi)\sin\left(\frac{n\pi\xi}{L}\right) d\xi$$

and hence

$$b_n = \frac{2}{n\pi c} \int_0^L \psi(\xi)\sin\left(\frac{n\pi\xi}{L}\right) d\xi.$$

With these choices of the $a_n's$ and $b_n's$, we claim that equation 4.28 is the solution of the initial-boundary value problem, assuming only that φ and ψ have Fourier sine representations on $[0, L]$.

4.9.1   Verification of the Solution

Solution 4.28 just found by the Fourier method requires verification. It is clear that $u(0, t) = u(L, t) = 0$ because all terms of the series vanish if $x = 0$ or $x = L$. However, even though each u_n satisfies the wave equation, this infinite sum might not because we are not justified in differentiating the series term by term (this series does not converge uniformly). This means that the process of computing $u_t(x, 0)$ by term by term differentiation was questionable.

We will verify equation 4.28 as the solution in a perhaps surprising way—by summing the series and showing that equation 4.28 is just the d'Alembert formula 4.21 in disguise.

First, write the proposed solution 4.28 as

$$
\begin{aligned}
u(x, t) &= \sum_{n=1}^{\infty} a_n \sin\left(\frac{n\pi x}{L}\right) \cos\left(\frac{n\pi ct}{L}\right) + \sum_{n=1}^{\infty} b \sin\left(\frac{n\pi x}{L}\right) \sin\left(\frac{n\pi ct}{L}\right) \\
&= \sum_{n=1}^{\infty} \frac{1}{2} a_n \left[\sin\left(\frac{n\pi(x - ct)}{L}\right) + \sin\left(\frac{n\pi(x + ct)}{L}\right)\right] \\
&\quad + \sum_{n=1}^{\infty} \frac{1}{2} b_n \left[\cos\left(\frac{n\pi(x - ct)}{L}\right) - \cos\left(\frac{n\pi(x + ct)}{L}\right)\right].
\end{aligned}
\tag{4.30}
$$

We will treat the two series on the right individually. For the first, recall that

$$\sum_{n=1}^{\infty} a_n \sin\left(\frac{n\pi x}{L}\right)$$

is the Fourier sine expression of $\varphi(x)$ on $[0, L]$. Suppose this series converges to $\varphi(x)$ for $0 \le x \le L$. This occurs if $\varphi(0) = \varphi(L) = 0$, φ is continuous on $[0, L]$, and φ' is piecewise continuous on $[0, L]$. This sine series is periodic of period $2L$, and so is defined for all x, while $\varphi(x)$ is defined only for $0 \le x \le L$. First extend φ to an odd function φ_o defined on $[-L, L]$, then extend to a periodic function φ_p of period $2L$ defined for all x. Then

$$\varphi_p(x) = \sum_{n=1}^{\infty} a_n \sin\left(\frac{n\pi x}{L}\right)$$

for all x, and

$$\sum_{n=1}^{\infty} \frac{1}{2} a_n \left[\sin\left(\frac{n\pi(x - ct)}{L}\right) + \sin\left(\frac{n\pi(x + ct)}{L}\right) \right] = \frac{1}{2} (\varphi_p(x - ct) + \varphi_p(x + ct)).$$

Thus the series of sine terms on the right side of equation 4.30 converges to the term

$$\frac{1}{2} (\varphi_p(x - ct) + \varphi_p(x + ct))$$

occurring in the d'Alembert formula 4.21.

For the series of cosine terms on the right side of equation 4.30, write

$$\sum_{n=1}^{\infty} \frac{1}{2} b_n \left[\cos\left(\frac{n\pi(x - ct)}{L}\right) - \cos\left(\frac{n\pi(x + ct)}{L}\right) \right]$$

$$= \sum_{n=1}^{\infty} \frac{1}{2} \frac{n\pi}{L} b_n \int_{x-ct}^{x+ct} \sin\left(\frac{n\pi s}{L}\right) ds$$

$$= \frac{1}{2c} \int_{x-ct}^{x+ct} \sum_{n=1}^{\infty} \frac{n\pi c}{L} b_n \sin\left(\frac{n\pi s}{L}\right) ds = \frac{1}{2c} \int_{x-ct}^{x+ct} \psi(s) \, ds$$

from equation 4.29. We have assumed here that the sum and integral can be interchanged, and that $0 \le x - ct < x + ct \le L$. Extend ψ to an odd function ψ_o defined on $[-L, L]$, then extend ψ_o to a periodic function ψ_p of period $2L$ defined on the entire real line. Then

$$\sum_{n=1}^{\infty} \frac{1}{2} b_n \left[\cos\left(\frac{n\pi(x - ct)}{L}\right) - \cos\left(\frac{n\pi(x + ct)}{L}\right) \right] = \frac{1}{2c} \int_{x-ct}^{x+ct} \psi_p(s) \, ds$$

for all x and $t > 0$. The series of cosine terms on the right side of equation 4.30 therefore corresponds to the integral term on the right side of equation 4.21.

We now have

$$u(x, t) = \frac{1}{2} (\varphi_p(x - ct) + \varphi_p(x + ct)) + \frac{1}{2c} \int_{x-ct}^{x+ct} \psi_p(s) \, ds.$$

Equation 4.28, obtained by separation of variables, and the d'Alembert formula 4.21, are different forms of the solution of the initial-boundary value problem.

Example 29   Here is an example in which a string of length 1 unit is raised to an initial position and released from rest. Consider the problem

$$u_{tt} = 4u_{xx} \text{ for } 0 < x < 1, t > 0$$

$$u(x, 0) = \begin{cases} x \text{ for } 0 \le x \le 1/3 \\ 1/3 \text{ for } 1/3 \le x \le 2/3 \\ 1 - x \text{ for } 2/3 \le x \le 1 \end{cases}$$

$$u_t(x, 0) = 0 \text{ for } 0 \le x \le 1$$

$$u(0, t) = u(1, t) = 0 \text{ for } t \ge 0.$$

With $\psi(x) = 0$, each $b_n = 0$ in solution 4.28. Compute

$$a_n = 2 \int_0^1 \varphi(\xi)\sin(n\pi\xi) \, d\xi$$

$$= 2 \left(\int_0^{1/3} \xi \sin(n\pi\xi) \, d\xi + \int_{1/3}^{2/3} \frac{1}{3} \sin(n\pi\xi) \, d\xi + \int_{2/3}^1 (1 - \xi)\sin(n\pi\xi) \, d\xi \right)$$

$$= \frac{2}{n^2\pi^2} (\sin(n\pi/3) + \sin(2n\pi/3)).$$

The solution is

$$u(x, t) = \frac{2}{\pi^2} \sum_{n=1}^{\infty} \left(\frac{\sin(n\pi/3) + \sin(2n\pi/3)}{n^2} \right) \sin(n\pi x)\cos(2n\pi t).$$

Figure 4.21(a) shows the initial position of the string, and Figures 4.21(b) through (m) provide snapshots of the string at different times. ∎

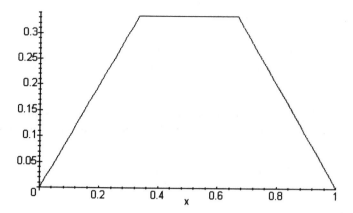

FIGURE 4.21(a)   Initial position of the string in Example 29.

Example 30   To contrast with the preceding example, here is a problem in which the initial position is zero (the string is not initially displaced), but it is given an initial velocity:

$$u_{tt} = 36u_{xx} \text{ for } 0 < x < 1, t > 0$$

$$u(x, 0) = 0 \text{ for } 0 \leq x \leq 1$$

$$u_t(x, 0) = 4 \text{ for } 0 < x < 1$$

$$u(0, t) = u(1, t) = 0 \text{ for } t \geq 0.$$

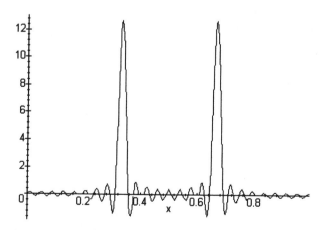

FIGURE 4.21(b)   String position at $t = 0.0037$.

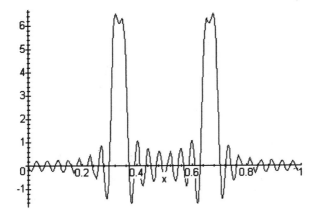

FIGURE 4.21(c) $t = 0.0068$.

With $\varphi(x) \equiv 0$, each $a_n = 0$ in solution 4.28, while

$$b_n = \frac{1}{3n\pi} \int_0^1 4\,\sin(n\pi\xi)\,d\xi = \frac{4}{3n^2\pi^2}\,(1 - (-1)^n).$$

The solution is

$$u(x,\,t) = \sum_{n=1}^{\infty} \frac{4}{3n^2\pi^2}\,(1 - (-1)^n)\sin(n\pi x)\sin(6n\pi t).$$

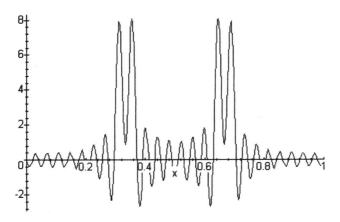

FIGURE 4.21(d) $t = 0.0093$.

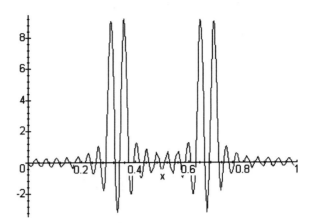

FIGURE 4.21(e) $t = 0.0126$.

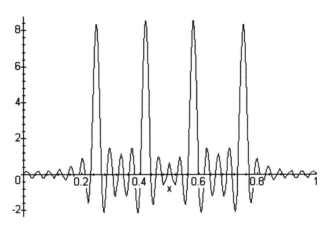

FIGURE 4.21(f) $t = 0.0427$.

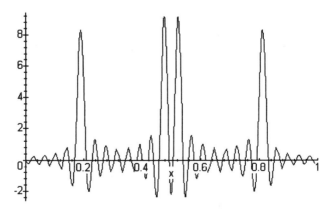

FIGURE 4.21(g) $t = 0.072$.

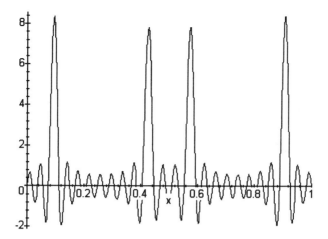

FIGURE 4.21(h) $t = 0.12$.

Figures 4.22(a) through (f) show the shape of the string at selected times. Some care must be exercised in reading these graphs. For example, the shape of the graphs in Figures 4.22(c) and (d) may not appear very different at a casual glance. However, notice that the height of the wave in Figure 4.22(d) is greater than that in Figure 4.22(c). Scales on the vertical axis may vary from graph to graph, even though the graphs are drawn about the same size.   ■

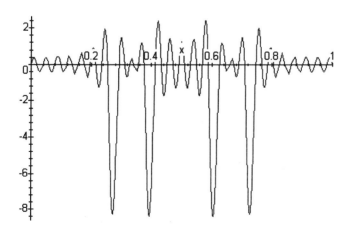

FIGURE 4.21(i) $t = 0.47$.

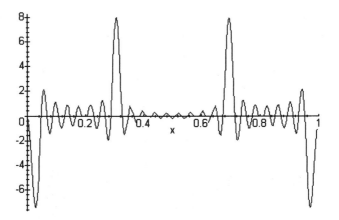

FIGURE 4.21(j) $t = 0.82$.

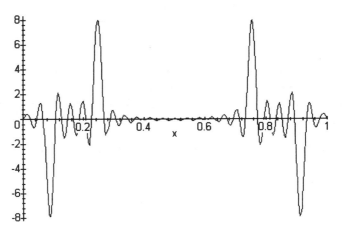

FIGURE 4.21(k) $t = 1.21$.

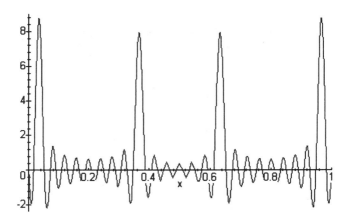

FIGURE 4.21(l) $t = 3.85$.

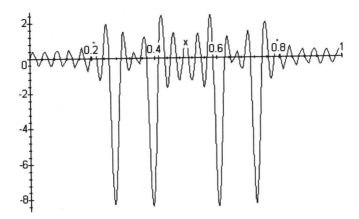

FIGURE 4.21(m) $t = 4.47$.

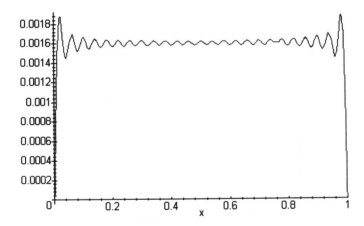

FIGURE 4.22(a)   Shape of the string of Example 30 at time $t = 0.0004$.

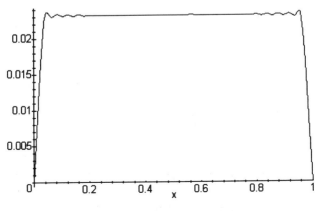

FIGURE 4.22(b) $t = 0.0058$.

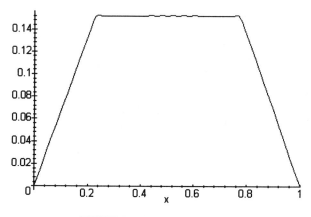

FIGURE 4.22(c) $t = 0.038$.

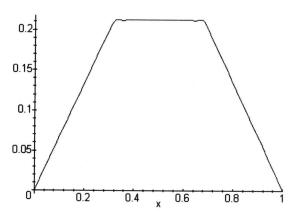

FIGURE 4.22(d) $t = 0.053$.

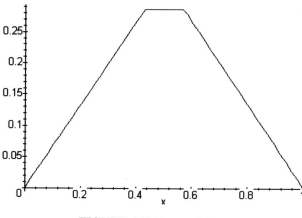

FIGURE 4.22(e) $t = 0.095$.

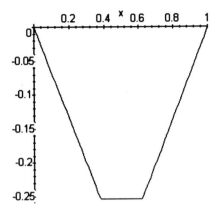

FIGURE 4.22(f) $t = 1.27$.

Example 31 This example has the string put in a nonzero initial configuration, and released with a nonzero initial velocity. Solve

$$u_{tt} = 9u_{xx} \text{ for } 0 < x < \pi, t > 0$$

$$u(x, 0) = x^2(\pi - x), u_t(x, 0) = \sin(x) \text{ for } 0 \le x \le \pi$$

$$u(0, t) = u(\pi, t) = 0 \text{ for } t \ge 0.$$

To allow comparison, we will write the solution obtained by the Fourier method, and also that obtained by making odd periodic extensions of φ and ψ.

Method 1 Extend φ and ψ to odd periodic functions φ_p and ψ_p of period 2π. Graphs of φ, its odd extension to $(-\pi, \pi]$, and its odd periodic extension φ_p are shown in Figures 4.23(a) through (c), while the odd periodic extension of ψ is just the familiar sine function defined on the entire real line. The solution is

$$u(x, t) = \frac{1}{2}(\varphi_p(x - 3t) + \varphi_p(x + 3t)) + \frac{1}{6}\int_{x-3t}^{x+3t} \psi_p(s) \, ds.$$

In this case $\psi_p(x) = \sin(x)$ for all x, since $\sin(x)$ is an odd function having period 2π. Therefore

$$u(x, t) = \frac{1}{2}(\varphi_p(x - 3t) + \varphi_p(x + 3t)) + \frac{1}{6}[\cos(x - 3t) - \cos(x + 3t)]$$

$$= \frac{1}{2}(\varphi_p(x - 3t) + \varphi_p(x + 3t)) + \frac{1}{3}\sin(x)\sin(3t).$$

This solution appears simple enough, but we must be careful in evaluating

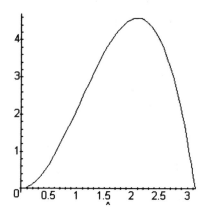

FIGURE 4.23(a)   Initial position function φ in Example 31.

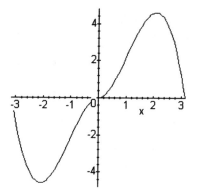

FIGURE 4.23(b)   Odd extension of φ to $(-\pi, \pi)$.

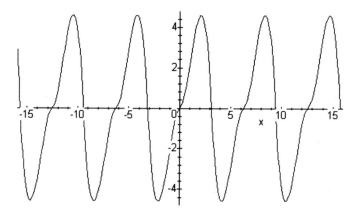

FIGURE 4.23(c)   Period extension of the function of Figure 4.23(b).

$\varphi_p(x - 3t)$ and $\varphi_p(x + 3t)$ for particular x and t because $\varphi_p(x)$ is not given directly for all x, but is constructed as an odd period extension of φ.

Method 2 We will write solution 4.28 obtained by the Fourier method. The coefficients are

$$a_n = \frac{2}{\pi} \int_0^\pi s^2(\pi - s)\sin(ns) \, ds = -\frac{4}{n^3} [1 + 2(-1)^n]$$

and

$$b_n = \frac{2}{3n\pi} \int_0^\pi \sin(s)\sin(ns) \, ds = \begin{cases} 0 \text{ if } n = 2, 3, \ldots \\ \dfrac{1}{3} \text{ if } n = 1. \end{cases}$$

Solution 4.28 is

$$u(x, t) = 4 \sin(x)\cos(3t) - \sum_{n=2}^{\infty} -\frac{4}{n^3} [1 + 2(-1)^n]\sin(nx)\cos(3nt) + \frac{1}{3} \sin(x)\sin(3t).$$

The first two terms in this expression form the Fourier sine expansion of

$$\frac{1}{2} (\varphi_p(x - 3t) + \varphi_p(x + 3t))$$

and the solutions obtained by d'Alembert's formula, and by separation of variables, agree.

The solution by Fourier method is convenient for viewing snapshots of the motion at different times. Figures 4.24(a) through (n) show graphs of this solution at different times. ■

EXERCISE 140 In each of problems 1 through 10, write the solution of the initial-boundary value problem 4.20 for the given information, using the Fourier method. Generate graphs of the string at different times, as in the examples.

1. $\varphi(x) = 1 - \cos(x)$, $\psi(x) = 0$, $c = 3$, $L = 2\pi$
2. $\varphi(x) = 0$, $\psi(x) = x$, $c = 2$, $L = 1$
3. $\varphi(x) = x(1 - x)$, $\psi(x) = 0$, $c = 6$, $L = 1$
4. $\varphi(x) = \sin(x)$, $\psi(x) = 0$, $c = 4$, $L = \pi$
5. $\varphi(x) = 0$, $\psi(x) = e^{-x}$, $c = 3$, $L = 2$
6. $\varphi(x) = 1 - \cos(x)$, $\psi(x) = 0$, $c = 4$, $L = 2\pi$
7. $\varphi(x) = \sin(x)$, $\psi(x) = x$, $c = 3$, $L = \pi$

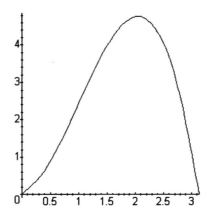

FIGURE 4.24(a) Position of the string in Example 31 at time $t = 0.1$.

8. $\varphi(x) = x^2(2 - x)$, $\psi(x) = x^2$, $c = 5$, $L = 2$
9. $\varphi(x) = \sin^2(x)$, $\psi(x) = 1$, $c = 1$, $L = \pi$
10. $\varphi(x) = x(1 - x^2)$, $\psi(x) = \cos(x)$, $c = 3$, $L = 1$

EXERCISE 141   In interpreting solutions of the wave equation as describing a vibrating string, it is interesting to study the effect c has on the motion. Take the solution found in Example 29 and adjust it for c equal to 0.2, 0.8, 3, 6, and then 10. For each choice of c, graph the solution on the same x, u-axis system for $t = 0.3$, then for $t = 0.5$, $t = 0.8$, $t = 1.2$, $t = 2$, and $t = 3$ seconds. This gives, for each of these times, a comparison of the position of the string for different choices of c. Experiment with different times and choices for c.

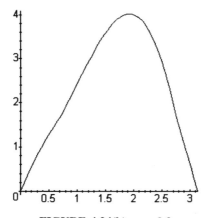

FIGURE 4.24(b) $t = 0.2$.

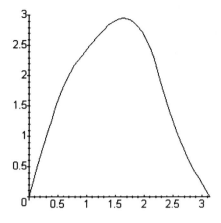

FIGURE 4.24(c) $t = 0.3$.

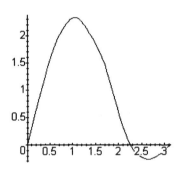

FIGURE 4.24(d) $t = 0.4$.

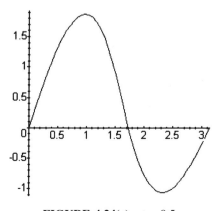

FIGURE 4.24(e) $t = 0.5$.

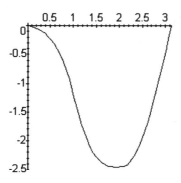

FIGURE 4.24(f) $t = 0.7$.

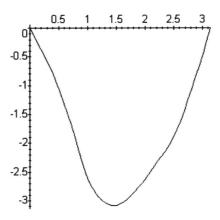

FIGURE 4.24(g) $t = 0.8$.

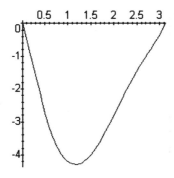

FIGURE 4.24(h) $t = 0.9$.

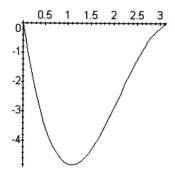

FIGURE 4.24(i) $t = 1.1$.

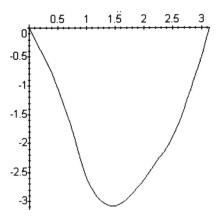

FIGURE 4.24(j) $t = 1.3$.

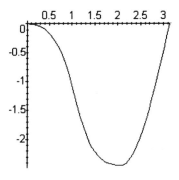

FIGURE 4.24(k) $t = 1.4$.

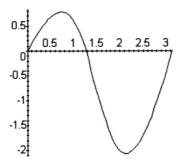

FIGURE 4.24(l) $t = 1.5$.

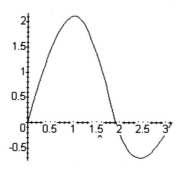

FIGURE 4.24(m) $t = 1.65$.

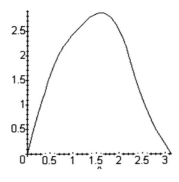

FIGURE 4.24(n) $t = 1.8$.

How would you describe the effect of increasing c on the ensuing motion at specific times?

EXERCISE 142   Repeat the program of Exercise 141 for the solution in Example 30.

EXERCISE 143   Repeat the program of Exercise 141 for the solution in Example 31.

EXERCISE 144   Solve the telegraph equation

$$u_{tt} + Au_t + Bu = c^2 u_{xx} \text{ for } 0 < x < L, \, t > 0$$

in which A, B, and a are positive constants. The boundary conditions are

$$u(0, t) = u(L, t) = 0 \text{ for } t > 0$$

and the initial conditions are

$$u(x, 0) = \varphi(x), \, u_t(x, 0) = 0 \text{ for } 0 < x < L.$$

Assume that $A^2 L^2 < 4(BL^2 + c^2 \pi^2)$. *Hint:* Proceed as in the derivation of equation 4.28, except now the differential equation for $T(t)$ is more complicated.

EXERCISE 145   The current $i(x, t)$ and voltage $v(x, t)$ at time t and distance x from one end of a transmission line satisfy the system

$$-v_x = Ri + Li_t$$

$$-i_x = Sv + Kv_t,$$

where R is the resistance, L the inductance, S the leakage conductance, and K the capacitance to ground all per unit length and all assumed constant. By differentiating appropriately and eliminating i, show that v satisfies the telegraph equation (Exercise 144). Similarly, show that i also satisfies the telegraph equation.

EXERCISE 146   Let $\theta(x, t)$ be the angular displacement at time t of the cross-section at x of a homogeneous cylindrical shell about the x-axis. It can be shown that

$$\theta_{tt} = a^2 \theta_{xx} \text{ for } 0 < x < L, \, t > 0.$$

Solve this equation subject to the conditions

$$\theta(x, 0) = \varphi(x), \; \theta_t(x, 0) = 0 \text{ for } 0 < x < L$$

if the ends of the shell are fixed elastically, which means that

$$\theta_x(0, t) - \alpha\theta(0, t) = 0, \; \theta_x(L, t) + \alpha\theta(L, t) = 0$$

for $t > 0$ and some positive constant α.

4.10 A NONHOMOGENEOUS PROBLEM ON A BOUNDED INTERVAL

Consider the initial-boundary value problem

$$u_{tt} = u_{xx} + Ax \text{ for } 0 < x < L, \; t > 0$$

$$u(x, 0) = u_t(x, 0) = 0 \text{ for } 0 \le x \le L$$

$$u(0, t) = u(L, t) = 0 \text{ for } t \ge 0.$$

This problem models the motion of a string pegged at its ends, with zero initial velocity and horizontal initial position, but with a forcing term acting parallel to the u-axis in the plane of motion and proportional to the distance from the left end of the string. A is a positive constant.

If we attempt a separation of variables by letting $u(x, t) = X(x)T(t)$, we get

$$XT'' = X''T + Ax$$

and we cannot isolate all terms involving x on one side of an equation, and all terms involving t on the other.

In such a case it is sometimes possible to transform the problem into one to which separation of variables applies. Let

$$u(x, t) = U(x, t) + f(x).$$

The idea is to choose f so that U satisfies an initial-boundary value problem we can solve. Substituting u into the wave equation yields

$$U_{tt} = U_{xx} + f''(x) + Ax$$

and this is just the homogeneous wave equation $U_{tt} = U_{xx}$ for U if we choose f so that

$$f''(x) + Ax = 0.$$

Next consider the boundary conditions. First

$$U(0, t) = u(0, t) - f(0) = -f(0)$$

and

$$U(L, t) = u(L, t) - f(L) = -f(L)$$

for $t \geq 0$. These conditions will reduce to $U(0, t) = 0$ and $U(L, t) = 0$ if we choose f so that

$$f(0) = f(L) = 0.$$

We therefore want to choose f so that

$$f''(x) + Ax = 0; f(0) = f(L) = 0.$$

First integrate $f''(x) = -Ax$ twice to obtain

$$f(x) = -\frac{A}{6} x^3 + Bx + C.$$

Then

$$f(0) = C = 0$$

and

$$f(L) = -\frac{A}{6} L^3 + BL = 0.$$

Thus choose $B = (A/6)L^2$ to obtain

$$f(x) = -\frac{1}{6} Ax^3 + \frac{1}{6} AL^2 x = \frac{A}{6} x(L^2 - x^2).$$

Finally, consider the initial conditions in the problem for U. We need

$$U(x, 0) = u(x, 0) - f(x) = -f(x)$$

and

$$U_t(x, 0) = u_t(x, 0) = 0.$$

In summary, the initial-boundary value problem for U is

$$U_{tt} = U_{xx} \text{ for } 0 < x < L, \, t > 0$$

$$U(x, 0) = -\frac{A}{6} x(L^2 - x^2) \text{ for } 0 \leq x \leq L$$

$$U_t(x, 0) = 0 \text{ for } 0 \leq x \leq L$$

$$U(0, t) = U(L, t) = 0 \text{ for } t \geq 0.$$

This problem was solved by the Fourier method in Section 4.9, except here we have U in place of u, $\varphi(x) = -(A/6)x(L^2 - x^2)$, $\psi(x) = 0$, and $c = 1$. From equation 4.28 the solution is

$$U(x, t) = \sum_{n=1}^{\infty} a_n \sin\left(\frac{n\pi x}{L}\right) \cos\left(\frac{n\pi t}{L}\right)$$

in which

$$a_n = \frac{2}{L} \int_0^L -\frac{A}{6} s(L^2 - s^2) \sin\left(\frac{n\pi s}{L}\right) ds = \frac{2AL^3(-1)^n}{n^3 \pi^3}.$$

Therefore

$$u(x, t) = \frac{2AL^3}{\pi^3} \sum_{n=1}^{\infty} \frac{(-1)^n}{n^3} \sin\left(\frac{n\pi x}{L}\right) \cos\left(\frac{n\pi t}{L}\right) + \frac{A}{6} x(L^2 - x^2).$$

Figures 4.25(a) through (i) show graphs of this solution for a succession of times, with $A = 1$ and $L = \pi$. Think of these graphs as photographs of the string at these times.

We can also use d'Alembert's formula to write the solution of this problem, using equation 4.21 and an odd periodic extension of φ. In this problem $\psi = 0$ so the integral term in equation 4.21 does not appear.

Example 32   Consider the initial-boundary value problem

$$u_{tt} = c^2 u_{xx} + K \text{ for } 0 < x < \pi, \, t > 0$$

$$u(0, t) = u(\pi, t) = 0 \text{ for } t > 0$$

$$u(x, 0) = \varphi(x) = x(\pi - x), \, u_t(x, 0) = 0 \text{ for } 0 \leq x \leq \pi,$$

in which K is a positive constant.

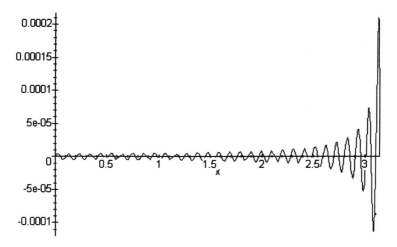

FIGURE 4.25(a) Position of the string at $t = 0.0006$.

Let $u(x, t) = U(x, t) + f(x)$. As in the above discussion, the idea is to choose f so that the problem for U is solvable by standard techniques. Substitute this expression into the partial differential equation to get

$$U_{tt} = c^2(U_{xx} + f''(x)) + K,$$

which is the homogeneous wave equation if we choose f so that $c^2 f''(x) + K = 0$, or

$$f''(x) = -\frac{K}{c^2}.$$

Next,

$$u(0, t) = U(0, t) + f(0) = U(0, t) = 0$$

if $f(0) = 0$, and

$$u(\pi, t) = U(\pi, t) + f(\pi) = U(\pi, t) = 0$$

if $f(\pi) = 0$. Thus choose f to be the solution of

$$f''(x) = -\frac{K}{c^2}; f(0) = f(\pi) = 0.$$

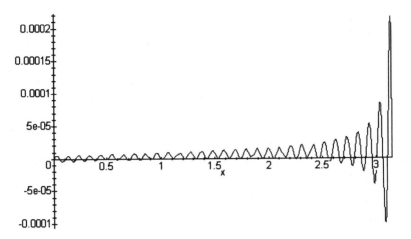

FIGURE 4.25(b) $t = 0.0028$.

By integrating twice and solving for the constants of integration, we obtain

$$f(x) = -\frac{K}{2c^2} x^2 + \frac{K\pi}{2c^2} x.$$

Finally,

$$u_t(x, 0) = U_t(x, 0) = 0 \text{ for } t \geq 0$$

and

$$u(x, 0) = U(x, 0) + f(x) = \varphi(x)$$

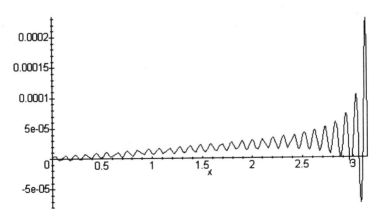

FIGURE 4.25(c) $t = 0.0047$.

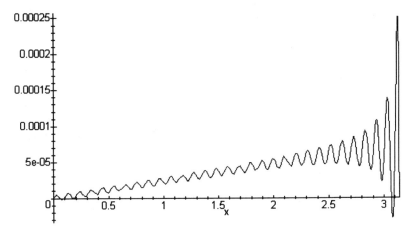

FIGURE 4.25(d) $t = 0.0069$.

if

$$U(x, 0) = \varphi(x) - f(x) \text{ for } 0 \leq x \leq L.$$

The problem for U is therefore the standard problem:

$$U_t = c^2 U_{xx} \text{ for } 0 < x < L, t > 0$$

$$U(x, 0) = x(\pi - x) + \frac{K}{2c^2} x(x - \pi), \ U_t(x, 0) = 0 \text{ for } 0 \leq x \leq \pi$$

$$U(0, t) = U(\pi, t) = 0 \text{ for } t \geq 0.$$

From equation 4.28 this problem has solution

$$U(x, t) = \sum_{n=1}^{\infty} a_n \sin(nx)\cos(nct)$$

in which

$$a_n = \frac{2}{\pi} \int_0^{\pi} \left(\xi(\pi - \xi) + \frac{K}{2c^2} \xi(\xi - \pi) \right) \sin(n\xi) \, d\xi = \frac{(2K - 4c^2)((-1)^n - 1)}{n^3 c^2 \pi}.$$

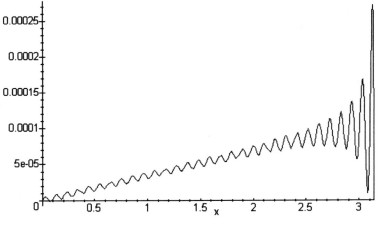

FIGURE 4.25(e) $t = 0.0083$.

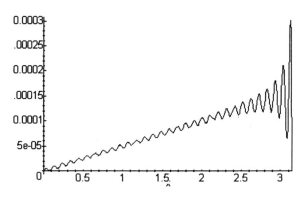

FIGURE 4.25(f) $t = 0.010$.

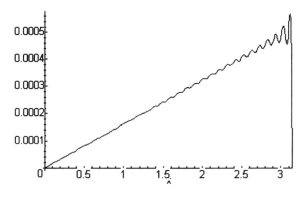

FIGURE 4.25(g) $t = 0.018$.

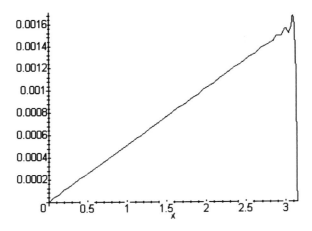

FIGURE 4.25(h) $t = 0.032$.

Therefore

$$u(x, t) = U(x, t) + f(x)$$

$$= \sum_{n=1}^{\infty} \left(\frac{(2K - 4c^2)((-1)^n - 1)}{n^3 c^2 \pi} \right) \sin(nx)\cos(nct) + \frac{Kx}{2c^2} (\pi - x).$$

We may think of this initial-boundary value problem as modeling the motion of a string initially placed in the configuration of Figure 4.26(a), released from rest, and having a constant vertical driving force acting on it. If the driving force has magnitude k per unit length, and ρ is the constant density of the

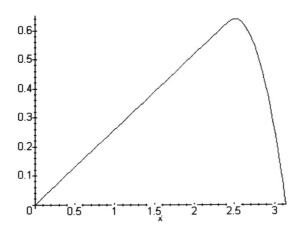

FIGURE 4.25(i) $t = 0.72$.

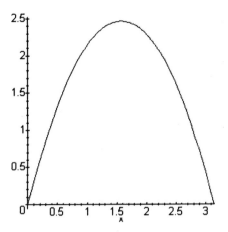

FIGURE 4.26(a)   Initial position of the string in Example 32.

string, then $K = k/\rho$. Figures 4.26(b) through (i) show string profiles at a succession of times, with $c = 2$ and $K = 3$.   ∎

EXERCISE 147   In Example 32, write the solution for the following values of K: 0.4, 2.5, 5.6, 10.2, and 15.1. Let $c = 2$, as in the graphs generated in the example. For each value of K, plot the graph of the solutions for the same value of t on the same u, t-axes, using the time values, in Figures 4.26(b) through (i). This will suggest, for each time, the effect of the magnitude of the forcing term K on the resulting motion. How would you describe this effect?

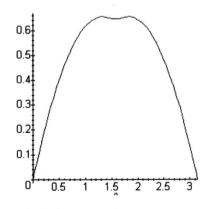

FIGURE 4.26(b)   Position at time $t = 0.86$.

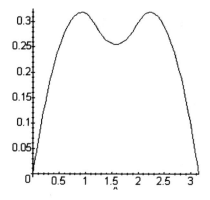

FIGURE 4.26(c) $t = 0.98$.

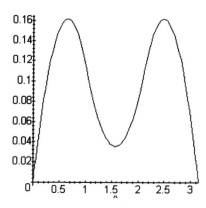

FIGURE 4.26(d) $t = 1.06$.

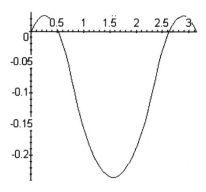

FIGURE 4.26(e) $t = 1.18$.

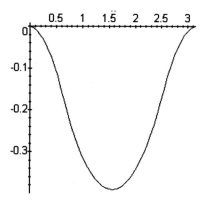

FIGURE 4.26(f) $t = 1.27$.

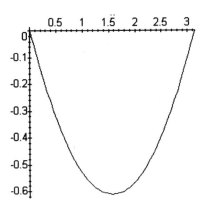

FIGURE 4.26(g) $t = 1.62$.

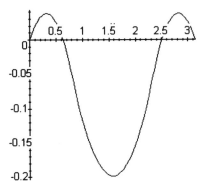

FIGURE 4.26(h) $t = 1.98$.

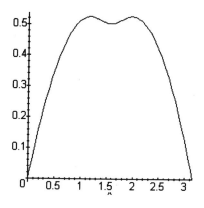

FIGURE 4.26(i) $t = 2.24$.

EXERCISE 148   Solve

$$u_{tt} = 9u_{xx} + Ax^2 \text{ for } 0 < x < 1, t > 0$$

$$u(x, 0) = u_t(x, 0) = 0 \text{ for } 0 \leq x \leq 1$$

$$u(0, t) = u(1, t) = 0 \text{ for } t \geq 0,$$

in which A is a positive constant. Generate sets of graphs of the string at different times, first with $A = 1.5$, then with $A = 5.7$, and finally with $A = 15.3$. This provides some feeling for the influence of the forcing term whose magnitude is dictated by A. How would you describe the effect of increasing A on the motion of the string at a particular time?

EXERCISE 149   Solve

$$u_{tt} = 4u_{xx} + \cos(x) \text{ for } 0 < x < \pi, t > 0$$

$$u(x, 0) = u_t(x, 0) = 0 \text{ for } 0 \leq x \leq \pi$$

$$u(0, t) = u(\pi, t) = 0 \text{ for } t \geq 0.$$

Generate graphs of the string at different times.

EXERCISE 150   Write the solution of the problem of Exercise 149 with $\cos(\alpha x)$ in place of $\cos(x)$. Choose a value of t and graph the solutions corresponding to $\alpha = 0.5, 1.2, 1.6, 2.1,$ and 2.8 on the same set of axes. Repeat this experiment for different times. What is the effect on the subsequent motion of increasing the frequency factor α in the forcing term?

EXERCISE 151   Solve

$$u_{tt} = u_{xx} - 3e^{-x} \text{ for } 0 < x < 2, \, t > 0$$

$$u(x, 0) = u_t(x, 0) = 0 \text{ for } 0 \le x \le 2$$

$$u(0, t) = u(2, t) = 0 \text{ for } t \ge 0.$$

Draw graphs of the string showing its position at different times.

EXERCISE 152   Solve

$$u_{tt} = 16u_{xx} - e^{-x} \text{ for } 0 < x < 3, \, t > 0$$

$$u(x, 0) = u_t(x, 0) = 0 \text{ for } 0 \le x \le 3$$

$$u(0, t) = u(3, t) = 0 \text{ for } t \ge 0.$$

Draw graphs of the string at various times.

EXERCISE 153   Solve

$$u_{tt} = 9u_{xx} + 4x \text{ for } 0 < x < 1, \, t > 0$$

$$u(0, t) = u(1, t) = 0 \text{ for } t > 0$$

$$u(x, 0) = 0, \, u_t(x, 0) = 1 \text{ for } 0 < x < 1.$$

EXERCISE 154   Solve

$$u_{tt} = 4u_{xx} \text{ for } 0 < x < 9, \, t > 0$$

$$u(0, t) = 0, \, u(9, t) = 1 \text{ for } t > 0$$

$$u(x, 0) = 0, \, u_t(x, 0) = x \text{ for } 0 < x < 9.$$

Draw graphs showing the shape of the string at different times.

EXERCISE 155   Try to adapt the method of this action to solve

$$u_{tt} = 4u_{xx} + t^2 \text{ for } 0 < x < 1, t > 0$$

$$u(x, 0) = u_t(x, 0) = 0 \text{ for } 0 \le x \le 1$$

$$u(0, t) = u(1, t) = 0 \text{ for } t \ge 0.$$

EXERCISE 156   Consider the general wave equation

$$u_{tt} = c^2 u_{xx} + g(x, t) \text{ for } 0 < x < L, t > 0.$$

Suppose u is a solution which is continuous with continuous first and second partial derivatives. Prove that

$$\frac{d}{dt} \int_a^b \frac{1}{2} [u_t^2 + c^2 u_x^2] \, dx = c^2 u_t u_x]_a^b + \int_a^b g u_t \, dx$$

for $0 < a \le x \le b < L$ any $t > 0$. *Hint:* Multiply the wave equation by u_t, integrate the resulting equation from a to b, and use integration by parts on one of the resulting integrals. It will also be useful to note that

$$u_t u_{tt} = \frac{1}{2} \frac{\partial}{\partial t} u_t^2,$$

with a similar relationship for $u_x u_{xt}$.

4.11   FOURIER INTEGRAL SOLUTION OF THE CAUCHY PROBLEM

Consider again the Cauchy problem

$$u_{tt} = c^2 u_{xx} \text{ for } -\infty < x < \infty, t > 0$$

$$u(x, 0) = \varphi(x), u_t(x, 0) = \psi(x) \text{ for } -\infty < x < \infty. \tag{4.31}$$

Previously we wrote a solution by d'Alembert's formula. We will now illustrate the Fourier method on this unbounded interval, considering the special case that the initial velocity $\psi(x)$ is identically zero.

Let $u(x, t) = X(x)T(t)$ in the wave equation to obtain

$$XT'' = c^2 X''T$$

or, as with the bounded interval,

$$X'' + \lambda X = 0, \; T'' + \lambda c^2 T = 0$$

for some separation constant λ. Consider cases on λ, noting that the ordinary differential equation for X has no boundary conditions to place limitations on solutions.

If $\lambda = 0$ then $X = ax + b$. There are no boundary conditions to restrict a or b. However, we will impose the condition that the solution must be bounded. Then $a = 0$ and $X =$ constant. This means that $\lambda = 0$ is an eigenvalue with constant eigenfunctions.

If $\lambda < 0$, write $\lambda = -\omega^2$ with $\omega > 0$. Then $X = ae^{\omega x} + be^{-\omega x}$. But $e^{\omega x} \to \infty$ as $x \to \infty$, so we choose $a = 0$ to have a bounded solution. And $e^{-\omega x} \to \infty$ as $x \to -\infty$ so we must choose $b = 0$ also. The case $\lambda < 0$ allows no nontrivial bounded solutions for X. This problem has no negative eigenvalue.

If $\lambda > 0$, write $\lambda = \omega^2$ with $\omega > 0$. In this case $X = a\cos(\omega x) + b\sin(\omega x)$. This is a bounded solution for X for any choices of the constants a and b. Every positive number $\lambda = \omega^2$ is an eigenvalue, with eigenfunctions

$$a\cos(\omega x) + b\sin(\omega x).$$

Since this eigenfunction is constant if $\omega = 0$, we can combine the cases $\lambda = 0$ and $\lambda > 0$ to write

$$X_\omega(x) = a_\omega \cos(\omega x) + b_\omega \sin(\omega x)$$

for $\lambda = \omega^2 \geq 0$.

Next consider the equation for T. In view of what we have found about λ this equation is

$$T'' + c^2\omega^2 T = 0$$

with $\omega \geq 0$. If $\omega = 0$ this has solution $T =$ constant. If $\omega > 0$ then

$$T = d\cos(\omega ct) + k\sin(\omega ct).$$

However, since the initial velocity is zero,

$$u_t(x, 0) = X(x)T'(0) = 0,$$

implying that $T'(0) = 0$. This forces us to choose $k = 0$.

For $\omega \geq 0$, we now have functions of the form

$$u_\omega(x, t) = [a_\omega \cos(\omega x) + b_\omega \sin(\omega x)]\cos(\omega ct)$$

which satisfy the wave equation and the condition that the initial velocity is zero. To satisfy the condition that the initial position is given by $\varphi(x)$, we must generally attempt a superposition, which in this case is over all $\omega \geq 0$. Such a superposition is accomplished by integrating $u_\omega(x, t)$ over $\omega \geq 0$, with $\int_0^\infty \cdots d\omega$ replacing the infinite series $\Sigma_{n=1}^\infty$ superposition used in the case of a bounded interval. There we had an eigenvalue $\lambda_n = n^2\pi^2/L^2$ corresponding to each positive integer (*discrete spectrum*), while in the present circumstance the eigenvalues cover the right half-line $\omega \geq 0$ (*continuous spectrum*). The totality of eigenvalues is called the *spectrum* of the partial differential equation.

We are therefore led to attempt a solution in the form of an integral superposition

$$u(x, t) = \int_0^\infty u_\omega(x, t) \, d\omega = \int_0^\infty [a_\omega \cos(\omega x) + b_\omega \sin(\omega x)]\cos(\omega ct) \, d\omega. \quad (4.32)$$

We require that

$$u(x, 0) = \int_0^\infty [a_\omega \cos(\omega x) + b_\omega \sin(\omega x)] \, dx = \varphi(x).$$

This is the Fourier integral representation of φ, so choose

$$a_\omega = \frac{1}{\pi} \int_{-\infty}^\infty \varphi(s) \cos(\omega s) \, ds \text{ and } b_\omega = \frac{1}{\pi} \int_{-\infty}^\infty \varphi(s)\sin(\omega s) \, ds.$$

If φ is absolutely integrable and continuous, and φ' is piecewise continuous, the Fourier integral presentation of φ converges to $\varphi(x)$ for all x and equation 4.32 is the solution of the Cauchy problem, with these choices of a_ω and b_ω.

Example 33   Let

$$\varphi(x) = \begin{cases} \cos(x) \text{ for } -\pi/2 \leq x \leq \pi/2 \\ 0 \text{ for } |x| > \pi/2. \end{cases}$$

Then

$$a_\omega = \frac{1}{\pi} \int_{-\pi/2}^{\pi/2} \cos(s)\cos(\omega s) \, ds = \begin{cases} \dfrac{2}{\pi} \dfrac{\cos(\pi\omega/2)}{1 - \omega^2} \text{ for } \omega \neq 1 \\ \dfrac{1}{2} \text{ for } \omega = 1. \end{cases}$$

Notice that a_ω is actually a continuous function of ω, since

$$\lim_{\omega \to 1} a_\omega = \frac{1}{2} = a_1.$$

Next,

$$b_\omega = \frac{1}{\pi} \int_{-\pi/2}^{\pi/2} \cos(s)\sin(\omega s) \, ds = 0.$$

The solution for this initial displacement, and zero initial velocity, is

$$u(x, t) = \frac{2}{\pi} \int_0^\infty \frac{\cos(\pi\omega/2)}{1 - \omega^2} \cos(\omega x)\cos(\omega ct) \, d\omega. \qquad \blacksquare$$

4.11.1 Solution by Fourier Transform

We will illustrate the use of the Fourier transform for the Cauchy problem just solved (again with $\psi = 0$). Since $-\infty < x < \infty$ we can attempt a Fourier transform $\mathscr{F}$ in x, carrying t through the transform as a parameter. Let $\mathscr{F}[u(x, t)](\omega) = \hat{u}(\omega, t)$.

Apply $\mathscr{F}$ to the wave equation:

$$\mathscr{F}[u_{tt}] = c^2 \mathscr{F}[u_{xx}].$$

Now

$$\mathscr{F}[u_{tt}(x, t)](\omega) = \int_{-\infty}^\infty u_{tt}(x, t)e^{-i\omega x} \, dx = \frac{\partial^2}{\partial t^2} \int_{-\infty}^\infty u(x, t)e^{-i\omega x} \, dx = \hat{u}_{tt}(\omega, t).$$

The operation $\partial/\partial t$ of differentiation with respect to time passes through the transform because the transform is with respect to x, and x and t are independent.

For the transform of u_{xx} use the operational formula for the Fourier transform (equation 3.28 in Section 3.7):

$$\mathscr{F}[u_{xx}(x, t)](\omega) = (i\omega)^2 \hat{u}(\omega, t) = -\omega^2 \hat{u}(\omega, t).$$

Therefore, application of the transform in the space variable to the wave equation yields:

$$\hat{u}_{tt}(\omega, t) = -c^2\omega^2\hat{u}(\omega, t).$$

Think of this as an ordinary differential equation in t, with ω carried along as a parameter. The general solution is

$$\hat{u}(\omega, t) = \alpha_\omega \cos(\omega ct) + \beta_\omega \sin(\omega ct).$$

Now,

$$\alpha_\omega = \hat{u}(\omega, 0) = \mathscr{F}[u(x, 0)](\omega) = \mathscr{F}[\varphi(x)](\omega) = \hat{\varphi}(\omega),$$

the Fourier transform of the initial position function. And

$$\omega c \beta_\omega = \hat{u}_t(\omega, 0) = \mathcal{F}[u_t(x, 0)](\omega) = 0$$

because we are considering the case that the initial velocity is zero. Therefore $\beta_\omega = 0$ and

$$\hat{u}(\omega, t) = \hat{\varphi}(\omega)\cos(\omega ct).$$

This is the Fourier transform of the solution of the Cauchy problem. Apply the inverse Fourier transform (equation 3.27, Section 3.7):

$$u(x, t) = \mathcal{F}^{-1}[\hat{u}(\omega, t)] = \frac{1}{2\pi} \int_{-\infty}^{\infty} \hat{\varphi}(\omega)\cos(\omega ct)e^{i\omega x} \, d\omega. \qquad (4.33)$$

We will show that this solution equals the solution 4.32 obtained using the Fourier integral. For this discussion only, let $u_{tr}(x, t)$ denote the solution 4.33 obtained by transform, and $u_{fi}(x, t)$ the solution 4.32 by Fourier integral. Begin by inserting the formula for $\hat{\varphi}(\omega)$ into the solution by transform 4.33:

$$u_{tr}(x, t) = \frac{1}{2\pi} \int_{-\infty}^{\infty} \left(\int_{-\infty}^{\infty} \varphi(s)e^{-i\omega s} \, ds \right) \cos(\omega ct)e^{i\omega x} \, d\omega$$

$$= \frac{1}{2\pi} \int_{-\infty}^{\infty} \int_{-\infty}^{\infty} e^{-i\omega(s-x)} \cos(\omega ct) \, d\omega \varphi(s) \, ds$$

$$= \frac{1}{2\pi} \int_{-\infty}^{\infty} \int_{-\infty}^{\infty} (\cos(\omega(x - s)) - i \sin(\omega(s - x)))\cos(\omega ct) \, d\omega \varphi(s) \, ds$$

$$= \frac{1}{2\pi} \int_{-\infty}^{\infty} \int_{-\infty}^{\infty} \cos(\omega(s - x))\cos(\omega ct)\varphi(s) \, d\omega \, ds,$$

the coefficient of i being omitted because $u_{tr}(x, t)$ is real. The integrand in the last integral is an even function in ω, being unchanged by a replacement of ω by $-\omega$. Therefore

$$u_{tr}(x, t) = \frac{1}{\pi} \int_{-\infty}^{\infty} \int_{0}^{\infty} \cos(\omega(s - x))\cos(\omega ct)\varphi(s) \, d\omega \, ds.$$

Now go back to solution 4.32 by Fourier integral and insert the formulas for the coefficients to obtain

$$u_{fi}(x, t) = \frac{1}{\pi} \int_0^\infty \left[\left(\int_{-\infty}^\infty \varphi(s)\cos(\omega s) \, ds \right) \cos(\omega x) \right.$$

$$+ \left(\int_{-\infty}^\infty \varphi(s)\sin(\omega s) \, ds \right) \sin(\omega x) \left. \right] \cos(\omega c t) \, d\omega$$

$$= \frac{1}{\pi} \int_{-\infty}^\infty \left(\int_0^\infty [\cos(\omega s)\cos(\omega x) + \sin(\omega s)\sin(\omega x)]\cos(\omega c t) \, d\omega \right) \varphi(s) \, ds$$

$$= \frac{1}{\pi} \int_{-\infty}^\infty \int_0^\infty \cos(\omega(s - x))\cos(\omega c t)\varphi(s) \, d\omega \, ds = u_{tr}(x, t).$$

EXERCISE 157   In each of problems 1 through 6, solve the boundary value problem 4.31 for the given initial position, assuming zero initial velocity, using the Fourier integral approach through separation of variables.

1. $\varphi(x) = e^{-|x|}$

2. $\varphi(x) = \begin{cases} 1 - x^2 & \text{for } -1 \le x \le 1 \\ 0 & \text{for } |x| > 1 \end{cases}$

3. $\varphi(x) = \begin{cases} \sin(x) & \text{for } -\pi \le x \le \pi \\ 0 & \text{for } |x| > \pi \end{cases}$

4. $\varphi(x) = \begin{cases} 4 - |x| & \text{for } -4 \le x \le 4 \\ 0 & \text{for } |x| > 4 \end{cases}$

5. $\varphi(x) = \begin{cases} e^{-x} & \text{for } x \ge 0 \\ 0 & \text{for } x < 0 \end{cases}$

6. $\varphi(x) = \begin{cases} \sin^2(x) & \text{for } -4\pi \le x \le 4\pi \\ 0 & \text{for } |x| > 4\pi \end{cases}$

EXERCISE 158   Use the Fourier method to write an integral solution of

$$u_{tt} = c^2 u_{xx} \text{ for } -\infty < x < \infty, \, t > 0$$

$$u(x, 0) = 0, \, u_t(x, 0) = \psi(x) \text{ for } -\infty < x < \infty.$$

EXERCISE 159   Solve the boundary value problem of Exercise 158 using the Fourier transform. Show that the solution obtained by transform agrees with that obtained by the separation of variables.

EXERCISE 160   In each of problems 1 through 4, use the Fourier method to solve the boundary value problem 4.31 with zero initial displacement, and the given initial velocity function.

1. $\psi(x) = e^{-|x|}$

2. $\psi(x) = \begin{cases} k \text{ for } -\alpha \leq x \leq \alpha \\ 0 \text{ for } |x| > \alpha \end{cases}$, in which k and α are positive constants

3. $\psi(x) = \begin{cases} \cos(x) \text{ for } -\pi/2 \leq x \leq \pi/2 \\ 0 \text{ for } |x| > \pi/2 \end{cases}$

4. $\psi(x) = \begin{cases} x \text{ for } -5 \leq x \leq 5 \\ 0 \text{ for } |x| > 5 \end{cases}$

4.12 A WAVE EQUATION IN TWO SPACE DIMENSIONS

Consider the initial-boundary value problem

$$u_{tt} = c^2(u_{xx} + u_{yy}) \text{ for } 0 < x < a, \ 0 < y < b$$

$$u(x, 0, t) = u(x, b, t) = 0 \text{ for } 0 \leq x \leq a, \ t \geq 0$$

$$u(0, y, t) = u(a, y, t) = 0 \text{ for } 0 \leq y \leq b, \ t \geq 0$$

$$u(x, y, 0) = \varphi(x, y), \ u_t(x, y, 0) = 0 \text{ for } 0 \leq x \leq a, \ 0 \leq y \leq b.$$

This problem models vibrations of a membrane fastened on a rectangular frame, given an initial displacement $\varphi(x, y)$ at (x, y), but no initial velocity.

To attempt a separation of variables with this problem, put $u(x, y, t) = X(x)Y(y)T(t)$ into the wave equation to obtain

$$XYT'' = c^2(X''YT + XY''T)$$

or

$$\frac{T''}{c^2T} - \frac{Y''}{Y} = \frac{X''}{X}.$$

The left side of this equation can be fixed by choosing specific values of y and t. Therefore the right side is constant for $0 < x < a$. For some constant λ,

$$\frac{T''}{c^2T} - \frac{Y''}{Y} = \frac{X''}{X} = -\lambda.$$

Now

$$\frac{T''}{c^2T} + \lambda = \frac{Y''}{Y} \text{ and } X'' + \lambda X = 0.$$

In the equation for T and Y, the left side depends only on t and the right side only on y, and these variables are independent. Therefore for some constant μ,

$$\frac{T''}{c^2 T} + \lambda = \frac{Y''}{Y} = -\mu.$$

Now we have

$$Y'' + \mu Y = 0 \text{ and } T'' + (\lambda + \mu)c^2 T = 0.$$

With two space variables we need two separation constants because it is impossible to isolate three variables independently on opposite sides of one equation.

Now consider the boundary conditions. First,

$$u(0, y, t) = X(0)Y(y)T(t) = 0$$

implies that $X(0) = 0$. Similarly, $u(a, y, t) = 0$ implies that $X(a) = 0$. The problem for X is

$$X'' + \lambda X = 0; \ X(0) = X(a) = 0.$$

We solved this problem in Section 4.9, finding eigenvalues and eigenfunctions

$$\lambda_n = \frac{n^2 \pi^2}{a^2}, \ X_n(x) = \sin\left(\frac{n\pi x}{a}\right) \text{ for } n = 1, 2, \ldots.$$

Since $u(x, 0, t) = u(x, b, t) = 0$, then $Y(0) = Y(b) = 0$ and the problem for Y is

$$Y'' + \mu Y = 0; \ Y(0) = Y(b) = 0.$$

This problem has eigenvalues and eigenfunctions

$$\mu_m = \frac{m^2 \pi^2}{b^2}, \ Y_m(y) = \sin\left(\frac{m\pi y}{b}\right) \text{ for } m = 1, 2, \ldots.$$

Notice that n and m independently take on all positive integer values.

Finally, since $u_t(x, y, 0) = X(x)Y(y)T'(0) = 0$, then $T'(0) = 0$. The problem for T is

$$T'' + \left(\frac{n^2 \pi^2}{a^2} + \frac{m^2 \pi^2}{b^2}\right) c^2 T = 0; \ T'(0) = 0.$$

This has as solution any constant multiple of

$$T_{nm}(y) = \cos(\alpha_{nm}ct)$$

in which

$$\alpha_{nm} = \sqrt{\frac{n^2\pi^2}{a^2} + \frac{m^2\pi^2}{b^2}}.$$

For each positive integer n and m we now have a function

$$u_{nm}(x, y, t) = b_{nm} \sin\left(\frac{n\pi x}{a}\right) \sin\left(\frac{m\pi y}{b}\right) \cos(\alpha_{nm}ct)$$

that satisfies the wave equation, the boundary conditions, and the initial condition $u_t(x, y, 0) = 0$. There remains to satisfy the initial position condition, and this generally requires a superposition of these functions. This is a double sum, since n and m independently take on all positive integer values. Let

$$u(x, y, t) = \sum_{n=1}^{\infty} \sum_{m=1}^{\infty} b_{nm} \sin\left(\frac{n\pi x}{a}\right) \sin\left(\frac{m\pi y}{b}\right) \cos(\alpha_{nm}ct). \qquad (4.34)$$

Now we need to choose the coefficients to satisfy

$$u(x, y, 0) = \sum_{n=1}^{\infty} \sum_{m=1}^{\infty} b_{nm} \sin\left(\frac{n\pi x}{a}\right) \sin\left(\frac{m\pi y}{b}\right) = \varphi(x, y). \qquad (4.35)$$

This is a *double Fourier sine expansion* of φ on the rectangle. We will reason informally to obtain the coefficients as follows. Imagine that y is fixed. Then $\varphi(x, y)$ is a function of x on $[0, a]$ (for each y in $[0, b]$). Think of

$$\sum_{n=1}^{\infty} \left(\sum_{m=1}^{\infty} b_{nm} \sin\left(\frac{m\pi y}{b}\right)\right) \sin\left(\frac{n\pi x}{a}\right) = \varphi(x, y) \qquad (4.36)$$

as the Fourier sine expansion of $\varphi(x, y)$, which is a function of x for fixed y. This means that the coefficient (the summation in large brackets in equation 4.36) is the Fourier sine coefficient in this expansion. Therefore, for each $n = 1, 2, \ldots$,

$$\sum_{m=1}^{\infty} b_{nm} \sin\left(\frac{m\pi y}{b}\right) = \frac{2}{a} \int_0^a \varphi(\xi, y) \sin\left(\frac{n\pi \xi}{a}\right) d\xi \equiv h_n(y). \qquad (4.37)$$

The integral in equation 4.37 defines a function $h_n(y)$ for $0 \le y \le b$. We can think of

$$h_n(y) = \sum_{m=1}^{\infty} b_{nm} \sin\left(\frac{m\pi y}{b}\right)$$

as the Fourier sine expansion of h_n on $[0, b]$, for $n = 1, 2, 3, \ldots$. Therefore b_{nm} is the mth Fourier sine coefficient in this expansion:

$$b_{nm} = \frac{2}{b}\int_0^b h_n(\zeta)\sin\left(\frac{m\pi\zeta}{b}\right) d\zeta$$

$$= \frac{2}{b}\int_0^b \int_0^a \left(\frac{2}{a}\varphi(\xi, \zeta)\sin\left(\frac{n\pi\xi}{a}\right) d\xi\right) \sin\left(\frac{m\pi\zeta}{b}\right) d\zeta$$

$$= \frac{4}{ab}\int_0^b \int_0^a \varphi(\xi, \zeta)\sin\left(\frac{n\pi\xi}{a}\right) \sin\left(\frac{m\pi\zeta}{b}\right) d\xi \, d\zeta.$$

By tracking back through this derivation of b_{nm} and using the theorem on convergence of Fourier sine series, it is possible to derive sufficient conditions on φ for the double sine series 4.36 to converge to $\varphi(x, y)$ for $0 \le x \le a$, $0 \le y \le b$. When this occurs, equation 4.34 defines the solution of the initial-boundary value problem.

Example 34   Consider this problem for the case that $a = b = \pi$ and

$$\varphi(x, y) = xy(\pi - x)(\pi - y).$$

Compute

$$b_{nm} = \frac{4}{\pi^2}\int_0^\pi \int_0^\pi \xi(\pi - \xi)\sin(n\xi)\zeta(\pi - \zeta)\sin(m\zeta) \, d\xi \, d\zeta$$

$$\frac{16}{n^3 m^3 \pi^2} [(-1)^n - 1][(-1)^m - 1].$$

The solution is

$$u(x, y, t) = \frac{16}{\pi^2}\sum_{n=1}^{\infty} \sum_{m=1}^{\infty} \frac{[(-1)^n - 1][(-1)^m - 1]}{n^3 m^3} \sin(nx)\sin(my)\cos(\alpha_{nm}ct),$$

in which

$$\alpha_{nm} = \sqrt{n^2 + m^2}. \qquad\blacksquare$$

EXERCISE 161   Solve

$$u_t = 9(u_{xx} + u_{yy}) \text{ for } 0 < x < 3, \, 0 < y < 6, \, t > 0$$

$$u(x, 0, t) = u(x, 6, t) = 0 \text{ for } 0 \le x \le 3, \, t \ge 0$$

$$u(0, y, t) = u(3, y, t) = 0 \text{ for } 0 \le y \le 6, \, t \ge 0$$

$$u(x, y, 0) = \sin\left(\frac{\pi x}{3}\right) y(6 - y), \, u_t(x, y, 0) = 0 \text{ for } 0 \le x \le 3, \, 0 \le y \le 6.$$

EXERCISE 162   Solve

$$u_{tt} = 4(u_{xx} + u_{yy}) \text{ for } 0 < x < \pi, \, 0 < y < 2\pi, \, t > 0$$

$$u(x, 0, t) = u(x, 2\pi, t) = 0 \text{ for } 0 \le x \le \pi, \, t \ge 0$$

$$u(0, y, t) = u(\pi, y, t) = 0 \text{ for } 0 \le y \le 2\pi, \, t \ge 0$$

$$u(x, y, 0) = x^2(\pi - x)y^2(2\pi - y) \text{ for } 0 \le x \le \pi, \, 0 \le y \le 2\pi,$$

$$u_t(x, y, 0) = 0 \text{ for } 0 \le x \le \pi, \, 0 \le y \le 2\pi.$$

EXERCISE 163   Solve

$$u_{tt} = 16(u_{xx} + u_{yy}) \text{ for } 0 < x < 1, \, 0 < y < 1, \, t > 0$$

$$u(x, 0, t) = u(x, 1, t) = 0 \text{ for } 0 \le x \le 1, \, t \ge 0$$

$$u(0, y, t) = u(1, y, t) = 0 \text{ for } 0 \le y \le 1, \, t \ge 0$$

$$u(x, y, 0) = 0, \, u_t(x, y, 0) = \sin(\pi x)\sin(\pi y) \text{ for } 0 \le x \le 1, \, 0 \le y \le 1.$$

EXERCISE 164   Solve

$$u_{tt} = 9(u_{xx} + u_{yy}) \text{ for } 0 < x < \pi, \, 0 < y < \pi, \, t > 0$$

$$u(x, 0, t) = u(x, \pi, t) = 0 \text{ for } 0 \le x \le \pi, \, t \ge 0$$

$$u(0, y, t) = u(\pi, y, t) = 0 \text{ for } 0 \le y \le \pi, \, t \ge 0$$

$$u(x, y, 0) = \sin(x)\sin(y), \, u_t(x, y, 0) = xy \text{ for } 0 \le x \le \pi, \, 0 \le y \le \pi.$$

EXERCISE 165   Solve

$$u_{tt} = u_{xx} + u_{yy} \text{ for } 0 \le x \le 1, 0 \le y \le 1$$

$$u(x, 0, t) = u(x, 1, t) = 0 \text{ for } 0 \le x \le 1, t \ge 0$$

$$u(0, y, t) = u(1, y, t) = 0 \text{ for } 0 \le y \le 1, t \ge 0$$

$$u(x, y, 0) = x(x - 1)^2 y(y - 1) \text{ for } 0 \le y \le 1$$

$$u_t(x, y, 0) = xy^2 \text{ for } 0 \le x \le 1, 0 \le y \le 1.$$

EXERCISE 166   Solve

$$u_{tt} = u_{xx} + u_{yy} \text{ for } 0 < x < 1, 0 < y < \pi, t > 0$$

$$u(x, 0, t) = u(x, \pi, t) = 0 \text{ for } 0 \le x \le 1, t \ge 0$$

$$u(0, y, t) = u(1, y, t) = 0 \text{ for } 0 \le y \le \pi, t \ge 0$$

$$u(x, y, 0) = x \cos(\pi x/2) \text{ for } 0 \le x \le 1, 0 \le y \le \pi$$

$$u_t(x, y, 0) = x + y \text{ for } 0 \le x \le 1, 0 \le y \le \pi.$$

4.13   KIRCHHOFF/POISSON SOLUTION IN 3-SPACE AND HUYGENS' PRINCIPLE

We will derive an integral solution of the Cauchy problem for the wave equation in three space dimensions. Let R^3 denote 3-dimensional space consisting of all points (x, y, z).

We want to solve the Cauchy problem

$$u_{tt} = u_{xx} + u_{yy} + u_{zz}$$

$$u(x, y, z, 0) = \varphi(x, y, z) \text{ and } u_t(x, y, z, 0) = \psi(x, y, z)$$

for (x, y, z) in R^3 and $t \ge 0$. We will refer to this problem as CP. We have let $c = 1$ in this wave equation as a convenience.

Let VCP denote the special Cauchy problem

$$u_{tt} = u_{xx} + u_{yy} + u_{zz}$$

$$u(x, y, z, 0) = 0, u_t(x, y, z, 0) = \psi(x, y, z)$$

for (x, y, z) in R^3.

We will show that if we can solve the apparently simpler VCP for any reasonable ψ, then we can solve CP as well. Two observations are needed.

Observation One If w is the solution of CP having initial position φ and initial velocity zero, and v is the solution of CP having initial position zero and initial velocity ψ, then $w + v$ is the solution of CP having initial position φ and initial velocity ψ.

To verify this, let $h = w + v$. It is routine to check that h satisfies the wave equation. Next,

$$h(x, y, z, 0) = w(x, y, z, 0) + v(x, y, z, 0) = \varphi(x, y, z)$$

and

$$h_t(x, y, z, 0) = w_t(x, y, z, 0) + v_t(x, y, z, 0) = \psi(x, y, z).$$

We can therefore solve CP by solving two simpler problems, one with zero initial velocity, the other with zero initial position.

Observation Two If v is the solution of VCP with initial velocity $\eta(x, y, z)$, then v_t is the solution of CP with initial position $\eta(x, y, z)$ and zero initial velocity.

To verify this, let $u = v_t$. We must first show that u satisfies the wave equation. Compute

$$u_{tt} = (v_t)_{tt} = (v_{tt})_t = (v_{xx} + v_{yy} + v_{zz})_t = (v_t)_{xx} + (v_t)_{yy} + (v_t)_{zz}$$

$$= (u_{xx} + u_{yy} + u_{zz}).$$

Next,

$$u(x, y, z, 0) = v_t(x, y, z, 0) = \eta(x, y, z).$$

Finally,

$$u_t(x, y, z, 0) = v_{tt}(x, y, z, 0) = [v_{xx} + v_{yy} + v_{zz}]_{t=0} = 0$$

because $v(x, y, z, 0) = 0$ for all (x, y, z).

Using these observations, we can solve CP if we can solve VCP. First write the solution u_ψ of VCP having initial velocity ψ. Then write the solution u_φ of VCP having initial velocity φ. By Observation two, $\partial u_\varphi / \partial t$ is the solution of

CP with initial position φ and zero initial velocity. By Observation one, the solution of CP is

$$u = \frac{\partial u_\varphi}{\partial t} + u_\psi. \tag{4.38}$$

Now here is the point. We claim that there is an integral formula for the solution of VCP. Hence, using equation 4.38, we will be able to write a formula for the solution of CP in terms of integrals involving the initial position and velocity functions.

Here is the promised integral formula for the solution of VCP.

Theorem 13 Let ψ by continuous with continuous first and second partial derivatives for all (x, y, z). Then for $t > 0$ and (x, y, z) in R^3, the solution of VCP is

$$u(x, y, z, t) = \frac{1}{4\pi t} \iint\limits_{S(x,y,z;t)} \psi(\xi, \eta, \tau) \, d\sigma_t \tag{4.39}$$

in which $S(x, y, z; t)$ is the sphere of radius t centered at (x, y, z).   ■

$S(x, y, z; t)$ consists of all (ξ, η, ζ) satisfying

$$(\xi - x)^2 + (\eta - y)^2 + (\zeta - z)^2 = t^2.$$

The solution of VCP, at any point (x, y, z) and time $t > 0$, is therefore $1/4\pi t$ times the surface integral of the initial velocity, over the sphere of radius t about the point. In this surface integral $d\sigma_t$ is the differential element of surface area on $S(x, y, z; t)$. This solution is known as *Kirchhoff's formula*.

Proof: We must first show that u as defined by equation 4.39 satisfies the wave equation. It will be convenient to think of points in R^3 as vectors and to denote vectors by boldface letters. Write $\mathbf{x} = (x, y, z)$ as an arbitrary point in 3-space, so $u(x, y, z, t)$ can be written $u(\mathbf{x}, t)$. We will denote $S(x, y, z; t)$ as S_t. Let U be the sphere of radius 1 about the origin and let $d\sigma$ be the differential element of surface area on U. Then $d\sigma_t = t^2 \, d\sigma$. Let $\mathbf{n}_t$ be the unit outer normal vector on S_t, and $\mathbf{n}$ the unit outer normal on U.

The vector notation, equation 4.39 is

$$u(\mathbf{x}, t) = \frac{1}{4\pi t} \iint\limits_{S_t} \psi(\boldsymbol{\xi}) \, d\sigma_t \tag{4.40}$$

with $\boldsymbol{\xi} = (\xi, \eta, \tau)$ as the variable of integration in this surface integral over S_t. Transform this surface integral over S_t into one over U, in which the variable of integration is $\boldsymbol{\zeta}$, by putting

$$\boldsymbol{\xi} = \mathbf{x} + t\boldsymbol{\zeta}.$$

This transformation is shown in Figure 4.27, in which a point on S_t is the vector sum of $\mathbf{x}$ and $t\boldsymbol{\zeta}$, with $\boldsymbol{\zeta}$ a unit vector. We obtain

$$u(\mathbf{x}, t) = \frac{t}{4\pi} \iint_U \psi(\mathbf{x}, t\boldsymbol{\zeta}) \, d\sigma, \tag{4.41}$$

with $\boldsymbol{\zeta}$ the variable of integration over the unit sphere U. From equation 4.41,

$$u_{xx} + u_{yy} + u_{zz} = \frac{t}{4\pi} \iint_U [\psi_{xx} + \psi_{yy} + \psi_{zz}]_{\mathbf{x}+t\boldsymbol{\zeta}} \, d\sigma$$

$$= \frac{1}{4\pi t} \iint_{S_t} (\psi_{xx}(\boldsymbol{\xi}) + \psi_{yy}(\boldsymbol{\xi}) + \psi_{zz}(\boldsymbol{\xi})) \, d\sigma_t. \tag{4.42}$$

Again using equation 4.41,

$$u_t(\mathbf{x}, t) = \frac{1}{4\pi} \iint_U \psi(\mathbf{x}, t\boldsymbol{\zeta}) \, d\sigma + \frac{t}{4\pi} \iint_U \nabla\psi(\mathbf{x} + t\boldsymbol{\zeta}) \cdot \mathbf{n} \, d\sigma$$

$$= \frac{1}{t} u(\mathbf{x}, t) + \frac{1}{4\pi t} \iint_{S_t} \nabla\psi(\boldsymbol{\xi}) \cdot \mathbf{n}_t \, d\sigma_t, \tag{4.43}$$

in which ∇ is the gradient operator. Apply the divergence theorem of Gauss to the last integral:

$$\iint_{S_t} \nabla\psi(\boldsymbol{\xi}) \cdot \mathbf{n}_t \, d\sigma_t = \iiint_B div(\nabla\psi) \, dV,$$

where B is the solid ball consisting of all points on or interior to S_t. It is routine to verify that

$$div(\nabla\psi) = \psi_{xx} + \psi_{yy} + \psi_{zz}.$$

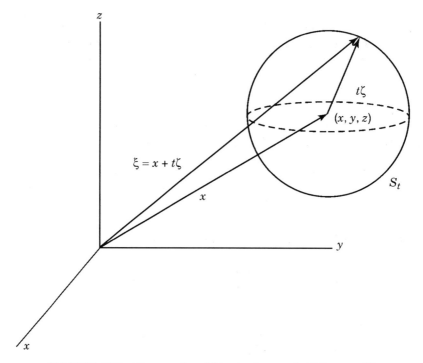

FIGURE 4.27   Change of variables in the proof of Theorem 13.

Therefore equation 4.43 becomes

$$u_t(\mathbf{x},\ t) = \frac{1}{t}\ u(\mathbf{x},\ t) + \frac{1}{4\pi t} \iiint_B (\psi_{xx} + \psi_{yy} + \psi_{zz})\ dV.$$

Denoting this triple integral by I, the last equation is

$$u_t(\mathbf{x},\ t) = \frac{1}{t}\ u(\mathbf{x},\ t) + \frac{1}{4\pi t}\ I. \qquad (4.44)$$

Then

$$u_{tt}(\mathbf{x},\ t) = -\frac{1}{t^2}\ u(\mathbf{x},\ t) + \frac{1}{t}\ u_t(\mathbf{x},\ t) + \frac{1}{4\pi t}\ I_t - \frac{1}{4\pi t^2}\ I$$

$$= \frac{1}{t}\left(-\frac{1}{t}\ u(\mathbf{x},\ t) + u_t(\mathbf{x},\ t) - \frac{1}{4\pi t}\ I\right) + \frac{1}{4\pi t}\ I_t = \frac{1}{4\pi t}\ I_t, \qquad (4.45)$$

with the term in large parentheses vanishing in view of equation 4.44. Now,

$$I_t = \int\int_{S_t} (\psi_{xx}(\boldsymbol{\xi}) + \psi_{yy}(\boldsymbol{\xi}) + \psi_{zz}(\boldsymbol{\xi}))\, d\sigma_t. \tag{4.46}$$

By equations 4.45, 4.46, and 4.42,

$$u_{tt} = \frac{1}{4\pi t} \int\int_{S_t} (\psi_{xx}(\boldsymbol{\xi}) + \psi_{yy}(\boldsymbol{\xi}) + \psi_{zz}(\boldsymbol{\xi}))\, d\sigma_t = u_{xx} + u_{yy} + u_{zz}.$$

Therefore the function u as defined by equation 4.39 satisfies the wave equation.

There remains to show that u satisfies the initial conditions. First,

$$u(\mathbf{x}, 0) = 0$$

by putting $t = 0$ into equation 4.41. Finally, from the first line of equation 4.43,

$$u_t(\mathbf{x}, 0) = \frac{1}{4\pi} \int\int_U \psi(\mathbf{x})\, d\sigma = \psi(\mathbf{x}) \frac{1}{4\pi} \int\int_U d\sigma = \psi(\mathbf{x})$$

since

$$\int\int_U d\sigma = \text{area of } U = 4\pi.$$

This completes the proof of the theorem.   ∎

Corollary 1   The solution of CP is

$$u(\mathbf{x}, t) = \frac{1}{4\pi} \frac{\partial}{\partial t} \left[\frac{1}{t} \int\int_{S_t} \varphi(\boldsymbol{\xi})\, d\sigma_t \right] + \frac{1}{4\pi t} \int\int_{S_t} \psi(\boldsymbol{\xi})\, d\sigma_t. \quad \blacksquare \tag{4.47}$$

The conclusion follows immediately from Theorem 13 and equation 4.38. This integral expression is known as *Poisson's formula* for the solution of the Cauchy problem for the wave equation in three dimensions.

Poisson's formula has an important consequence for wave motion in 3-space. For a given point (x, y, z) and time t, $u(x, y, z, t)$ depends only on t and data given on the sphere S_t of radius t about the point. This was noticed by the Dutch natural philosopher Christian Huygens (1629–1695), a contemporary of

Newton and Leibniz. S_t is the *domain of influence* of the initial conditions of time t.

To see the physical effect of this observation, imagine an initial disturbance that vanishes outside a solid ball B_R of radius R about the origin, and picture yourself standing at (x, y, z) outside this ball. Let ρ be the distance from the origin to (x, y, z). Notice that S_t intersects B_R if and only if

$$\rho - R \le t \le \rho + R$$

(Figure 4.28). This means that you feel the disturbance first at time $\rho - R$, but after time $\rho + R$ the effect is suddenly gone, because the initial disturbance vanishes outside B_R and the domain of influence no longer intersects any non-zero data. This effect is known as *Huygens' Principle*.

If we think of the solution as a sound wave, this means that at first you hear nothing at (x, y, z). Then at time $\rho - R$ you hear the sound, continuing to

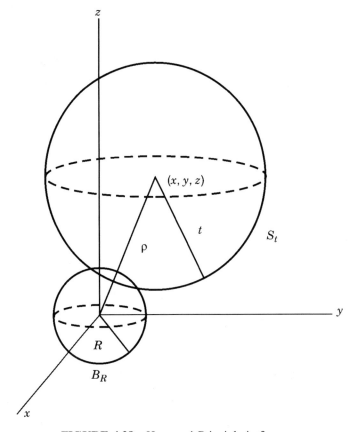

FIGURE 4.28 Huygens' Principle in 3-space.

experience it until time $\rho + R$, at which time the sound suddenly vanishes. This is what happens when someone standing some distance away gives a short blast on a whistle. At first you hear nothing, then the whistle sound, then nothing again.

EXERCISE 167   Suppose in Problem CP the initial position function is identically zero, and the initial velocity is constant. Show that the solution is independent of **x**. Does this make sense from a physical point of view, thinking of the solution as a sound wave?

EXERCISE 168   State a version of Duhamel's Principle (Exercise 122, Section 4.1) for the three-dimensional wave problem

$$\frac{\partial^2 u}{\partial x^2} + \frac{\partial^2 u}{\partial y^2} + \frac{\partial^2 u}{\partial z^2} - \frac{\partial^2 u}{\partial t^2} = f(x, y, z, t)$$

for $(x, y, z) \in R^3$, $t > 0$,

$$u(x, y, z, 0) = u_t(x, y, z, 0) = 0.$$

Call this initial value problem Problem K. Use the requested version of Duhamel's Principle and Kirchhoff's formula 4.39 for the solution of problem VCP to show that the solution of Problem K is given by

$$u(\mathbf{x}, t) = -\frac{1}{4\pi} \iiint_{\bar{B}(\mathbf{x},t)} \frac{1}{r} f(\mathbf{y}, t - r) \, d\mathbf{y},$$

in which $r = |\mathbf{x} - \mathbf{y}|$. Assume that f is continuous for all $(\mathbf{x}, t)$ with $\mathbf{x}$ in R^3, $t \geq 0$.

EXERCISE 169   Write an integral formula for the solution of

$$\frac{\partial^2 u}{\partial x^2} + \frac{\partial^2 u}{\partial y^2} + \frac{\partial^2 u}{\partial z^2} - \frac{\partial^2 u}{\partial t^2} = f(\mathbf{x}, t) \text{ for } \mathbf{x} \text{ in } R^3, t > 0$$

$$u(\mathbf{x}, 0) = \varphi(\mathbf{x}), \, u_t(\mathbf{x}, 0) = \psi(\mathbf{x}).$$

4.14   HADAMARD'S METHOD OF DESCENT

We will discuss a method for solving the Cauchy problem for the wave equation in two space dimensions, using Poisson's formula for the solution of the corresponding problem in 3-space. The method is due to the French mathematician

Jacques Hadamard (1865–1963) and is called the *method of descent* because it descends from three to two space dimensions.

Consider the Cauchy problem in two space variables:

$$u_{tt} = u_{xx} + u_{yy}$$

$$u(x, y, 0) = \varphi(x, y), \ u_t(x, y, 0) = \psi(x, y)$$

for all (x, y) in the plane for $t \geq 0$. Again we have let $c = 1$.

The idea is to think of the plane R^2 as consisting of points $(x, y, 0)$ in R^3, and of this Cauchy problem in the plane as a Cauchy problem in R^3 in which the initial data and partial differential equation are independent of z.

We may use Poisson's formula 4.47 to write the solution of this Cauchy problem in R^2 as

$$u(x, y, t) = \frac{1}{4\pi} \frac{\partial}{\partial t} \left[\frac{1}{t} \int \int_{S_t} \varphi(\xi, \eta) \, d\sigma_t \right] + \frac{1}{4\pi t} \int \int_{S_t} \psi(\xi, \eta) \, d\sigma_t.$$

Since φ and ψ are functions of just x and y, we can project these surface integrals to double integrals over a region in the plane. Recall that $\mathbf{n}_t$ denotes the unit outer normal vector to S_t. Consider a typical surface element having area $d\sigma$ (Figure 4.29). This projects onto a $d\xi$ by $d\eta$ rectangle centered at $(x, y, 0)$ in the (ξ, η) plane. The area of this rectangle is related to the area of the surface element by

$$d\xi \, d\eta = \mathbf{n}_t \cdot \mathbf{k} d\sigma = \frac{1}{t} \sqrt{t^2 - (\xi - x)^2 - (\eta - y)^2} \, d\sigma.$$

Here we have used the fact that the sphere circle of radius t about $(x, y, 0)$ has equation

$$(\xi - x)^2 + (\eta - y)^2 + \zeta^2 = t^2,$$

with (ξ, η, ζ) an arbitrary point on S_t. Since the sphere has both an upper and lower hemisphere we introduce a factor of 2 and write

$$\int \int_{S_t} \varphi(\xi, \eta) \, d\sigma_t = 2t \int \int_{D_t} \frac{\varphi(\xi, \eta)}{\sqrt{t^2 - (\xi - x)^2 - (\eta - y)^2}} \, d\xi \, d\eta$$

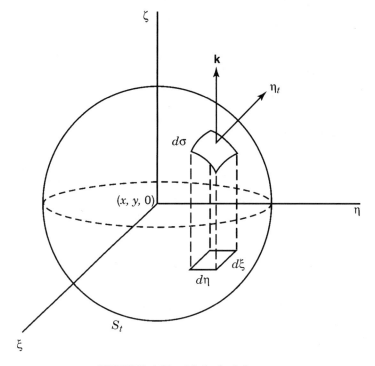

FIGURE 4.29 Method of descent.

in which D_t is the disk of radius t about (x, y) in the plane, and consists of all (ξ, η) which

$$(\xi - x)^2 + (\eta - y)^2 \leq t^2.$$

A similar formula holds for the integral of ψ over S_t. The solution of the Cauchy problem in the plane is therefore

$$u(x, y, t) = \frac{1}{2\pi} \frac{\partial}{\partial t} \left[\int \int_{D_t} \frac{\varphi(\xi, \eta)}{\sqrt{t^2 - (\xi - x) - (\eta - y)^2}} \, d\xi \, d\eta \right]$$

$$+ \frac{1}{2\pi} \int \int_{D_t} \frac{\psi(\xi, \eta)}{\sqrt{t^2 - (\xi - x)^2 - (\eta - y)^2}} \, d\xi \, d\eta. \quad (4.48)$$

Notice a significant difference between the solution 4.47 in 3-space, and the solution 4.48 in the plane. The domain of influence of the initial data at a point in 3-space at time t is the surface of a sphere of radius t about the point, and Huygens' Principle holds. In the plane, however, the domain of influence at (x, y) at time t is the *solid disk* of radius t about the point, not just its boundary circle. A disturbance (say in a small disk D about the origin) will eventually be felt at (x, y) and will then be felt at all later times because once D_t intersects D at some time, it will do so at all later times (Figure 4.30). Huygens' Principle does not hold in the plane. A wave disturbance in a plane will eventually be felt at any point, and then at that point at all later times, although with decreasing magnitude. This can be seen by dropping a stone in a smooth pond. Waves will spread out from the point of impact and continue indefinitely.

It is an interesting exercise to descend one more dimension, going from equation 4.48 to the solution of the Cauchy problem on the real line. This results in d'Alembert's formula, which should be no surprise.

EXERCISE 170 Use the method of descent to derive d'Alembert's formula for the one-dimensional Cauchy problem, from Poisson's formula for two space dimensions (equation 4.48).

EXERCISE 171 Using d'Alembert's formula, explain why Huygens' Principle fails to hold in one space dimension. Give a physical model for this (consider waves along a vibrating string).

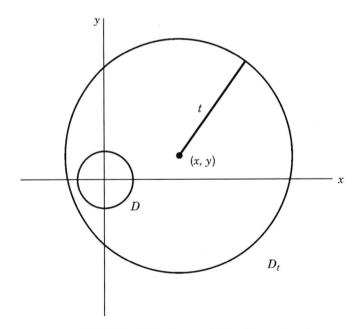

FIGURE 4.30 Huygens' Principle in the plane.

5

THE HEAT EQUATION

We now turn to the constant coefficient, second order, linear parabolic partial differential equation, which we will also study in its canonical form.

As with the hyperbolic wave equation, this parabolic equation has both mathematical interest and physical relevance. In the parabolic case the physical context is in modeling heat conduction and diffusion phenomena. For this reason the parabolic equation is often referred to as the heat equation, or diffusion equation.

We will begin with the Cauchy problem for the heat equation.

5.1 THE CAUCHY PROBLEM FOR THE HEAT EQUATION IS ILL POSED

We will show that the Cauchy problem for the heat equation is not well posed.
Consider the Cauchy problem

$$u_t = u_{xx} \text{ for } -\infty < x < \infty, \, t > 0$$

$$u(0, t) = \varphi(t), \, u_x(0, t) = \psi(t) \text{ for } t \geq 0.$$

Initial conditions are given on the line $x = 0$ in the x, t-plane.

The heat equation is parabolic and the characteristics are the lines $t = $ constant. This Cauchy problem therefore specifies data about the function and its normal derivative on a line that is not characteristic. For this problem to be well posed, it is required that it have a unique solution, and that the solution depend continuously on the data. We will show by an example that the last criterion is not met.

Let $\psi(t) \equiv 0$ and $\varphi_n(t) = (2/n)\sin(2n^2 t)$ for any positive integer n. It is easy to check that this Cauchy problem has solution

$$u_n(x, t) = \frac{1}{n}(e^{nx}\sin(2n^2 t + nx) + e^{-nx}\sin(2n^2 t - nx)).$$

Now,

$$\frac{2}{n}\sin(2n^2 t) \rightarrow 0 \text{ as } n \rightarrow \infty.$$

We can therefore make the initial data as small in magnitude as we like by choosing n large enough.

However, by choosing n large enough, we can, for certain values of x and t, make the solution as large in magnitude as we like.

The solution therefore does not continuously depend on the initial data, and the problem is ill posed.

EXERCISE 172 Verify the assertion that $u_n(x, t)$ can be made as large in magnitude as we like by choosing n, x, and t appropriately.

5.2 INITIAL AND BOUNDARY CONDITIONS

For the time being we will restrict ourselves to the one-dimensional heat equation 2.22 in Section 2.5.2. We think of this equation as modeling the temperature distribution in a wire or bar of material, with $u(x, t)$ the temperature at time t in the cross section at x.

Generally t will range over the nonnegative real numbers. The space variable x may vary over the entire real line (infinite medium), the half line $x \geq 0$ (semi-infinite medium), or may be restricted to some finite interval, say $[0, L]$.

In the latter case there are certain kinds of boundary conditions that are commonly encountered. For example, conditions of the form

$$u(0, t) = g(t), u(L, t) = h(t) \text{ for } t \geq 0$$

specify that the temperature at the left end of the bar at time t is $g(t)$, and that at the right end is $h(t)$.

Instead of giving the temperature at each end at all times, we might specify conditions of the form

$$u_x(0, t) = g(t), u_x(L, t) = h(t) \text{ for } t \geq 0.$$

These boundary conditions give information about radiation of energy from the ends of the bar at times $t \geq 0$. In the case that $h(t) = g(t) = 0$ for $t \geq 0$, these boundary conditions are called *insulation conditions*, and they state that there is no transfer of heat energy across the ends of the bar into the surrounding medium.

We can also specify *free radiation*, or *convection*, boundary conditions, in which the bar loses heat by radiation from its ends into the surrounding medium. If we assume that the medium is kept at a constant temperature zero, then these conditions take the form

$$u_x(0, t) + Au(0, t) = 0, \; u_x(L, t) + Au(L, t) = 0 \text{ for } t \geq 0,$$

in which A is constant.

An *initial-boundary value problem* for the heat equation on a finite interval consists of the heat equation, together with an initial condition and boundary conditions at the ends of the bar.

For example, the initial-boundary value problem

$$u_t = ku_{xx} \text{ for } 0 < x < L, \, t > 0$$

$$u(x, 0) = f(x) \text{ for } 0 \leq x \leq L$$

$$u(0, t) = u(L, t) = 0 \text{ for } t \geq 0$$

models heat conduction in a bar of length L, with initial temperature in the cross section at x given by $f(x)$, and with ends kept at temperature zero for all times $t \geq 0$. Under certain conditions on f, this problem will be seen to be well posed—it has a unique solution that depends continuously on the initial data. This is consistent with the intuition that the heat equation governing the energy distribution, together with the initial temperature throughout the bar, and the temperature at the ends of the bar at all later times, should be sufficient to uniquely determine the temperature throughout the bar at any time. Our intuition also tells us that small changes in the initial temperature should cause small changes in the resulting temperature distribution.

As another example of an initial-boundary value problem, consider

$$u_t = ku_{xx} \text{ for } 0 < x < L, \, t > 0$$

$$u(x, 0) = f(x) \text{ for } 0 \leq x \leq L$$

$$u(0, t) = T_1, \; u_x(L, t) = -A[u(L, t) - T_2] \text{ for } t \geq 0.$$

This problem models the temperature distribution in a bar of length L, with initial temperature in the cross section at x equal to $f(x)$, temperature at the left end maintained at the constant value T_1, and the right end radiating heat energy into a medium which is kept at constant temperature T_2.

In the case of a semi-infinite medium, $0 \leq x < \infty$, we will still require an initial condition

$$u(x, 0) = f(x) \text{ for } x \geq 0.$$

Now, however, there is only one end point of the bar, and we specify a boundary condition only at $x = 0$. This might give the temperature at the left end at all times $t \geq 0$, or be in the form of an insulation condition, or a free radiation condition. The absence of a boundary condition at a "right end" of the bar is compensated for by insisting that the solution be a bounded function as $x \rightarrow \infty$. This is consistent with the physical interpretation of $u(x, t)$ as a temperature distribution, which we expect to be a bounded function.

For an infinite medium $-\infty < x < \infty$ there are no boundary conditions, only an initial condition

$$u(x, 0) = f(x) \text{ for all } x.$$

Again, we will seek bounded solutions for this initial value problem.

EXERCISE 173   Formulate an initial-boundary value problem modeling heat conduction in a homogeneous bar of length L and uniform cross section, if the left end is kept at temperature zero and the right end is insulated. The initial temperature distribution in the cross section at x is $f(x)$.

EXERCISE 174   Formulate an initial-boundary value problem modeling heat conduction in a homogeneous bar of length L and uniform cross section if the left end is kept at temperature $\alpha(t)$, the right end at constant temperature K, and the initial temperature in the cross section at x is $f(x)$.

EXERCISE 175   Suppose u is a solution of $u_t = k^2 u_{xx}$ for $0 < x < L$, $t > 0$. Prove that

$$\frac{\partial}{\partial t} \left(\frac{1}{2} \int_0^L u^2(x, t) \, dx \right) = k^2 [u(x, t) u_x(x, t)]_{x=0}^{x=L} - k^2 \int_0^L u_x^2(x, t) \, dx.$$

Hint: Multiply the heat equation by u and integrate.
      Use this result to show that the problem

$$u_t = k^2 u_{xx} \text{ for } 0 < x < L, \, t > 0$$

$$u(0, t) = u(L, t) = 0 \text{ for } t \geq 0$$

$$u(x, 0) = f(x) \text{ for } 0 \leq x \leq L$$

can have only one solution.

EXERCISE 176   Prove the following version of Duhamel's Principle for the heat equation (see Exercise 122, Section 4.1 for a version related to the wave equation). Let u satisfy

$$u_t = u_{xx} + f(x, t) \text{ for } 0 < x < L, t > 0$$

$$u(0, t) = u(L, t) = 0 \text{ for } t \geq 0$$

$$u(x, 0) = 0 \text{ for } 0 \leq x \leq L.$$

For $T > 0$, let $v(x, t, T)$ satisfy

$$v_t = v_{xx} \text{ for } 0 < x < L, t > T$$

$$v(x, T, T) = f(x, T) \text{ for } 0 \leq x \leq L$$

$$v(0, t, T) = v(L, t, T) = 0 \text{ for } t \geq T.$$

Prove that

$$u(x, t) = \int_0^t v(x, t, T) \, dT.$$

5.3   THE WEAK MAXIMUM PRINCIPLE

We will begin with the one-dimensional heat equation and derive a property enjoyed by any continuous solution, independent of initial and boundary conditions.

Consider the rectangle in the x, t-plane defined by $0 \leq x \leq L$ and $0 \leq t \leq T$. We know that any function that is continuous on this compact set must achieve a maximum on it. We claim that, if the function is a solution of $u_t = ku_{xx}$, then this maximum is achieved at a point on the lower horizontal side or a vertical side of the rectangle.

Theorem 14 (Weak Maximum Principle)   Let v be a solution of $u_t = ku_{xx}$ that is continuous on the closed rectangle R consisting of all (x, t) with $0 \leq x \leq L, 0 \leq t \leq T$. Then $v(x, t)$ assumes its maximum value on R at a point on the base $t = 0$ or on a vertical side $x = 0$ or $x = L$.   ∎

Here is a physically based rationale for this conclusion. A solution of the heat equation on $[0, L]$ may be interpreted as the temperature distribution in a bar of length L. Because heat conduction is a dissipative process, in the absence of new sources of heat energy, we would not expect the temperature $v(x, t)$ in the bar at any later time to exceed either the initial temperature ($t = 0$) throughout the bar, or the temperature at the ends of the bar (at $x = 0$ and at $x = L$).

Now here is an analytical argument.

Proof: Let C consist of points $(x, 0)$ with $0 \le x \le L$, and points $(0, t)$ and (L, t) with $0 \le t \le T$. C consists of the lower and vertical sides of the rectangle (Figure 5.1).

We know that $v(x, t)$ achieves a maximum value M on the closed rectangle R. $v(x, t)$ must also achieve a maximum value M_C on the closed set C. We will prove the theorem by showing that $M = M_C$.

Clearly $M_C \le M$, since every point of C is also in R. Suppose $M - M_C = \epsilon > 0$. Choose any point (x_0, t_0) in R such that $v(x_0, t_0) = M$. Since $\epsilon > 0$, (x_0, t_0) is not in C, so $0 < x_0 < L$ and $0 < t_0 \le T$. Define

$$w(x, t) = v(x, t) + \frac{\epsilon}{4L^2} (x - x_0)^2.$$

Consider $w(x, t)$ at points in C. First, for $0 \le x \le L$,

$$w(x, 0) = v(x, 0) + \frac{\epsilon}{4L^2} (x - x_0)^2 \le M - \epsilon + \frac{\epsilon}{4L^2} L^2 = M - \frac{3\epsilon}{4} < M.$$

In similar fashion,

$$w(0, t) < M \text{ and } w(L, t) < M \text{ for } 0 \le t \le T.$$

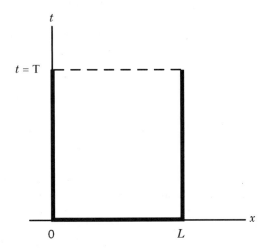

FIGURE 5.1 Weak maximum principle for the heat equation (Theorem 14).

But $w(x_0, t_0) = v(x_0, t_0) = M$. Therefore the maximum of $w(x, t)$ on R is at least M, and is achieved at a point (x_1, t_1) of R not on C. Because $0 < x_1 < L$ and $0 < t_1 \le T$, then

$$w_t(x_1, t_1) \ge 0 \text{ and } w_{xx}(x_1, t_1) \le 0.$$

Then

$$w_t(x_1, t_1) - kw_{xx}(x_1, t_1) \ge 0.$$

But

$$w_t(x_1, t_1) - kw_{xx}(x_1, t_1) = v_t(x_1, t_1) - kv_{xx}(x_1, t_1) - k\frac{\epsilon}{2L^2} = -\frac{k\epsilon}{2L^2} < 0.$$

This is a contradiction, and we conclude that $M = M_C$, completing the proof. ∎

We have proved that a continuous solution achieves its maximum at a point on the lower or vertical sides of R. We have *not* proved that this solution cannot *also* achieve its maximum at another point of R, as certainly occurs with a constant solution. For this reason the theorem is referred to as the *weak maximum principle*. This version is sufficient for our purpose.

By considering $-v$, we can adapt the proof to show that a continuous solution also achieves its minimum at a point on the lower or vertical sides of R.

We will derive two important consequences of the weak maximum principle. Consider the general initial-boundary value problem

$$u_t = ku_{xx} + F(x, t) \text{ for } 0 < x < L, t > 0$$

$$u(x, 0) = f(x) \text{ for } 0 \le x \le L$$

$$u(0, t) = g(t), u(L, t) = h(t) \text{ for } t \ge 0. \tag{5.1}$$

It is assumed that f, g, and h are continuous and that F is continuous for $0 \le x \le L$ and $t \ge 0$. Our first consequence is a uniqueness result for this problem.

Theorem 15 (Uniqueness) The initial-boundary value problem 5.1 can have only one continuous solution. ∎

Proof: Suppose u_1 and u_2 are continuous solutions. Let $w = u_1 - u_2$. It is routine to check that

$$w_t - kw_{xx} = 0 \text{ for } 0 < x < L, t > 0$$

$$w(x, 0) = 0 \text{ for } 0 \le x \le L$$

$$w(0, t) = w(L, t) = 0 \text{ for } t \ge 0.$$

Now let $T > 0$ and let R be the rectangle defined by $0 \le x \le L, 0 \le t \le T$. The initial and boundary conditions mean that $w(x, t) = 0$ on the lower and vertical sides of R. Since w achieves its maximum on R on these sides, then $w(x, t) \le 0$ for $0 \le x \le L$ and $t \ge 0$.

But $w(x, t)$ also achieves its minimum on R on these sides, so $0 \le w(x, t)$ for $0 \le x \le L$ and $T \ge 0$.

We conclude that $w(x, t) = 0$ for $0 \le x \le L$ and $0 \le t \le T$, hence $u_1 = u_2$ on R. Since T is any positive number, then $u_1(x, t) = u_2(x, t)$ for $0 \le x \le L$ and $t \ge 0$. ∎

We can also use the weak maximum principle to show that solutions of the problem 5.1 depend continuously on the initial and boundary data. The problem 5.1 is therefore well posed.

Theorem 16   For $j = 1, 2$, let u_j be the continuous solution of

$$u_t = ku_{xx} + F(x, t) \text{ for } 0 < x < L, t > 0$$

$$u(x, 0) = f_j(x) \text{ for } 0 \le x \le L$$

$$u(0, t) = g_j(t), u(L, t) = h_j(t) \text{ for } t \ge 0.$$

Let ϵ be a positive number such that

$$|f_1(x) - f_2(x)| \le \epsilon \text{ for } 0 \le x \le L$$

and

$$|g_1(t) - g_2(t)| \le \epsilon \text{ and } |h_1(t) - h_2(t)| \le \epsilon$$

for $t \ge 0$. Then,

$$|u_1(x, t) - u_2(x, t)| \le \epsilon$$

for $0 \le x \le L$ and $t \ge 0$. ∎

Proof: Let $w = u_1 - u_2$. Then

$$w_t - kw_{xx} = 0 \text{ for } 0 < x < L, t > 0$$

$$w(x, 0) = f_1(x) - f_2(x) \text{ for } 0 \le x \le L$$

$$w(0, t) = g_1(t) - g_2(t), w(L, t) = h_1(t) - h_2(t) \text{ for } t \ge 0.$$

Now let T be any positive number and consider values of $w(x, t)$ on the lower and vertical sides of the rectangle $R: 0 \le x \le L, 0 \le t \le T$. On the lower side of R,

$$|w(x, 0)| = |f_1(x) - f_2(x)| \le \epsilon.$$

On the left vertical side,

$$|w(0, t)| = |g_1(t) - g_2(t)| \le \epsilon.$$

And on the right vertical side,

$$|w(L, t)| = |h_1(t) - h_2(t)| \le \epsilon.$$

By the weak maximum principle, $|w(x, t)| \le \epsilon$ for $0 \le x \le L$ and $0 \le t \le T$. Since T can be any positive number, this proves the theorem. ∎

EXERCISE 177 Fill in the details of this alternative proof of the weak maximum principle. Let

$$M = \max_{(x,t) \text{ in } R} v(x, y).$$

There is at least one point (x_0, t_0) in R at which $v(x_0, t_0) = M$. If $M_C < M$, then this point is not on C. Define $w(x, t) = v(x, t) - \epsilon(t - t_0)$, with ϵ any positive number. Then w is continuous on R and $w(x_0, t_0) = v(x_0, t_0) = M > M_C$. Choose ϵ sufficiently small that $w(x_0, t_0)$ is greater than the maximum value achieved by $w(x, t)$ on C. Then the maximum of $w(x, t)$ on R is achieved at a point (x_1, t_1) not on C. Now reason as in the proof given in the text.

EXERCISE 178 Prove the following theorem. Suppose

$$u_t = ku_{xx} \text{ for } -\infty < x < \infty, t > 0$$

$$u(x, 0) = f(x) \text{ for } -\infty < x < \infty.$$

Suppose f is continuous on the entire real line. Let $u(x, t) \to 0$ uniformly in t as $x \to \pm\infty$. Then

$$|u(x, t)| \le \max|f(x)|.$$

Hint: Apply the Weak Maximum Principle on $-a \le x \le a$, $t \ge 0$, and then let $a \to \infty$.

5.4   SOLUTIONS ON BOUNDED INTERVALS

In this section we will consider several initial-boundary value problems involving the one-dimensional heat equation, for a bounded interval $[0, L]$.

5.4.1   Ends of the Bar at Temperature Zero

Consider the temperature distribution $u(x, t)$ within a homogeneous bar of length L and uniform cross section, with given initial temperature $f(x)$ at (the cross section at) x and whose ends are kept at temperature zero. This temperature distribution is modeled by the initial-boundary value problem:

$$u_t = ku_{xx} \text{ for } 0 < x < L, \, t > 0$$
$$u(0, t) = u(L, t) = 0 \text{ for } t \ge 0$$
$$u(x, 0) = f(x) \text{ for } 0 \le x \le L. \tag{5.2}$$

In Section 3.1 we started to solve this problem by separation of variables, in the case $L = \pi$. Let $u(x, t) = X(x)T(x)$ and substitute into the heat equation to obtain

$$\frac{X''}{X} = \frac{T'}{kT} = -\lambda,$$

in which λ is the separation constant. Then

$$X'' + \lambda X = 0 \text{ and } T' + \lambda kT = 0.$$

Following the reasoning used in Section 3.1, with L in place of π, the problem for X is:

$$X'' + \lambda X = 0; \, X(0) = X(L) = 0.$$

It is intriguing that separation of variables applied to the wave and heat equations has led to exactly the same boundary value problem for the function

X containing the space variable, since this is problem 4.26 of Section 4.9. The eigenvalues are

$$\lambda_n = \frac{n^2\pi^2}{L^2} \text{ for } n = 1, 2, \ldots$$

and corresponding eigenfunctions are constant multiples of

$$X_n(x) = \sin\left(\frac{n\pi x}{L}\right).$$

The problem for T now has the form

$$T' + \frac{kn^2\pi^2}{L^2} T = 0,$$

with solutions constant multiples of

$$e^{-kn^2\pi^2 t/L^2}.$$

Thus, for each positive integer n, we have a function

$$u_n(x, t) = b_n \sin(n\pi x/L)e^{-kn^2\pi^2 t/L^2}$$

which satisfies the heat equation and the boundary conditions. As we found with the initial-boundary value problem for the wave equation in Section 4.9, there are functions f for which it is impossible to choose a particular n and b_n so that this function satisfies the initial condition. In similar fashion, any finite sum of the u_ns will fail to provide a solution for some choices of f. We are thus led, as in the case of the wave equation, to try an infinite superposition

$$u(x, t) = \sum_{n=1}^{\infty} b_n \sin(n\pi x/L)e^{-kn^2\pi^2 t/L^2}.$$

We need to choose the b_ns so that

$$u(x, 0) = f(x) = \sum_{n=1}^{\infty} b_n \sin(n\pi x/L).$$

Recognize this as the Fourier sine expansion of the initial temperature function
f. Hence choose

$$b_n = \frac{2}{L} \int_0^L f(\xi)\sin\left(\frac{n\pi\xi}{L}\right) d\xi.$$

We therefore propose the function

$$u(x, t) = \frac{2}{L} \sum_{n=1}^{\infty} \left(\int_0^L f(\xi)\sin\left(\frac{n\pi\xi}{L}\right) d\xi\right) \sin\left(\frac{n\pi x}{L}\right) e^{-n^2\pi^2 kt/L^2} \qquad (5.3)$$

as the solution.

Verification of the Solution   How do we know that the expression of equation
5.3 is indeed the solution?

It is clear that u satisfies the boundary conditions, since each term of the
series vanishes at $x = 0$ and at $x = L$.

The initial condition is not so obvious, since we have placed no conditions
on f. In order for $u(x, 0) = f(x)$ to be satisfied for $0 \le x \le L$, we need the
Fourier sine series for f to converge to $f(x)$ on this interval. For this to occur,
it is sufficient that $f(0) = f(L) = 0$ and that f is continuous and piecewise smooth
on $[0, L]$. These are certainly reasonable conditions for an initial temperature
function, and the conditions that $f(0) = f(L) = 0$ are consistent with the model
under consideration, in which the ends of the bar are kept at temperature zero.

Finally, does u satisfy the heat equation? Even though each term in the
infinite series 5.3 does, it is conceivable that this infinite series does not. To
check on this, let T be any positive number. Then

$$\left| b_n \sin\left(\frac{n\pi x}{L}\right) e^{-n^2\pi^2 kt/L^2} \right| \le |b_n| e^{-n^2\pi^2 kT/L^2}$$

for $t \ge T$. We know from Bessel's inequality that $b_n \to 0$ as $n \to \infty$. Therefore

$$\sum_{n=1}^{\infty} |b_n| e^{-n^2\pi^2 kT/L^2}$$

is a convergent series of constants. By a theorem of Weierstrass, the series 5.3
converges absolutely and uniformly for $0 \le x \le L$ and $t \ge T > 0$.

By similar reasoning, the series obtained by partially differentiating equation
5.3 term by term, once with respect to t, or twice with respect to x, also
converges uniformly for $0 \le x \le L$ and $t \ge T$. Since T can be chosen as close
to zero as we like, this is enough to be able to compute u_t and u_{xx} by differentiating the series term by term, for $0 \le x \le L$ and $t > 0$. Since each $u_n(x, t)$
satisfies the heat equation, $u(x, t)$ does also.

We conclude that, with certain conditions on f, equation 5.3 defines the solution of the initial-boundary value problem.

Example 35 Solve

$$u_t = ku_{xx} \text{ for } 0 < x < L, t > 0$$

$$u(0, t) = u(L, t) = 0 \text{ for } t > 0$$

$$u(x, 0) = f(x) = \begin{cases} x & \text{for } 0 \le x \le L/2 \\ L - x & \text{for } L/2 \le x \le L. \end{cases}$$

Compute

$$\int_0^L f(\xi)\sin\left(\frac{n\pi\xi}{L}\right) d\xi = \frac{2L^2}{n^2\pi^2} \sin\left(\frac{n\pi}{2}\right).$$

The solution of initial-boundary value problem is

$$u(x, t) = \frac{4L}{\pi^2} \sum_{n=1}^{\infty} \frac{1}{n^2} \sin\left(\frac{n\pi}{2}\right) \sin\left(\frac{n\pi x}{L}\right) e^{-n^2\pi^2 kt/L^2}.$$

We may interpret this as the temperature throughout a homogeneous bar of length L, with the given initial temperature and the temperature at the ends kept at zero. Figures 5.2(a) through (d) show the evolution of the temperature distribution (with $k = 1$ and $L = \pi$) from the triangular-shaped initial temperature to the more parabolic distribution we might intuitively expect. Figure 5.2(e) compares the distributions for times $t = 0.351, 0.951, 1.87, 3.87$ and 8.87, with the temperature decreasing with time, as we should expect.

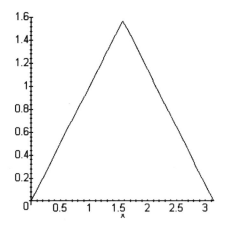

FIGURE 5.2(a)   Initial temperature distribution in Example 35.

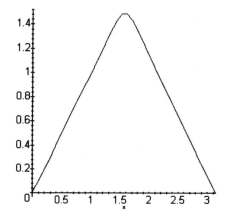

FIGURE 5.2(b)   Temperature distribution at time $t = 0.005$.

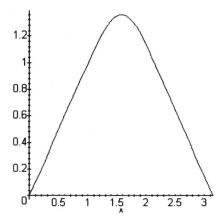

FIGURE 5.2(c) $t = 0.035$.

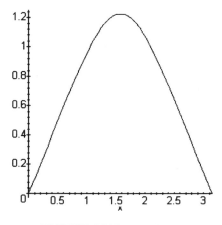

FIGURE 5.2(d) $t = 0.094$.

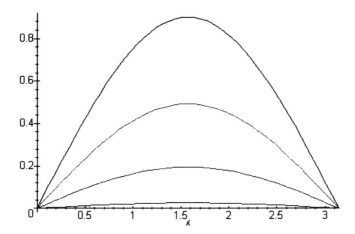

FIGURE 5.2(e) Comparison of the temperature at times $t = 0.351, 0.951, 1.87, 3.87$, and 8.87 (the latter temperature being essentially zero).

It is interesting to use different values of k and compare the effects on the temperature function at selected times. We leave this experiment to the student. ∎

Example 36 Suppose the initial temperature is given on different segments of a homogeneous metal bar of length 1 by

$$f(x) = \begin{cases} 1 \text{ for } 0 \le x \le 1/3 \\ 100 \text{ for } 1/3 \le x \le 2/3 \\ 0 \text{ for } 2/3 \le x \le 1. \end{cases}$$

Suppose $k = 1$ for the material of this bar. The temperature function is given by equation 5.3. Compute

$$\int_0^1 f(\xi)\sin(n\pi\xi)\, d\xi = \int_0^{1/3} \sin(n\pi\xi)\, d\xi + \int_{1/3}^{2/3} 100 \sin(n\pi\xi)\, d\xi$$

$$= \frac{99 \cos(n\pi/3) + 1 - 100 \cos(2n\pi/3)}{n\pi}.$$

Thus the temperature distribution is

$$u(x, t) = 2 \sum_{n=1}^{\infty} \left(\frac{99 \cos(n\pi/3) + 1 - 100 \cos(2n\pi/3)}{n\pi} \right) \sin(n\pi x)e^{-n^2\pi^2 t}.$$

Figures 5.3(a) through (c) show the initial temperature function and then the

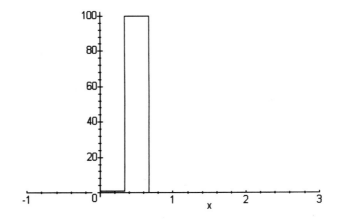

FIGURE 5.3(a)   Initial temperature distribution in Example 36.

temperature function at different times, while Figure 5.3(d) shows the temperature distribution at two later times for comparison. ■

5.4.2   Temperature in a Bar With Insulated Ends

Model the temperature distribution in a bar with insulated ends by:

$$u_t = ku_{xx} \text{ for } 0 < x < L, \, t > 0$$

$$u_x(0, \, t) = u_x(L, \, t) = 0 \text{ for } t \geq 0$$

$$u(x, \, 0) = f(x) \text{ for } 0 \leq x \leq L.$$

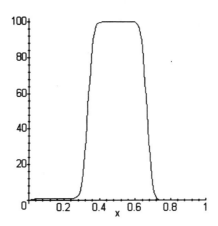

FIGURE 5.3(b)   Temperature at time $t = 0.0003$.

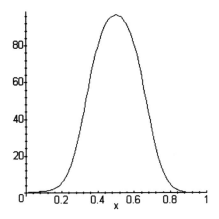

FIGURE 5.3(c) $t = 0.003$.

We can also solve this problem by separation of variables. Let $u(x, t) = X(x)T(t)$. Proceeding as in Section 5.4.1, we obtain

$$X'' + \lambda X = 0, \ T' + \lambda k T = 0.$$

Now $u_x(0, t) = X'(0)T(t) = 0$ implies that $X'(0) = 0$; and $u_x(L, t) = 0$ implies that $X'(L) = 0$. The boundary value problem for X is therefore

$$X'' + \lambda X = 0; \ X'(0) = X'(L) = 0.$$

As in the case of boundary conditions $X(0) = X(L) = 0$, values of λ for which this problem has nontrivial solutions are called eigenvalues, and a nontrivial solution corresponding to an eigenvalue is an eigenfunction. We proceed to

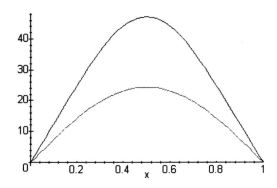

FIGURE 5.3(d)   Comparison of the temperature at times $t = 0.035$ (top curve) and $t = 0.098$.

find the eigenvalues and eigenfunctions of this problem as we did for the problem in Section 4.9. Consider cases on λ.

If $\lambda = 0$ then $X'' = 0$ so $X(x) = ax + b$. Now $X'(0) = X'(L) = a = 0$ and b can be any constant. Therefore 0 is an eigenvalue of this problem, with (non-zero) constant eigenfunction.

If $\lambda < 0$, write $\lambda = -k^2$ with $k > 0$. Now $X'' - k^2 X = 0$ has general solution

$$X(x) = ae^{kx} + be^{-kx}.$$

The condition

$$X'(0) = ka - kb = 0$$

implies that $a = b$, so $u(x) = 2a \cosh(kx)$. But then

$$X'(L) = 2ka \sinh(kL) = 0$$

implies that $a = 0$ and the solution is trivial. This problem has no negative eigenvalue.

If $\lambda > 0$, write $\lambda = k^2$ for $k > 0$. Now $X'' + k^2 X = 0$ has general solution

$$X(x) = a \cos(kx) + b \sin(kx).$$

Since

$$X'(0) = bk = 0$$

we conclude that $b = 0$. Then $X(x) = a \cos(kx)$. Now

$$X'(L) = -ak \sin(kL) = 0$$

implies that kL must be an integer multiple of π, hence

$$k = \frac{n\pi}{L} \text{ for } n = 1, 2, \ldots$$

and

$$\lambda_n = \frac{n^2 \pi^2}{L^2}.$$

These numbers are eigenvalues of the problem. Corresponding to each such eigenvalue is an eigenfunction

$$X_n(x) = \cos\left(\frac{n\pi x}{L}\right).$$

In summary, the eigenvalues and corresponding eigenfunctions for the present problem are

$$\lambda_0 = 0;\ X_0(x) = 1$$

and

$$\lambda_n = \frac{n^2\pi^2}{L^2};\ X_n(x) = \cos\left(\frac{n\pi x}{L}\right)\ \text{for } n = 1, 2, \ldots,$$

with the understanding that we may also take constant multiples of these eigenfunctions.

Now recall that the differential equation for T is

$$T' + \lambda kT = 0.$$

Corresponding to $\lambda = 0$ this equation has solution

$$T_0(t) = \text{constant.}$$

Corresponding to $\lambda = n^2\pi^2/L^2$ for $n = 1, 2, \ldots$, $T_n(t)$ can be any constant multiple of

$$e^{-n^2\pi^2 kt/L^2}.$$

We therefore have functions

$$u_0(x, t) = \text{constant,}$$

and

$$u_n(x, t) = a_n \cos\left(\frac{n\pi x}{L}\right) e^{-n^2\pi^2 kt/L^2}\ \text{for } n = 1, 2, \ldots$$

which satisfy the heat equation and the insulation conditions for any choices of the constant coefficients. To satisfy the initial condition, we attempt a superposition

$$u(x, t) = \sum_{n=0}^{\infty} u_n(x, t) = \frac{1}{2}a_0 + \sum_{n=1}^{\infty} a_n \cos\left(\frac{n\pi x}{L}\right) e^{-n^2\pi^2 kt/L^2}. \tag{5.4}$$

The constant term is denoted $a_0/2$ in anticipation of a Fourier cosine expansion. We require that

$$u(x, 0) = f(x) = \frac{1}{2} a_0 + \sum_{n=1}^{\infty} a_n \cos\left(\frac{n\pi x}{L}\right).$$

This is the Fourier cosine expansion of f, hence choose

$$a_n = \frac{2}{L} \int_0^L f(\xi)\cos\left(\frac{n\pi\xi}{L}\right) d\xi \text{ for } n = 0, 1, 2, \ldots$$

in Equation 5.4. This gives the solution of the heat equation with insulation boundary conditions.

We normally assume that f is continuous and piecewise smooth on $[0, L]$. Because of the decaying exponential factor in the series 5.4 this solution can be verified by reasoning similar to that used when the ends were kept at zero temperature.

Example 37   Consider a homogeneous bar with insulated ends, of length π and with material having $k = 4$. Suppose the initial temperature if given by

$$f(x) = \begin{cases} 0 \text{ for } 0 \le x < \pi/2 \\ 50 \text{ for } \pi/2 \le x \le \pi. \end{cases}$$

Compute the numbers

$$a_n = \frac{2}{\pi} \int_0^{\pi} f(\xi)\cos(n\xi) \, d\xi.$$

We get

$$a_0 = \frac{2}{\pi} \int_{\pi/2}^{\pi} 50 \, d\xi = 50$$

and, for $n = 1, 2, \ldots$,

$$a_n = \frac{2}{\pi} \int_{\pi/2}^{\pi} 50 \cos(n\xi) \, d\xi = -\frac{100}{n\pi} \sin(n\pi/2).$$

The solution is

$$u(x, t) = 25 - \sum_{n=1}^{\infty} \left(\frac{100}{n\pi} \sin(n\pi/2)\right) \cos(nx)e^{-4n^2 t}.$$

Figures 5.4(a) through (e) show the initial temperature distribution and then the evolution of this temperature distribution over selected increasing times. Note that the temperature at the right end has substantially decreased from Figure 5.4(d), in which $t = 0.097$, to Figure 5.4(e), where $t = 3.6$. ■

5.4.3 Ends of the Bar at Different Temperatures

Consider the initial-boundary value problem

$$u_t = ku_{xx} \text{ for } 0 < x < L, \, t > 0$$

$$u(0, \, t) = T_1, \, u(L, \, t) = T_2 \text{ for } t \geq 0$$

$$u(x, \, 0) = f(x) \text{ for } 0 \leq x \leq L$$

in which T_1 and T_2 are distinct positive numbers. This problem models temperature distribution in a bar in which the ends are kept at different constant temperatures.

If we attempt a separation of variables $u(x, \, t) = X(x)T(t)$, there is a rude surprise. We require that

$$u(0, \, t) = T_1 = X(0)T(t).$$

When $T_1 = 0$ this implies that $X(0) = 0$. However, if $T_1 \neq 0$ then $T(t) = T_1/X(0)$ for $t \geq 0$, hence

$$u(x, \, t) = \frac{T_1}{X(0)} X(x),$$

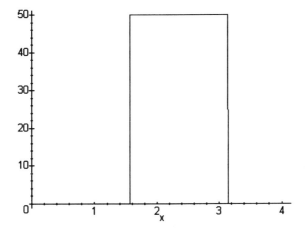

FIGURE 5.4(a) Initial temperature distribution in Example 37.

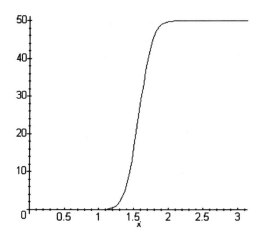

FIGURE 5.4(b)   Temperature at time $t = 0.003$.

independent of t. This is an unrealistic conclusion. Similarly, if we apply the boundary condition at $x = L$ we have to conclude that

$$u(x, t) = \frac{T_2}{X(L)} X(x).$$

We are led to this state of affairs by attempting to apply separation of variables when the boundary conditions are nonhomogeneous, specifying nonzero temperatures at the ends.

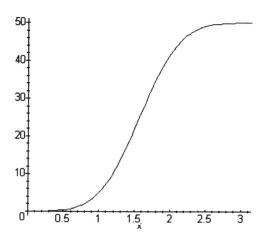

FIGURE 5.4(c) $t = 0.025$.

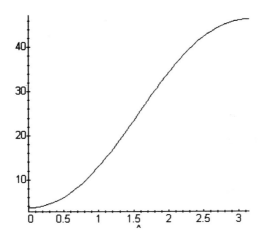

FIGURE 5.4(d) $t = 0.097$.

We can overcome this difficulty by perturbing the temperature function to define a new initial-boundary value problem having homogeneous boundary conditions, which we know how to solve. To do this, begin by putting

$$u(x, t) = U(x, t) + \psi(x).$$

Substitute u into the heat equation to obtain

$$U_t = k(U_{xx} + \psi''(x))$$

and this is a standard heat equation for U if $\psi''(x) = 0$. Thus choose $\psi(x) = Cx + D$. Now consider the boundary conditions. First,

$$u(0, t) = T_1 = U(0, t) + \psi(0)$$

becomes the homogeneous condition $U(0, t) = 0$ if

$$\psi(0) = T_1.$$

This leads us to choose $D = T_1$. So far $\psi(x) = Cx + T_1$.
  Next, $u(L, t) = U(L, t) + \psi(L) = T_2$ becomes $U(L, t) = 0$ if

$$\psi(L) = CL + T_1 = T_2.$$

This suggests that we set

$$C = \frac{1}{L}(T_2 - T_1).$$

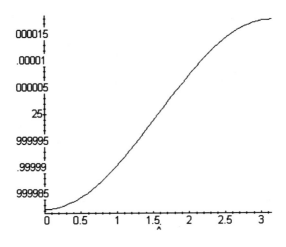

FIGURE 5.4(e) $t = 3.6$.

Now

$$\psi(x) = \frac{1}{L} (T_2 - T_1)x + T_1.$$

With this choice of ψ, the initial-boundary value problem for U is

$$U_t = kU_{xx} \text{ for } 0 < x < L, t > 0$$

$$U(0, t) = U(L, t) = 0$$

$$U(x, 0) = u(x, 0) - \psi(x) = f(x) - \frac{1}{L} (T_2 - T_1)x - T_1 \text{ for } 0 \le x \le L.$$

We solved this initial-boundary value problem in Section 5.4.1. By equation 5.3,

$$U(x, t) = \sum_{n=1}^{\infty} b_n \sin \left(\frac{n\pi x}{L} \right) e^{-n^2\pi^2 kt/L^2}$$

with

$$b_n = \frac{2}{L} \int_0^L \left(f(\xi) - \frac{1}{L} (T_2 - T_1)\xi - T_1 \right) \sin \left(\frac{n\pi\xi}{L} \right) d\xi$$

$$= \frac{2}{L} \int_0^L f(\xi)\sin \left(\frac{n\pi\xi}{L} \right) d\xi + 2 \frac{(-1)^n T_2 - T_1}{n\pi}.$$

Finally, the solution of the original initial-boundary value problem is

$$u(x, t) = U(x, t) + \frac{1}{L}(T_2 - T_1)x + T_1.$$

Physically, we may interpret this solution as a decomposition of the temperature distribution into a transient part $U(x, t)$, which decays to zero as $t \to \infty$, and a steady-state part $\psi(x)$ which is independent of time. Indeed,

$$\psi(x) = \lim_{t \to \infty} u(x, t).$$

Such decompositions are commonly seen in physical systems. For example, the current in a circuit can be written as the sum of a transient part which decays to zero as $t \to \infty$, and a steady-state part which is the limit of the solution as $t \to \infty$.

Example 38   We will solve the initial-boundary value problem

$$u_t = 7u_{xx} \text{ for } 0 < x < 5, t > 0$$

$$u(0, t) = 1, u(5, t) = 4 \text{ for } t > 0$$

$$u(x, 0) = f(x) = \begin{cases} 3 - x \text{ for } 0 < x < 3 \\ 10(x - 3) \text{ for } 3 < x < 5. \end{cases}$$

Following the above discussion, let

$$u(x, t) = U(x, t) + \frac{3}{5}x + 1,$$

where U is the solution of the initial-boundary value problem

$$U_t = 7U_{xx} \text{ for } 0 < x < 5, t > 0$$

$$U(0, t) = U(5, t) = 0 \text{ for } t > 0$$

$$U(x, 0) = u(x, 0) - \frac{3}{5}x - 1 = \begin{cases} 2 - \frac{8}{5}x \text{ for } 0 < x \le 3 \\ -31 + \frac{47}{5}x \text{ for } 3 \le x < 5. \end{cases}$$

This problem has solution

$$U(x, t) = \sum_{n=1}^{\infty} b_n \sin(n\pi x/5)e^{-7n^2\pi^2 t/25},$$

in which

$$b_n = \frac{2}{5} \int_0^5 U(\xi, 0)\sin(n\pi\xi/5)\, d\xi$$

$$= \frac{2}{5}\left[\int_0^3 \left(2 - \frac{8}{5}\xi\right) \sin(n\pi\xi/5)\, d\xi + \int_3^5 \left(-31 + \frac{47}{5}\xi\right) \sin(n\pi\xi/5)\, d\xi \right]$$

$$= -110\,\frac{\sin(3n\pi/5)}{n^2\pi^2} + \frac{4}{n\pi} - \frac{32}{n\pi}(-1)^n.$$

Then

$$u(x, t) = \sum_{n=1}^{\infty} \left(-110\,\frac{\sin(3n\pi/5)}{n^2\pi^2} + \frac{4}{n\pi} - \frac{32}{n\pi}(-1)^n\right) \sin(n\pi x/5)e^{-7n^2\pi^2 t/25} + \frac{3}{5}x + 1.$$

Figures 5.5(a) through (f) show the initial temperature function and then the temperature distribution throughout the bar at a succession of times. Notice that, by the time $t = 0.86$ in Figure 5.5(f), the temperature distribution is nearly the straight line $y = (3/5)x + 1$, which is the steady state temperature distribution for this problem (the limit of $u(x, t)$ as $t \to \infty$). ∎

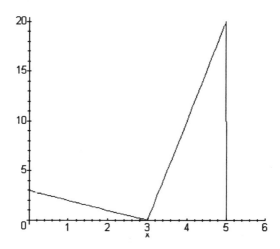

FIGURE 5.5(a)   Initial temperature distribution in Example 38.

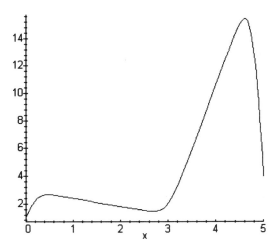

FIGURE 5.5(b) Temperature at time $t = 0.003$.

5.4.4 Diffusion of Charges in a Transistor

The heat equation is so named because it models heat conduction in various settings. It can, however, model other diffusion phenomena. Let $h(x, t)$ be the concentration of positive charge carriers at time t and position x in a transistor occupying the interval $[0, L]$. It has been shown that h is modeled by the initial-boundary value problem:

$$h_t = k \left(h_{xx} - \frac{\eta}{L} h_x \right) \text{ for } 0 < x < L, \, t > 0$$

$$h(0, t) = h(L, t) = 0 \text{ for } t \geq 0$$

$$h(x, 0) = \frac{KL}{k\eta} (1 - e^{-\eta(1 - x/L)}) \text{ for } 0 \leq x \leq L.$$

Here η and K are positive constants.

We can attempt the separation of variables method with the problem as it stands, or we can try a transformation of the partial differential equation into an equation with no first derivative term with respect to x. We will develop the second approach and leave the separation of variables experiment to the student.

Experience has shown that it is fruitful to try

$$h(x, t) = e^{\alpha x + \beta t} u(x, t)$$

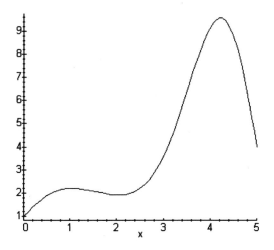

FIGURE 5.5(c) $t = 0.026$.

(see Exercise 181 in this section). We will choose α and β to obtain a problem for u that we have already solved. Substitute $h(x, t)$ into the partial differential equation to get

$$\beta e^{\alpha x + \beta t} u + e^{\alpha x + \beta t} u_t = k(\alpha^2 e^{\alpha x + \beta t} u + 2\alpha e^{\alpha x + \beta t} u_x + e^{\alpha x + \beta t} u_{xx})$$

$$- k \frac{\eta}{L} (\alpha e^{\alpha x + \beta t} u + e^{\alpha x + \beta t} u_x).$$

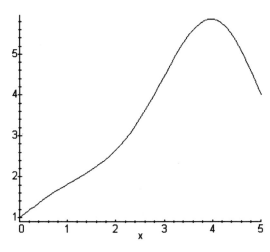

FIGURE 5.5(d) $t = 0.087$.

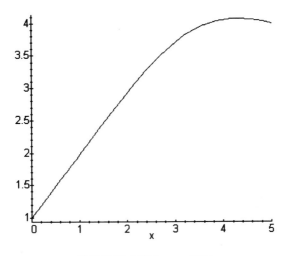

FIGURE 5.5(e) $t = 0.27$.

This equation can be written

$$u_t = ku_{xx} + u_x \left(2\alpha k - \frac{k\eta}{L} \right) + u \left(k\alpha^2 - \frac{\alpha k\eta}{L} - \beta \right).$$

This is the canonical heat equation for u if we choose α and β so that the terms in parentheses vanish. Thus choose

$$\alpha = \frac{\eta}{2L} \text{ and } \beta = -\frac{k\eta^2}{4L^2}.$$

Next,

$$h(0, t) = e^{\beta t} u(0, t) = 0$$

implies that $u(0, t) = 0$. And $h(L, t) = 0$ implies in the same way that $u(L, t) = 0$. The initial-boundary value problem for u is

$$u_t = ku_{xx} \text{ for } 0 < x < L, t > 0$$

$$u(0, t) = u(L, t) = 0 \text{ for } t \geq 0$$

$$u(x, 0) = \frac{KL}{k\eta} e^{-\eta x/2L} (1 - e^{-\eta(1-x/L)}) \text{ for } 0 \leq x \leq L.$$

By equation 5.3 the solution of this problem for u is

$$u(x, t) = \sum_{n=1}^{\infty} b_n \sin \left(\frac{n\pi x}{L} \right) e^{-n^2 \pi^2 kt/L^2}$$

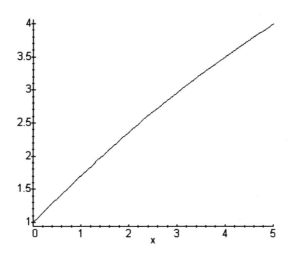

FIGURE 5.5(f) $t = 0.86$.

in which

$$b_n = \frac{2K}{k\eta} \int_0^L e^{-\eta\xi/2L}(1 - e^{-\eta(1-\xi/L)})\sin\left(\frac{n\pi\xi}{L}\right) d\xi = \frac{8KL}{k\eta} \frac{n\pi(1 - e^{-\eta})}{4n^2\pi^2 + \eta^2}.$$

Therefore

$$u(x, t) = \frac{8KL}{k\eta} (1 - e^{-\eta}) \sum_{n=1}^{\infty} \frac{n\pi}{4n^2\pi^2 + \eta^2} \sin\left(\frac{n\pi x}{L}\right) e^{-n^2\pi^2 kt/L^2}$$

and

$$h(x, t) = e^{\eta x/2L} e^{-k\eta^2 t/4L^2} u(x, t).$$

EXERCISE 179   In each of problems 1 through 7, solve $u_t = ku_{xx}$ with the given boundary and initial conditions. Interpreting the solution as a heat distribution, graph the temperature at different times.

1. $u(0, t) = u(1, t) = 0$ for $t \geq 0$; $u(x, 0) = \sin(\pi x)$ for $0 \leq x \leq 1$
2. $u(0, t) = u(3, t) = T$ for $t \geq 0$; $u(x, 0) = T$ for $0 \leq x \leq 3$
3. $u_x(0, t) = u_x(4, t) = 0$ for $t \geq 0$; $u(x, 0) = x^2$ for $0 \leq x \leq 4$
4. $u(0, t) = u(2, t) = 0$ for $t \geq 0$; $u(x, 0) = \sin(\pi x)$ for $0 \leq x \leq 2$
5. $u_x(0, t) = u_x(6, t) = 0$ for $t \geq 0$; $u(x, 0) = e^{-x}$ for $0 \leq x \leq 6$
6. $u(0, t) = 3, u(5, t) = \sqrt{7}$ for $t \geq 0$; $u(x, 0) = x^2$ for $0 \leq x \leq 5$
7. $u_x(0, t) = u_x(L, t) = 0$ for $t \geq 0$; $u(x, 0) = L - x^2$ for $0 \leq x \leq L$

EXERCISE 180   Consider the partial differential equation

$$w_t = kw_{xx} + hw,$$

with h constant. Determine a choice of α so that, if

$$w(x, t) = e^{\alpha t}u(x, t),$$

then u satisfies the one-dimensional heat equation.

EXERCISE 181   Let a and b be real numbers and $k > 0$. Consider the partial differential equation

$$w_t = k(w_{xx} + aw_x + bw).$$

Determine α and β so that, if

$$w(x, t) = e^{\alpha x + \beta t}u(x, t),$$

then u satisfies the standard heat equation.

EXERCISE 182   Let u be a solution of the heat equation

$$u_t = ku_{xx} \text{ for } 0 < x < L, \ t > 0.$$

Let $0 < a < b < L$. Prove that, for any positive t,

$$\frac{d}{dt} \int_a^b \frac{1}{2} u(x, t)^2 \, dx = kuu_x]_a^b - \int_a^b ku_x(x, t)^2 \, dx.$$

This is the analogue for the heat equation of the energy integral defined previously for the sine wave equation. *Hint:* Multiply the heat equation by u and integrate both sides from a to b, applying integration by parts where appropriate.

EXERCISE 183   Consider the problem

$$u_t = ku_{xx} \text{ for } 0 < x < L, \ t > 0$$
$$u(0, t) = u_x(L, t) = 0 \text{ for } t \geq 0$$
$$u(x, 0) = f(x).$$

This models heat conduction in a bar of length L with the left end kept at temperature zero, but with an insulation condition on the right end. Using separation of variables $u(x, t) = X(x)T(t)$, show that the problem to solve for X is

$$X'' + \lambda X = 0; \ X(0) = X'(L) = 0.$$

By considering cases on λ, show that this problem has eigenvalues

$$\lambda_n = \frac{(2n - 1)^2 \pi^2}{4L^2}$$

for $n = 1, 2, \ldots$, and corresponding eigenfunctions

$$X_n(x) = \sin \left(\frac{(2n - 1)\pi x}{2L} \right).$$

For each λ_n, show that corresponding solutions for T are constant multiples of

$$T_n(t) = e^{-(2n-1)^2 \pi^2 kt/4L^2}.$$

Show that, for $n = 1, 2, \ldots$, the functions

$$u_n(x, t) = b_n \sin \left(\frac{(2n - 1)\pi x}{2L} \right) e^{-(2n-1)^2 kt/4L^2}$$

are solutions of the heat equation satisfying both boundary conditions. To satisfy the initial condition, attempt a superposition

$$u(x, t) = \sum_{n=1}^{\infty} b_n \sin \left(\frac{(2n - 1)\pi x}{2L} \right) e^{-(2n-1)^2 \pi^2 kt/4L^2}.$$

We require that

$$u(x, 0) = f(x) = \sum_{n=1}^{\infty} b_n \sin \left(\frac{(2n - 1)\pi x}{2L} \right).$$

Explain why this is not a Fourier sine expansion of f on $[0, L]$. However, it is an expansion of f in a series of eigenfunctions of a boundary value problem for X. Show that

$$\int_0^L \sin \left(\frac{(2n - 1)\pi x}{2L} \right) \sin \left(\frac{(2m - 1)\pi x}{2L} \right) dx = 0$$

if n and m are distinct positive integers, and derive a formula for the $b_n's$ by reasoning informally as was done in originally motivating the choice of the Fourier coefficients in Section 3.2.

EXERCISE 184   Solve

$$u_t = ku_{xx} \text{ for } 0 < x < L, t > 0$$
$$u_x(0, t) = u(L, t) = 0 \text{ for } t \geq 0$$
$$u(x, 0) = f(x) \text{ for } 0 \leq x \leq L.$$

EXERCISE 185   Solve

$$u_t = ku_{xx} - hu_x \text{ for } 0 < x < L, t > 0$$
$$u(0, t) = u(L, t) = 0 \text{ for } t \geq 0$$
$$u(x, 0) = f(x) \text{ for } 0 \leq x \leq L,$$

in which h is a positive number.

EXERCISE 186   Solve

$$u_t = 4u_{xx} - 2u_x \text{ for } 0 < x < L, t > 0$$
$$u(0, t) = u(L, t) = 0 \text{ for } t > 0$$
$$u(x, 0) = x^2(L - x), \text{ for } 0 \leq x \leq L.$$

Graph the solution for some values of the time, with $L = \pi$.

EXERCISE 187   Solve

$$u_t = u_{xx} - 6u_x \text{ for } 0 < x < \pi, t > 0$$
$$u(0, t) = u(\pi, t) = 0 \text{ for } t > 0$$
$$u(x, 0) = \sin(x) \text{ for } 0 \leq x \leq \pi.$$

Graph the solution for selected values of t.

EXERCISE 188   Solve

$$u_t = ku_{xx} - hu \text{ for } 0 < x < L, t > 0$$
$$u_x(0, t) = u_x(L, t) = 0 \text{ for } t \geq 0$$
$$u(x, 0) = f(x) \text{ for } 0 \leq x \leq L,$$

in which h is a positive constant. *Hint:* Note Exercise 180.

EXERCISE 189   Solve

$$u_t = 4u_{xx} - 8u \text{ for } 0 < x < 2\pi, t > 0$$

$$u_x(0, t) = u_x(2\pi, t) = 0 \text{ for } t > 0$$

$$u(x, 0) = x(2\pi - x) \text{ for } 0 \le x \le 2\pi.$$

Generate some graphs of the solution for selected values of t.

EXERCISE 190   Solve

$$u_t = 7u_{xx} \text{ for } 0 < x < 5, t > 0$$

$$u(0, t) = 1, u(5, t) = 4 \text{ for } t > 0$$

$$u(x, 0) = e^{-x} \text{ for } 0 < x < 5.$$

Graph the solution for selected values of t.

EXERCISE 191   A thin, homogeneous bar of length L has insulated ends and initial temperature B, a constant. Find the temperature distribution in the bar.

EXERCISE 192   A thin, homogeneous bar of length L has initial temperature equal to a constant B, and the right end ($x = L$) is insulated while the left end is kept at temperature zero. Find the temperature distribution in the bar.

EXERCISE 193   A thin, homogeneous bar of thermal diffusivity 9, length 2 cm, and insulated sides has its left end maintained at temperature zero, while the right end is insulated. The bar has an initial temperature given by $f(x) = x^2$. Determine the temperature distribution in the bar. Calculate $\lim_{t \to \infty} u(x, t)$.

EXERCISE 194   Consider a long, thin, homogeneous bar of length L, with sides poorly insulated. Heat radiates freely from the bar along its length. Assuming a positive transfer coefficient A and a constant temperature T in the surrounding medium, the equation for the temperature distribution is

$$u_t = ku_{xx} - A(u - T).$$

Assume insulated ends and an initial temperature of $f(x)$. Solve for $u(x, t)$. *Hint:* Let $w = u - T$ and note Exercise 180.

EXERCISE 195   Solve

$$u_t = ku_{xx} \text{ for } -L < x < L, t > 0$$

$$u(-L, t) = u(L, t), u_x(-L, t) = u_x(L, t) \text{ for } t \ge 0$$

$$u(x, 0) = f(x) \text{ for } 0 \le x \le L.$$

This initial-boundary value problem models heat conduction in an insulated wire of length $2L$ which is thought of as bent into the form of a ring, with the ends joined. The boundary conditions identify the points $(-L, t)$ and (L, t) in the sense that the temperature and flux at these points are the same. This problem is known as Fourier's ring.

EXERCISE 196   Solve

$$u_t = k(u_{xx} - 3u_x + 2u) \text{ for } 0 \le x \le L, t > 0$$

$$u(0, t) = u(L, t) = 0 \text{ for } t > 0$$

$$u(x, 0) = f(x) \text{ for } 0 \le x \le L.$$

EXERCISE 197   In the analysis of the diffusion of charges in a transistor, define

$$q(t) = c \frac{k}{K} h_x(0, t)$$

for $t > 0$. c is a constant. The reclaimable charge on the transistor is the quantity Q defined by

$$Q = \int_0^\infty q(t) \, dt.$$

Show that

$$Q = c \frac{L^2}{k\eta} (1 - e^{-\eta}) \left[\frac{\dfrac{\sinh(\eta)}{\eta} - 1}{\cosh(\eta) - 1} \right].$$

Hint: Starting with the series expression for $h(x, t)$, obtain a series for $q(t)$. Integrate term by term to obtain a series for Q. Now sum this series by applying Parseval's equation.

EXERCISE 198   Solve the following special case of the Stefan problem:

$$u_t = u_{xx} \text{ for } 0 < x < f(t), t > 0$$

$$u(0, t) = A \text{ for } t \ge 0$$

$$u(f(t), t) = 0 \text{ for } t > 0$$

$$u_x(f(t), t) = -f'(t) \text{ for } t > 0,$$

in which both u and f are to be determined and A is a given constant. *Hint:* Assume a solution

$$u(x, t) = B + C \operatorname{erf}\left(\frac{x}{2\sqrt{t}}\right) \text{ for } 0 < x < f(t).$$

Here erf is the error function, which is defined by

$$\operatorname{erf}(x) = \frac{2}{\sqrt{\pi}} \int_0^x e^{-\xi^2} d\xi.$$

Show that this choice of u satisfies the heat equation. Now determine B, C, and f to satisfy the other conditions of the problem.

5.5 SOLUTIONS ON UNBOUNDED DOMAINS

We often use Fourier series to write the solution of an initial-boundary value problem on a bounded domain. On an unbounded domain we may turn to Fourier integrals and transforms, and sometimes to other transforms such as the Laplace transform. We will consider some examples.

5.5.1 Heat Conduction in an Infinite Bar

Consider the initial value problem

$$u_t = ku_{xx} \text{ for } -\infty < x < \infty, t > 0$$

$$u(x, 0) = f(x) \text{ for all } x. \tag{5.5}$$

This problem models heat flow in a homogeneous bar occupying the x-axis. The bar has no end points and therefore there are no boundary conditions. We compensate for this absence of information by seeking a bounded solution.

Substitute $u(x, t) = X(x)T(t)$ into the heat equation to obtain, as in the bounded interval case,

$$\frac{X''}{X} = \frac{T'}{kT} = -\lambda$$

or

$$X'' + \lambda X = 0, \ T + \lambda kT = 0.$$

There are no boundary conditions on X.

In Section 4.11 we found that the eigenvalues and (bounded) eigenfunctions for this problem for X are

$$\lambda = \omega^2 \geq 0; \quad X_\omega(x) = a_\omega \cos(\omega x) + b_\omega \sin(\omega x).$$

Now consider T. Corresponding to $\lambda = 0$ the equation for T is $T' = 0$, so $T = $ constant. If $\lambda = \omega^2 > 0$ then $T' + \omega^2 kT = 0$ with general solution $T = ce^{-\omega^2 kt}$.

We now have

$$u_\omega(x, t) = [a_\omega \cos(\omega x) + b_\omega \sin(\omega x)]e^{-\omega^2 kt}$$

as a bounded solution of the heat equation for $-\infty < x < \infty$, with a_ω and b_ω real numbers which are as yet arbitrary.

To satisfy the initial condition, attempt a superposition

$$u(x, t) = \int_0^\infty [a_\omega \cos(\omega x) + b_\omega \sin(\omega x)]e^{-\omega^2 kt} \, d\omega \tag{5.6}$$

as we did in solving the wave equation on the real line by separation of variables. We must choose the coefficients a_ω and b_ω so that

$$u(x, 0) = f(x) = \int_0^\infty [a_\omega \cos(\omega x) + b_\omega \sin(\omega x)] \, d\omega.$$

Since this is the Fourier integral expansion of f, choose

$$a_\omega = \frac{1}{\pi} \int_{-\infty}^\infty f(\xi)\cos(\omega\xi) \, d\xi$$

and

$$b_\omega = \frac{1}{\pi} \int_{-\infty}^\infty f(\xi)\sin(\omega\xi) \, d\xi.$$

This Fourier expansion converges to $f(x)$ (hence $u(x, 0) = f(x)$) if f is absolutely integrable and continuous, and f' is piecewise continuous, on the real line. In this event $u(x, t)$ as given by equation 5.6 is the solution of the problem.

Example 39 Suppose the initial temperature is given by $f(x) = e^{-|x|}$. Then

$$a_\omega = \frac{1}{\pi} \int_{-\infty}^\infty e^{-|\xi|} \cos(\omega\xi) \, d\xi = \frac{2}{\pi} \frac{1}{1 + \omega^2}$$

and

$$b_\omega = \frac{1}{\pi} \int_{-\infty}^{\infty} e^{-|\xi|} \sin(\omega\xi) \, d\xi = 0$$

(the last integral is zero by inspection because the integrand is an odd function). With this initial temperature function, the solution is

$$u(x, t) = \frac{2}{\pi} \int_{0}^{\infty} \frac{1}{1 + \omega^2} \cos(\omega x) e^{-\omega^2 kt} \, d\omega. \qquad \blacksquare$$

EXERCISE 199   Suppose the function f in the initial value problem 5.5 is an odd function. Show that the solution of the problem is odd (in the space variable). Show that, if f is an even function, then the solution is even in the space variable.

Solution by Fourier Transform   We will solve problem 5.5 using the Fourier transform as an illustration of this technique.

Since $-\infty < x < \infty$ we may attempt a Fourier transform of $u(x, t)$ in x, treating t as a parameter carried along in the process. Denote the Fourier transform of $u(x, t)$ in the x-variable by $\hat{u}(\omega, t)$. That is, $\mathfrak{F}[u(x, t)](\omega) = \hat{u}(\omega, t)$.

Apply $\mathfrak{F}$ to the heat equation:

$$\mathfrak{F}[u_t] = \mathfrak{F}[ku_{xx}]$$

or, since k is constant,

$$\mathfrak{F}[u_t] = k\mathfrak{F}[u_{xx}].$$

Since the transform is with respect to x, $\mathfrak{F}$ and the operation $\partial/\partial t$ of differentiating with respect to time commute:

$$\mathfrak{F}[u_t](\omega) = \int_{-\infty}^{\infty} \frac{\partial u(x, t)}{\partial t} e^{-i\omega x} \, dx = \frac{\partial}{\partial t} \int_{-\infty}^{\infty} u(x, t) e^{-i\omega x} \, dx = \frac{\partial}{\partial t} \hat{u}(\omega, t).$$

For the transform of u_{xx} apply the operational rule (equation 3.28 of Section 3.7) with $n = 2$:

$$\mathfrak{F}[u_{xx}](\omega) = (i\omega)^2 \hat{u}(\omega, t) = -\omega^2 \hat{u}(\omega, t).$$

The transform of the heat equation is

$$\hat{u}_t = -k\omega^2 \hat{u},$$

or

$$\hat{u}_t + k\omega^2 \hat{u} = 0.$$

This can be thought of as an ordinary differential equation in t, with ω as a parameter. The solution is

$$\hat{u}(\omega, t) = A_\omega e^{-\omega^2 kt}, \tag{5.7}$$

in which the coefficient A_ω may depend on ω. To determine A_ω, transform the initial condition $u(x, 0) = f(x)$ and use equation 5.7 at $\omega = 0$ to obtain

$$\mathfrak{F}[u(x, 0)](\omega) = \hat{u}(\omega, 0) = A_\omega = \mathfrak{F}[f](\omega) = \hat{f}(\omega).$$

Now we have

$$\hat{u}(\omega, t) = \hat{f}(\omega)e^{-\omega^2 kt}.$$

This is the Fourier transform of the solution. Apply the inverse Fourier transform (equation 3.27) to obtain

$$u(x, t) = \mathfrak{F}^{-1}[\hat{f}(\omega)e^{-\omega^2}kt] = \frac{1}{2\pi} \int_{-\infty}^{\infty} \hat{f}(\omega)e^{-\omega^2 kt}e^{i\omega x}\, d\omega.$$

To write this solution in terms of f (instead of $\hat{f}$), insert into this integral the definition of $\hat{f}(\omega)$ from equation 3.26:

$$u(x, t) = \frac{1}{2\pi} \int_{-\infty}^{\infty} \left(\int_{-\infty}^{\infty} f(\xi)e^{-i\omega\xi}\, d\xi \right) e^{-\omega^2 kt}e^{i\omega x}\, d\omega$$

$$= \frac{1}{2\pi} \int_{-\infty}^{\infty} \int_{-\infty}^{\infty} f(\xi)e^{-i\omega(\xi-x)}e^{-\omega^2 kt}\, d\xi\, d\omega$$

$$= \frac{1}{2\pi} \int_{-\infty}^{\infty} \int_{-\infty}^{\infty} f(\xi)[\cos(\omega(\xi - x)) - i\sin(\omega(\xi - x))]e^{-\omega^2 kt}\, d\xi\, d\omega$$

$$= \frac{1}{2\pi} \int_{-\infty}^{\infty} \int_{-\infty}^{\infty} f(\xi)\cos(\omega(\xi - x))e^{-\omega^2 kt}\, d\xi\, d\omega$$

$$- \frac{i}{2\pi} \int_{-\infty}^{\infty} \int_{-\infty}^{\infty} f(\xi)\sin(\omega(\xi - x))e^{-\omega^2 kt}\, d\xi\, d\omega.$$

Because $u(x, t)$ is real valued,

$$u(x, t) = \frac{1}{2\pi} \int_{-\infty}^{\infty} \int_{-\infty}^{\infty} f(\xi)\cos(\omega(\xi - x))e^{-\omega^2 kt} \, d\xi \, d\omega. \tag{5.8}$$

EXERCISE 200   Substitute the Fourier integral coefficients into the solution 5.6 obtained by Fourier integral to verify that the solutions obtained by Fourier integral and Fourier transform agree. *Hint:* This is like the argument carried out in Section 4.11.1 for solutions of the wave equation on the real line.

A Single Integral Expression for the Fourier Transform Solution   Equation 5.8 is the solution of the initial-boundary value problem 5.5 obtained by using the Fourier transform. It is possible to write this solution in a neater form by exploiting the integral

$$\int_{-\infty}^{\infty} e^{-\zeta^2} \cos\left(\frac{\alpha\zeta}{\beta}\right) d\zeta = \sqrt{\pi}e^{-\alpha^2/4\beta^2}. \tag{5.9}$$

Exercises 202 and 211 suggest derivations of equation 5.9. To use this integral, let

$$\zeta = \omega\sqrt{kt}, \quad \alpha = x - \xi \text{ and } \beta = \sqrt{kt}$$

in the integral with respect to ω in equation 5.8. Now,

$$d\zeta = \sqrt{kt} \, d\omega \text{ and } \frac{\alpha\zeta}{\beta} = \omega(x - \xi)$$

so

$$\int_{-\infty}^{\infty} e^{-\omega^2 kt} \cos(\omega(\xi - x)) \, d\omega = \int_{-\infty}^{\infty} e^{-\zeta^2} \cos\left(\frac{\alpha\zeta}{\beta}\right) \frac{1}{\sqrt{kt}} \, d\zeta = \frac{1}{\sqrt{kt}} \sqrt{\pi}e^{-(x-\xi)^2/4kt}.$$

Therefore equation 5.8 can be written

$$u(x, t) = \frac{1}{2\pi} \int_{-\infty}^{\infty} \sqrt{\frac{\pi}{kt}} e^{-(x-\xi)^2/4kt} f(\xi) \, d\xi,$$

and the solution of problem 5.5 is

$$u(x, t) = \frac{1}{2\sqrt{\pi kt}} \int_{-\infty}^{\infty} e^{-(x-\xi)^2/4kt} f(\xi) \, d\xi. \tag{5.10}$$

In this form it is not obvious that u satisfies the heat equation or the initial condition. To verify the former, we can proceed as follows. Suppose f is continuous and bounded on the real line, say $|f(\xi)| \leq M$ for all ξ. Choose any $L > 0$ and $t_0 > 0$. Then

$$0 \leq \left| e^{-(x-\xi)^2/4kt} f(\xi) \right| \leq M e^{-(L+\xi)^2/4kt_0}$$

for $-L \leq x \leq L$ and $t \geq t_0$. Since

$$\int_{-\infty}^{\infty} M e^{-(L+\xi)^2/4kt_0} \, d\xi$$

converges, the integral in equation 5.10 converges uniformly in x and t for $-L \leq x \leq L$ and $t \geq t_0$. In similar fashion we can check that the integrals obtained by differentiating under the integral sign, once with respect to t and twice with respect to x, also converge uniformly in this range of values of x and t. We can therefore verify that $u(x, t)$ as given by equation 5.10 satisfies the heat equation by directly computing u_t and u_{xx}, carrying out the differentiation under the integral sign for the factor in the solution that is defined by the improper integral. We leave this differentiation to the student. Since L is any positive number and t_0 can be chosen as close as we like to zero, this shows that $u(x, t)$ satisfies the heat equation for all x and for $t > 0$.

To show that $u(x, 0) = f(x)$, make the change of variable

$$\zeta = \frac{\xi - x}{\sqrt{4kt}}$$

in equation 5.10 to obtain

$$u(x, t) = \frac{1}{\sqrt{\pi}} \int_{-\infty}^{\infty} f(x + \sqrt{4kt}\,\zeta) e^{-\zeta^2} \, d\zeta.$$

Because f is bounded this integral converges uniformly in x and t (all x, $t \geq 0$), and the limit and integral can be interchanged to write

$$\lim_{t \to 0+} u(x, t) = \frac{1}{\sqrt{\pi}} \int_{-\infty}^{\infty} \left(\lim_{t \to 0+} f(x + \sqrt{4kt}\,\zeta) \right) e^{-\zeta^2} \, d\zeta.$$

Now

$$\lim_{t \to 0+} f(x + \sqrt{4kt}\,\zeta) = f(x)$$

by the continuity of f, and $f(x)$ passes through the integration with respect to ζ to yield

$$\lim_{t \to 0+} u(x, t) = u(x, 0) = f(x) \frac{1}{\sqrt{\pi}} \int_{-\infty}^{\infty} e^{-\zeta^2} \, d\zeta = f(x),$$

in which we have used the fact that

$$\int_{-\infty}^{\infty} e^{-\zeta^2} \, d\zeta = \sqrt{\pi}.$$

EXERCISE 201   Let a be a positive number and suppose $f(x) = 0$ for $|x| > a$. Suppose also that $f(x) > 0$ for $-a < x < a$. Let u be the solution of

$$u_t = ku_{xx} \text{ for } -\infty < x < \infty, \, t > 0$$

$$u(x, 0) = f(x) \text{ for } -\infty < x < \infty.$$

Prove that $u(x, t) > 0$ for $-\infty < x < \infty$ and for all $t > 0$. This conclusion can be interpreted to mean that the parabolic heat equation propagates disturbances with infinite velocity, in contrast to what we saw with the hyperbolic wave equation.

An Alternate Derivation Using Convolution   Consider again the initial value problem 5.5. We derived the solution 5.6 using the Fourier integral, and the solution 5.8 using the Fourier transform. These solutions are different expressions for the same function. We also derived the more compact expression 5.10 for this solution, involving only a single integral.

It is also possible to proceed directly from the boundary value problem 5.5 to the solution 5.10 by using the Fourier transform and the convolution theorem. By applying the Fourier transform in the space variable to the heat equation, we obtain, as in Section 5.6,

$$\hat{u}(\omega, t) = \hat{f}(\omega)e^{-k\omega^2 t}.$$

It is at this point that we diverge from the calculation done in that section. We would like to write $\hat{f}(\omega)e^{-k\omega^2 t}$ as the Fourier transform of some continuous function, because then $u(x, t)$ must be this function. To do this, it is enough to write just $e^{-k\omega^2 t}$ as the transform of some function, because then $\hat{u}(x, t)$ will be the product of two transformed functions, and this is the transform of the convolution of the individual functions.

To carry out this strategy we will use the following two results, derived previously (Lemma 4(1) and 4(2) of Section 3.7), but restated here for ease of reference.

1. If $m > 0$ and we set $g_m(x) = mg(mx)$, then $\widehat{g_m}(\omega) = \hat{g}(\omega/m)$.
2. If $g(x) = e^{-x^2/2}$, then $\hat{g}(\omega) = \sqrt{2\pi}e^{-\omega^2/2} = \sqrt{2\pi}g(\omega)$.

Use these as follows. Let $g(x) = e^{-x^2/2}$. With $m = 1/\sqrt{2kt}$,

$$e^{-k\omega^2 t} = g(\sqrt{2kt}\omega) = \frac{1}{\sqrt{2\pi}}\hat{g}(\sqrt{2kt}\omega) = \frac{1}{\sqrt{2\pi}}\hat{g}(\omega/m)$$

$$= \frac{1}{\sqrt{2\pi}}\widehat{g_m}(\omega) = \frac{1}{\sqrt{2\pi}}\widehat{g_{1/\sqrt{2kt}}}(\omega).$$

Thus

$$\hat{u}(\omega, t) = \hat{f}(\omega)\frac{1}{\sqrt{2\pi}}\widehat{g_{1/\sqrt{2kt}}}(\omega) = \frac{1}{\sqrt{2\pi}}\widehat{g_{1/\sqrt{2kt}}}\hat{f}(\omega).$$

We have written the factors in this order to obtain the particular form of the solution given in equation 5.10. By the convolution theorem (Theorem 10(2), Section 3.8),

$$u(x, t) = \frac{1}{\sqrt{2\pi}}(g_{1/\sqrt{2kt}} * f)(x) = \frac{1}{\sqrt{2\pi}}\int_{-\infty}^{\infty} g_{1/\sqrt{2kt}}(x - \xi)f(\xi)\, d\xi$$

$$= \frac{1}{\sqrt{2\pi}}\int_{-\infty}^{\infty}\frac{1}{\sqrt{2kt}}g((x - \xi)/\sqrt{2kt})f(\xi)\, d\xi = \frac{1}{2\sqrt{\pi kt}}\int_{-\infty}^{\infty} e^{-(x-\xi)^2/4kt}f(\xi)\, d\xi.$$

This is equation 5.10.

EXERCISE 202   Derive equation 5.9 as follows. Let

$$F(x) = \int_0^{\infty} e^{-\zeta^2}\cos(x\zeta)\, d\zeta.$$

(a) Show that this integral converges uniformly for all x, as does the integral obtained by interchanging d/dx and $\int_0^{\infty} \cdots d\zeta$.
(b) Compute $F'(x)$ by interchanging d/dx and $\int_0^{\infty} \cdots d\zeta$. By integrating by parts, show that

$$F'(x) = -\frac{x}{2}F(x).$$

(c) Solve this differential equation to obtain $F(x)$ to within a constant.

(d) Evaluate this constant by using the standard integral

$$\int_0^\infty e^{-\zeta^2} \, d\zeta = \frac{\sqrt{\pi}}{2}.$$

EXERCISE 203   For each of problems 1 through 5, obtain the solution of problem 5.5 by separation of variables (Fourier integral), then by Fourier transform, and finally by convolution. Is it obvious that each method yields the same solution?

1. $f(x) = e^{-4|x|}$

2. $f(x) = \begin{cases} \sin(x) \text{ for } |x| \le \pi \\ 0 \text{ for } |x| > \pi. \end{cases}$

3. $f(x) = \begin{cases} x \text{ for } 0 \le x \le 4 \\ 0 \text{ for } x < 0 \text{ and for } x > 4. \end{cases}$

4. $f(x) = \begin{cases} e^{-x} \text{ for } -1 \le x \le 1 \\ 0 \text{ for } |x| > 1. \end{cases}$

5. $f(x) = \begin{cases} 1 \text{ for } 0 \le x \le 1 \\ -1 \text{ for } -1 \le x < 0 \\ 0 \text{ for } |x| > 1. \end{cases}$

EXERCISE 204   Let

$$f(x) = \begin{cases} 1 \text{ for } -1 \le x \le 1 \\ 0 \text{ for } |x| > 1 \end{cases}$$

in the solution 5.10. Expand the integrand in a Maclaurin series and integrate term by term to obtain a series solution for $u(x, t)$.

EXERCISE 205   Carry out the program of Exercise 204, with

$$f(x) = \begin{cases} x \text{ for } -2 \le x \le 2 \\ 0 \text{ for } |x| > 2. \end{cases}$$

EXERCISE 206   Use separation of variables and then the Fourier transform to solve

$$u_t = ku_{xx} \text{ for } -\infty < x < \infty, \, t > 0$$

$$u_x(x, 0) = \begin{cases} 1 \text{ for } -h \le x \le h \\ 0 \text{ for } |x| > h. \end{cases}$$

Show that the same solution is obtained by both methods.

EXERCISE 207   Use separation of variables and then a Fourier transform to solve

$$u_t = ku_{xx} \text{ for } -\infty < x < \infty, \, t > 0$$

$$u_x(x, 0) = \begin{cases} \cos(\pi x) \text{ for } -1 \le x \le 1 \\ 0 \text{ for } |x| > 1. \end{cases}$$

Show that the same solution is obtained by both methods.

EXERCISE 208   Consider the problem

$$u_t = u_{xx} \text{ for } -\infty < x < \infty, \, t > 0$$

$$u(x, 0) = 0.$$

Define

$$s(x, t) = \begin{cases} xt^{-3/2}e^{-x^2/4t} \text{ for } -\infty < x < \infty, \, t > 0 \\ 0 \text{ for } t = 0. \end{cases}$$

Show that s satisfies the heat equation for $-\infty < x < \infty$ and $t > 0$, and that

$$\lim_{t \to 0+} s(x, t) = 0.$$

Hence conclude that s is continuous on $-\infty < x < \infty$, $t \ge 0$ and is a solution of the initial value problem. But $u(x, t) \equiv 0$ is also a solution. Does this show that this problem does not have a unique solution?

EXERCISE 209   For the problem

$$u_t = ku_{xx} \text{ for } -\infty < x < \infty, \, t > 0$$

$$u(x, 0) = f(x) \text{ for } -\infty < x < \infty,$$

determine a function $G(x, t)$ such that the solution can be written

$$u(x, t) = \int_{-\infty}^{\infty} G(x - \xi, t)f(\xi) \, d\xi.$$

G is called the Green's function for this initial-boundary value problem. Physically, it may be thought of as the temperature at x, at time t, due to a heat source concentrated at ξ at time zero. Show that

$$\int_{-\infty}^{\infty} G(x - \xi, t) \, d\xi = 1$$

for all x.

EXERCISE 210   Derive a Duhamel-type principle for the problem

$$u_t = u_{xx} + f(x, t) \text{ for } -\infty < x < \infty, t > 0$$

$$u(x, 0) = 0.$$

Using this principle, derive the solution

$$u(x, t) = \int_0^t \int_{-\infty}^{\infty} \frac{1}{2\sqrt{\pi(t - \tau)}} e^{-(x-\xi)^2/4(t-\tau)} f(\xi, \tau) \, d\xi \, d\tau.$$

5.5.2   Heat Conduction in a Semi-Infinite Bar

Consider the initial-boundary value problem

$$u_t = ku_{xx} \text{ for } x \geq 0$$

$$u(0, t) = 0 \text{ for } t \geq 0$$

$$u(x, 0) = f(x). \tag{5.11}$$

This problem is posed on $[0, \infty)$ and so has a boundary condition at the left end. We seek a bounded solution.

Let $u(x, t) = X(x)T(t)$ and proceed as with the Fourier integral solution on the entire real line to obtain

$$X'' + \lambda X = 0, \ T' + \lambda kT = 0.$$

Now $u(0, t) = X(0)T(t) = 0$ implies that

$$X(0) = 0.$$

Consider cases on λ. If $\lambda = 0$ then $X = ax + b$. For a bounded solution we need $a = 0$. But then $X(0) = b = 0$, so this case has only the trivial solution for X.

If $\lambda = -\omega^2$ with $\omega > 0$, then $X = ae^{\omega x} + be^{-\omega x}$. This is unbounded for $x \geq 0$ unless $a = 0$. Then $X = be^{-\omega x}$. But then $X(0) = b = 0$, so this case also yields only the trivial solution.

If $\lambda = \omega^2$ with $\omega > 0$, then $X = a \cos(\omega x) + b \sin(\omega x)$. This is bounded for $x \geq 0$. However, $X(0) = a = 0$, so we retain only the sine term.

With $\lambda = \omega^2$ the equation for T is $T' + \omega^2 kT = 0$ with solutions $T = ce^{-\omega^2 kt}$.

Thus form functions

$$u_\omega(x, t) = b_\omega \sin(\omega x)e^{-\omega^2 kt}.$$

To satisfy the initial condition form a superposition

$$u(x, t) = \int_0^\infty b_\omega \sin(\omega x)e^{-\omega^2 kt} \, d\omega.$$

We require that

$$u(x, 0) = f(x) = \int_0^\infty b_\omega \sin(\omega x) \, d\omega.$$

Choose b_ω as the coefficient in the Fourier sine integral expansion of f:

$$b_\omega = \frac{2}{\pi} \int_0^\infty f(\xi)\sin(\omega \xi) \, d\xi.$$

This expansion converges to $f(x)$ if f is continuous and absolutely integrable on $[0, \infty)$, and f' is piecewise continuous.

Solution by Fourier Sine Transform   We will illustrate another technique to solve this problem on $[0, \infty)$. Since $x \geq 0$, we can entertain the possibility of using a Fourier sine or cosine transform in x. The operational formula for the cosine transform (Theorem 11(2), Section 3.9) requires that we know $u_x(0, t)$, and we do not. The operational formula for the sine transform (Theorem 11(1)) uses $u(0, t)$, which we are given. We therefore choose the sine transform in x and apply it to the heat equation. As with the Fourier transform in the preceding section, $\partial/\partial t$ passes through $\mathfrak{F}_S$ and we use the operational formula to compute $\mathfrak{F}_S\{u_{xx}\}$. Letting $\mathfrak{F}_S[U(x, t)](\omega) = \hat{u}_S(\omega, t)$, we obtain

$$\frac{\partial}{\partial t} \hat{u}_S(\omega, t) = k[-\omega^2 \hat{u}_S(\omega, t) + \omega u(0, t)].$$

Since $u(0, t) = 0$,

$$\frac{\partial}{\partial t} \hat{u}_S(\omega, t) + \omega^2 k\hat{u}_S(\omega, t) = 0.$$

Think of this as a first order differential equation in t. The general solution is

$$\hat{u}_S(\omega, t) = b_\omega e^{-\omega^2 kt}.$$

Now

$$\hat{u}_S(\omega, 0) = b_\omega = \mathfrak{F}_S[u(x, 0)](\omega) = \mathfrak{F}_S[f(x)](\omega) = \hat{f}_S(\omega).$$

Therefore

$$\hat{u}_S(\omega, t) = \hat{f}_S(\omega)e^{-\omega^2 kt}.$$

This is the sine transform of the solution. Apply the inverse sine transform (equation 3.32) to obtain

$$u(x, t) = \frac{2}{\pi} \int_0^\infty \hat{f}_S(\omega)e^{-\omega^2 kt} \sin(\omega x) \, d\omega.$$

Upon inserting the definition of $\hat{f}_S(\omega)$ we obtain the solution

$$u(x, t) = \frac{2}{\pi} \int_0^\infty \int_0^\infty f(\xi)\sin(\omega\xi)\sin(\omega x)e^{-\omega^2 kt} \, d\xi \, d\omega.$$

This solution is often written in a different way. Use the identity

$$\sin(\omega\xi)\sin(\omega x) = \frac{1}{2} [\cos(\omega(x - \xi)) - \cos(\omega(x + \xi))]$$

and interchange the order of integration in the solution to write

$$u(x, t) = \frac{1}{\pi} \int_0^\infty \left(\int_0^\infty e^{-\omega^2 kt} \cos(\omega(x - \xi)) \, d\omega \right) f(\xi) \, d\xi$$

$$- \frac{1}{\pi} \int_0^\infty \left(\int_0^\infty e^{-\omega^2 kt} \cos(\omega(x + \xi)) \, d\omega \right) f(\xi) \, d\xi.$$

Now use the fact that

$$\frac{1}{\pi} \int_0^\infty e^{-\omega^2 kt} \cos(\alpha\omega) \, d\omega = \frac{1}{2\sqrt{\pi kt}} e^{-\alpha^2/4kt},$$

which can be obtained by a change of variables in the integral 5.9. This enables us to write the solution as

$$u(x, t) = \frac{1}{2\sqrt{\pi kt}} \int_0^\infty [e^{-(x-\xi)^2/4kt} - e^{-(x+\xi)^2/4kt}] f(\xi) \, d\xi. \tag{5.12}$$

Note the similarity in form between this solution on the half-line and the solution 5.10 on the entire line.

EXERCISE 211   Show that

$$\int_{-\infty}^\infty e^{-\zeta^2} \cos(2\beta\zeta) \, d\zeta = \sqrt{\pi} e^{-\beta^2}$$

by complex analytic methods as follows. First integrate e^{-z^2} about the closed rectangular path having vertices at $\pm R$ and $\pm R + \beta i$. Use Cauchy's theorem to conclude that this integral is zero. Set this integral equal to the sum of the integrals over the individual sides of the rectangle and take the limit as $R \to \infty$.

EXERCISE 212   In each of problems 1 through 5, solve the initial-boundary value problem 5.11 by separation of variables (Fourier integral) and also by Fourier sine transform.

1. $f(x) = e^{-\alpha x}$, with α a positive number.

2. $f(x) = xe^{-\alpha x}$, α positive

3. $f(x) = \begin{cases} 1 \text{ for } 0 \le x \le h \\ 0 \text{ for } x > h. \end{cases}$

4. $f(x) = \begin{cases} x \text{ for } 0 \le x \le 2 \\ 0 \text{ for } x > 2. \end{cases}$

5. $f(x) = \begin{cases} 1 \text{ for } 0 \le x \le h \\ -1 \text{ for } h < x \le 2h \\ 0 \text{ for } x > 2h. \end{cases}$

EXERCISE 213   Solve

$$u_t = ku_{xx} \text{ for } x > 0, \, t > 0$$

$$u(x, 0) = f(x) \text{ for } x > 0$$

$$u_x(0, t) = 0 \text{ for } t > 0.$$

EXERCISE 214 Solve the problem of Exercise 213 with

$$f(x) = \begin{cases} e^{-x} & \text{for } 0 \le x \le 1 \\ 0 & \text{for } x > 1. \end{cases}$$

EXERCISE 215 Use separation of variables to solve

$$u_t = ku_{xx} \text{ for } x > 0, \ t > 0$$

$$u_x(x, 0) = \begin{cases} A & \text{for } 0 \le x \le h \\ 0 & \text{for } x > h. \end{cases}$$

Solve the problem again using the Fourier transform. Show that the same solution is obtained by both methods.

EXERCISE 216 Carry out the program of Exercise 215 for

$$u_t = ku_{xx} \text{ for } x > 0, \ t > 0$$

$$u_x(x, 0) = \begin{cases} x & \text{for } 0 \le x \le 2 \\ 0 & \text{for } x > 2. \end{cases}$$

EXERCISE 217 Carry out the program of Exercise 215 for

$$u(x, 0) = \begin{cases} 1 - x & \text{for } 0 \le x \le 1 \\ 0 & \text{for } x > 1. \end{cases}$$

EXERCISE 218 Solve

$$u_t = u_{xx} - u \text{ for } x > 0, \ t > 0$$

$$u_x(0, t) = f(t) \text{ for } t > 0$$

$$u(x, 0) = 0 \text{ for } x > 0.$$

using the Fourier sine or cosine transform, as appropriate.

EXERCISE 219 Solve

$$u_t = u_{xx} - tu \text{ for } x > 0, \ t > 0$$

$$u(x, 0) = xe^{-x} \text{ for } x > 0$$

$$u(0, t) = 0 \text{ for } t > 0.$$

EXERCISE 220

(a) Solve the problem

$$u_t = ku_{xx} \text{ for } x > 0,\ t > 0$$
$$u(x, 0) = 0 \text{ for } x > 0$$
$$u(0, t) = B \text{ for } t > 0$$

in which B is constant. *Hint:* Transform the problem into one solved previously by setting $v = u - B$.

Show that the solution can be written in the form

$$u(x, t) = B(1 - \text{erf}(x/\sqrt{4kt}))$$

for $x > 0$, $t > 0$, in which erf is the error function defined in Exercise 198.

(b) Solve the problem

$$u_t = ku_{xx} \text{ for } x > 0,\ t > 0$$
$$u(x, 0) = 0 \text{ for } x > 0$$
$$u(0, t) = B(t) \text{ for } t > 0.$$

Show that the solution of this problem can be written in the form

$$u(x, t) = \int_0^t \frac{\partial}{\partial t} W(x, \tau, t - \tau)\, d\tau,$$

in which $W(x, \tau, t)$ is the solution of the problem

$$W_t = kW_{xx} \text{ for } x > 0,\ t > 0$$
$$W(x, 0) = 0 \text{ for } x > 0$$
$$W(0, t) = B(\tau) \text{ for } t > 0.$$

Note that, in the last condition, $W(0, t)$ is constant for each choice of τ.

(c) Use the result of part (a) to solve the problem for W in part (b) and write the solution for $u(x, t)$ in the problem for u in part (b). This solution can be simplified by an appropriate change of variables.

A Digression: The Great Debate Over the Age of the Earth   The solution 5.12 for the heat equation in a semi-infinite medium played an important role in one of the more significant and interesting scientific controversies ever to take place. The entire story is told in Joe D. Burchfield's book *Lord Kelvin and The Age of the Earth,* published by Macmillan in 1975. We will recall some of the background and then use the solution 5.12 to derive Lord Kelvin's estimate for the age of the earth.

In the eighteenth century the common belief, among natural philosophers and nonscientists alike, was that the earth and everything on it (plants, animals, mankind) had all been created at about the same time. This meant that the age of the earth was approximately the age of mankind, and this was estimated to be several thousand years. Some scholars estimated the age of the earth by forming a list of families in the Bible and guessing at the average age of each generation. One biblical chronology, constructed in the third century A.D. by Julius Africanus, made the underlying assumption that history is comprised of a cosmic week, with each week one thousand years. Estimating that Christ was born some time in the sixth day of this cosmic week, Africanus arrived at a creation date of about 5500 B.C. for the earth. This would make the earth about 7500 years old today.

In the nineteenth century, scientists began to suspect that certain geological processes, such as the formation of mountains and canyons, must have taken a considerable period of time. Estimates of the age of the earth were extended to tens of thousands of years.

Then, in the period 1830–1833, Sir Charles Lyell's monumental three-volume work *Principles of Geology* was published. In it Lyell argued that processes such as erosion and sedimentation in rivers had to occur over extended periods of time. Unlike his predecessors and colleagues, Lyell thought in terms of many millions of years, perhaps hundreds of millions, a bold step beyond anything previously imagined. This view became known as the *uniformitarian theory*, because it disputed the belief that early violent geological activity had caused the present day earth to form over a relatively short time, and maintained that the earth reached its present state over an extended period of gradual, nearly uniform change.

Charles Darwin, who in 1859 published his classic *On the Origin of Species*, was very much a supporter of Lyell's view because evolution required longer periods of time than the several thousand years the eighteenth century scientists had been willing to allow.

In the 1850s, Lord Kelvin (Sir William Thomson, 1824–1907) entered the arena with a completely different perspective, that of a physicist. Kelvin was instrumental in the development of the branch of physics known as thermodynamics, and was sensitive to arguments involving generation and conservation of energy. He began by considering the sun as the primary source of energy for the earth. Since nothing was known of nuclear processes at this time, it was at first believed that the sun must be a gigantic chemical reaction. But this notion was soon discarded for want of a sufficiently large source of fuel. Kelvin therefore turned to the idea that the sun generated heat energy converted from gravitational potential energy as the sun contracts. This led him to an initial estimate that the sun, and therefore the earth, could not be more than one hundred million years old.

While this approach was interesting enough, it was Kelvin's next argument that caused the revolution in geological thinking and methodology. Kelvin had observed from data taken from mines extending deep into the earth, that the temperature of the earth increases with depth. He reasoned that this must be caused by a continual loss of heat energy from the interior of the planet, by conduction into the earth's upper crust. However, it can also be observed that this upper crust does not increase in temperature on a year to year basis. Hence Kelvin reasoned that there must be a continual loss of heat energy from the earth. Further, Kelvin, and many geologists, believed that the earth is a solid ball that had solidified at a uniform temperature. Thus, if one could make an estimate of this initial temperature, and assuming that there is a continual loss of heat energy, it should be possible, by determining the rate of heat loss and current temperatures in the crust, to estimate the age of the earth.

Kelvin was prepared mathematically for this task. Unlike geologists, who at this time did not approach their science by any kind of mathematical modeling, Kelvin was quite expert at constructing models and solving the resulting equations. He began by assuming that the earth is an approximately homogeneous sphere of radius R. If $\psi(\mathbf{x}, t)$ is the temperature at a point $\mathbf{x}$ (a vector denoting a point in 3-space, with origin at the center of the sphere) and time t, then ψ satisfies the three-dimensional heat equation

$$\frac{\partial \psi}{\partial t}(\mathbf{x}, t) = k\nabla^2\psi(\mathbf{x}, t) \text{ for } |\mathbf{x}| < R, t > 0.$$

Kelvin had solved this equation while engaged in studies of heat conduction. However, he now became convinced that, very near the surface of the earth, the temperature has remained constant and is on average approximately its initial value. This led him to neglect the curvature of the earth and model heat conduction through the earth as being along the half-line from its center. Let $u(x, t)$ be the temperature of the earth at distance x from the center and time t, for $x > 0$, $t > 0$. The one-dimensional problem to be solved for u is

$$u_t(x, t) = ku_{xx}(x, t) \text{ for } x > 0, t > 0$$

$$u(x, 0) = c \text{ for } x > 0$$

$$u(0, t) = 0 \text{ for } t > 0.$$

Kelvin solved this problem, obtaining the integral expression given in equation 5.12. We repeat this expression here for ease of reference, noting that in the present context $f(x) = c$:

$$u(x, t) = \frac{c}{2\sqrt{\pi kt}} \int_0^\infty [e^{-(x-\xi)^2/4kt} - e^{-(x+\xi)^2/4kt}] \, d\xi.$$

Kelvin had no way of measuring the earth's temperature at large depths. However, by making measurements as he descended into mine tunnels, he could estimate the rate of increase of temperature as one moved from the surface toward the center of the earth. This suggested that he compute the rate of change of temperature with depth. From the solution, compute

$$\frac{\partial u}{\partial x} = \frac{c}{2\sqrt{\pi kt}} \int_0^\infty \left\{ -\frac{x - \xi}{2kt} e^{-(x-\xi)^2/4kt} + \frac{x + \xi}{2kt} e^{-(x+\xi)^2/4kt} \right\} d\xi.$$

Then

$$\frac{\partial u}{\partial x}(0, t) = \frac{c}{2\sqrt{\pi kt}} \int_0^\infty \frac{\xi}{kt} e^{-\xi^2/4kt} d\xi = \frac{c}{\sqrt{\pi kt}} [-e^{-\xi^2/4kt}]_0^\infty = \frac{c}{\sqrt{\pi kt}}.$$

Solve this equation for t to obtain

$$t = \frac{1}{\pi k} \left(\frac{c}{u_x(0, t)} \right)^2.$$

As mentioned, Kelvin had used measurements taken in mine shafts to estimate $u_x(0, t)$. He also took measurements of the temperature of rocks and molten lava to estimate c and k. These led him to estimate the age of the earth at between 100 million and 400 million years, with the variation to take into account possible errors in the estimates of c, k, and $u_x(0, t)$.

The effect of Kelvin's analysis was profound. Many geologists disagreed with his conclusions, but they were at a loss to contradict his mathematics, which they did not understand. The mathematical arguments did indeed look formidable. And Kelvin's deserved high status in the scientific community tended to lend further credence to his position. Despite his primary involvement with physics and thermodynamics, Kelvin retained an interest in geology and the age of the earth until his death early this century. Partly as a result of his influence, the science of geology became more attuned to the careful collection and analysis of data based on geological samples, and also, to a lesser extent, on mathematical modeling.

There is an amusing footnote to the story of Kelvin and the age of the earth. In 1904 Rutherford announced his discovery of heat generation by radioactive processes (he was studying radium). This introduced a source of heat energy beyond anything Kelvin could have imagined when he was making his estimates. Now it happened that Rutherford was about to give a lecture on his discovery, and he expected Kelvin to be present. Rutherford was worried about Kelvin's reaction to this introduction of a new factor in measuring heat energy within the earth, which of course negated Kelvin's estimates and conclusions. During the lecture Kelvin appeared to doze off, but he suddenly came awake when Rutherford approached the crucial point. Fortunately, Rutherford thought fast and said that Kelvin had estimated the age of the earth, assuming the absence of new and unanticipated sources of heat. Since no one could fault Kelvin for not foreseeing the phenomenon of radioactivity, this satisfied him and all was well.

5.5.3   A Problem Involving a Nonhomogeneous Boundary Condition

Consider the initial-boundary value problem

$$u_t = ku_{xx} \text{ for } x > 0, \ t > 0$$

$$u(x, 0) = A \text{ for } x > 0$$

$$u(0, t) = \begin{cases} B \text{ for } 0 < t < t_0 \\ 0 \text{ for } t > t_0. \end{cases}$$

A, B, and t_0 are positive constants. This models a homogeneous bar extending from 0 to infinity. At time zero the temperature throughout the bar for $x > 0$ is a constant A. The left end is kept at temperature B from time zero until time t_0, after which it is at temperature zero.

The boundary condition is more conveniently written using the Heaviside function, which is defined by

$$H(t) = \begin{cases} 1 \text{ for } t \geq 0 \\ 0 \text{ for } t < 0. \end{cases}$$

In terms of H,

$$u(0, t) = B[1 - H(t - t_0)].$$

We will solve this problem by using a Laplace transform in t. Let $\mathcal{L}$ denote the Laplace transform, defined by

$$\mathcal{L}[f](s) = \int_0^\infty f(t)e^{-st} \, dt.$$

Let $\mathcal{L}[u(x, t)](s) = U(x, s)$. The variable of the transformed function is s, and x is carried along as a parameter.

Since the transform is in the t variable, $\partial/\partial x$ and $\mathcal{L}$ commute, so

$$\mathcal{L}[u_{xx}(x, t)](s) = \frac{\partial^2}{\partial x^2} U(x, s).$$

For the transform of u_t, integrate by parts to obtain

$$\mathcal{L}[u_t(x, t)](s) = \int_0^\infty u_t(x, t)e^{-st} \, dt = u(x, t)e^{-st}\big]_0^\infty - \int_0^\infty u(x, t)(-s)e^{-st} \, dt$$

$$= -u(x, 0) + s \int_0^\infty u(x, t)e^{-st} \, dt = -A + sU(x, s).$$

Therefore

$$\mathcal{L}[u_t] = \mathcal{L}[ku_{xx}]$$

becomes

$$-A + sU(x, s) = k\,\frac{\partial^2 U(x, s)}{\partial x^2}$$

or

$$\frac{\partial^2 U}{\partial x^2} - \frac{s}{k}U = -\frac{A}{k}.$$

Think of this as a constant coefficient second order differential equation for U in terms of the variable x. Its general solution is

$$U(x, s) = a(s)e^{\sqrt{s/k}x} + b(s)e^{-\sqrt{s/k}x} + \frac{A}{s},$$

in which the coefficients may be functions of s. We will impose the condition that $a(s) = 0$ because $e^{\sqrt{s/k}x} \to \infty$ as $x \to \infty$ and we want a bounded solution. Then

$$U(x, s) = b(s)e^{-\sqrt{s/k}x} + \frac{A}{s}.$$

To solve for $b(s)$ use the condition that $u(0, t) = B[1 - H(t - t_0)]$. Then

$$\mathcal{L}[u(0, t)](s) = U(0, s) = b(s) + \frac{A}{s} = \mathcal{L}[B[1 - H(t - t_0)]](s) = \frac{B}{s} - \frac{B}{s}e^{-t_0 s}$$

in which the transform of $1 - H(t - t_0)$ can be computed by integration, by consulting a table, or with computer software. We conclude that

$$b(s) = \frac{B - A}{s} - \frac{B}{s}s^{-t_0 s}$$

and therefore

$$U(x, s) = \left[\frac{B - A}{s} - \frac{B}{s}e^{-t_0 s}\right]e^{-\sqrt{s/k}x} + \frac{A}{s}.$$

The solution is $u(x, t) = \mathcal{L}^{-1}[U(x, s)]$ and this inverse can be determined by consulting a table or using software. There is also a complex integral formula for the inverse Laplace transform of a function. We obtain

$$u(x, t) = \left[A \text{ erf} \left(\frac{x}{2\sqrt{kt}} \right) + B \text{ erf } c \left(\frac{x}{2\sqrt{kt}} \right) \right] (1 - H(t - t_0)) + \left[A \text{ erf} \left(\frac{x}{2\sqrt{kt}} \right) \right.$$

$$\left. + B \text{ erf } c \left(\frac{x}{2\sqrt{kt}} \right) - B \text{ erf } c \left(\frac{x}{2\sqrt{k(t - t_0)}} \right) \right] H(t - t_0).$$

Here erf is the error function defined in Exercise 198, and erf c is the complementary error function defined by

$$\text{erf } c(x) = 1 - \text{erf}(x) = \frac{2}{\sqrt{\pi}} \int_0^x e^{-\xi^2} \, d\xi.$$

5.6   THE NONHOMOGENEOUS HEAT EQUATION

We will consider the initial-boundary value problem

$$u_t = ku_{xx} + F(x, t) \text{ for } 0 < x < L, t > 0$$

$$u(0, t) = u(L, t) = 0 \text{ for } t \geq 0$$

$$u(x, 0) = f(x) \text{ for } 0 \leq x \leq L. \tag{5.13}$$

This is a nonhomogeneous problem because of the term $F(x, t)$ in the heat equation. This could, for example, represent a source of heat energy in the medium. We will place conditions on F as needed as we solve this problem.

In the case $F \equiv 0$ we obtained a solution (equation 5.3) of the form

$$\sum_{n=1}^{\infty} b_n \sin \left(\frac{n\pi x}{L} \right) e^{-n^2\pi^2kt/L^2},$$

in which the $b_n's$ are the coefficients in the Fourier sine expansion of f on $[0, L]$. For the current problem, we will take a cue from that case and attempt a solution

$$u(x, t) = \sum_{n=1}^{\infty} T_n(t) \sin \left(\frac{n\pi x}{L} \right). \tag{5.14}$$

The problem is to determine each T_n.

First observe that, for a given t, equation 5.14 can be thought of as the Fourier sine expansion of $u(x, t)$, considered a function of x, with $T_n(t)$ the nth

Fourier coefficient in this expansion. Of course we do not yet know $u(x, t)$, but on the basis of equation 5.14 we can formally write

$$T_n(t) = \frac{2}{L} \int_0^L u(\xi, t)\sin\left(\frac{n\pi\xi}{L}\right) d\xi. \tag{5.15}$$

Now assume that, for each $t \geq 0$, $F(x, t)$, as a function of x, can also be expanded in a Fourier sine series:

$$F(x, t) = \sum_{n=1}^{\infty} B_n(t)\sin\left(\frac{n\pi x}{L}\right) dx, \tag{5.16}$$

where

$$B_n(t) = \frac{2}{L} \int_0^L F(\xi, t)\sin\left(\frac{n\pi\xi}{L}\right) d\xi \tag{5.17}$$

is the coefficient in this expansion, and may of course depend on t.

Differentiate equation 5.15 to obtain

$$T_n'(t) = \frac{2}{L} \int_0^L u_t(\xi, t)\sin\left(\frac{n\pi\xi}{L}\right) d\xi. \tag{5.18}$$

Since $u_t = ku_{xx} + F(x, t)$, equation 5.18 becomes

$$T_n'(t) = \frac{2k}{L} \int_0^L u_{xx}(\xi, t)\sin\left(\frac{n\pi\xi}{L}\right) d\xi + \frac{2}{L} \int_0^L F(\xi, t)\sin\left(\frac{n\pi\xi}{L}\right) d\xi.$$

In view of equation 5.17,

$$T_n'(t) = \frac{2k}{L} \int_0^L u_{xx}(\xi, t)\sin\left(\frac{n\pi\xi}{L}\right) d\xi + B_n(t). \tag{5.19}$$

Apply integration by parts twice to the integral in equation 5.19, using the boundary conditions and, at the last step, equation 5.15:

$$\int_0^L u_{xx}(\xi, t)\sin\left(\frac{n\pi\xi}{L}\right) d\xi = \sin\left(\frac{n\pi\xi}{L}\right) u_x(\xi, t)\Bigg]_0^L$$

$$- \int_0^L u_x(\xi, t)\frac{n\pi}{L}\cos\left(\frac{n\pi\xi}{L}\right) d\xi = -\int_0^L \frac{n\pi}{L} u_x(\xi, t)\cos\left(\frac{n\pi\xi}{L}\right) d\xi$$

$$= -\frac{n\pi}{L} u(\xi, t)\cos\left(\frac{n\pi\xi}{L}\right)\Bigg]_0^L + \frac{n\pi}{L} \int_0^L u(\xi, t)\left(-\frac{n\pi}{L}\right)\sin\left(\frac{n\pi\xi}{L}\right) d\xi$$

$$= -\frac{n^2\pi^2}{L^2} \int_0^L u(\xi, t)\sin\left(\frac{n\pi\xi}{L}\right) d\xi = -\frac{n^2\pi^2}{L^2}\frac{L}{2} T_n(t) = -\frac{n^2\pi^2}{2L} T_n(t).$$

Substituting this into equation 5.19 yields

$$T_n'(t) = -k\frac{n^2\pi^2}{L^2} T_n(t) + B_n(t).$$

For $n = 1, 2, \ldots$, we now have an ordinary differential equation for $T_n(t)$:

$$T_n' + k\frac{n^2\pi^2}{L^2} T_n = B_n(t). \tag{5.20}$$

Next, use equation 5.15 to obtain the condition,

$$T_n(0) = \frac{2}{L}\int_0^L u(\xi, 0)\sin\left(\frac{n\pi\xi}{L}\right) d\xi = \frac{2}{L}\int_0^L f(\xi)\sin\left(\frac{n\pi\xi}{L}\right) d\xi = b_n,$$

the nth coefficient in the Fourier sine expansion of f on $[0, L]$.

Solve the ordinary differential equation 5.20 subject to the condition $T_n(0) = b_n$ to obtain the unique solution

$$T_n(t) = \int_0^t e^{-kn^2\pi^2(t-\tau)/L^2} B_n(\tau)\, d\tau + b_n e^{-kn^2\pi^2 t/L^2}.$$

Finally, substitute this into equation 5.14 to obtain

$$u(x, t) = \sum_{n=1}^{\infty} \left(\int_0^t e^{-kn^2\pi^2(t-\tau)/L^2} B_n(\tau)\, d\tau\right) \sin\left(\frac{n\pi x}{L}\right)$$

$$+ \sum_{n=1}^{\infty} b_n \sin\left(\frac{n\pi x}{L}\right) e^{-kn^2\pi^2 t/L^2} \tag{5.21}$$

in which $B_n(\tau)$ is determined by equation 5.17, and

$$b_n = \frac{2}{L}\int_0^L f(\xi)\sin\left(\frac{n\pi\xi}{L}\right) d\xi$$

for $n = 1, 2, \ldots$. We leave it for the student to verify that $u(x, t)$ as defined by equation 5.21 is the solution of problem 5.13.

Example 40   Solve the initial-boundary value problem

$$u_t = 4u_{xx} + t^2 \cos(x/2) \text{ for } 0 < x < \pi, \, t > 0$$

$$u(0, t) = u(\pi, t) = 0 \text{ for } t \geq 0$$

$$u(x, 0) = f(x) = \begin{cases} 0 \text{ for } 0 \leq x \leq \pi/2 \\ 25 \text{ for } \pi/2 < x < \pi. \end{cases}$$

In the notation of the preceding discussion, $F(x, t) = t^2 \cos(x/2)$ and $L = \pi$. First compute

$$B_n(t) = \frac{2}{\pi} \int_0^\pi t^2 \cos(\xi/2)\sin(n\xi) \, d\xi = \frac{8}{\pi} \frac{2n}{4n^2 - 1} t^2$$

from equation 5.17. Now we can evaluate the integral occurring in the first summation of equation 5.21:

$$\int_0^t \frac{8}{\pi} \frac{2n}{4n^2 - 1} \tau^2 e^{-4n^2(t-\tau)} \, d\tau = \frac{1}{2} \frac{-4n^2 t + 8n^4 t^2 + 1 - e^{-4n^2 t}}{n^5 \pi (4n^2 - 1)}.$$

Finally, compute

$$b_n = \frac{2}{\pi} \int_0^\pi f(\xi)\sin(n\xi) \, d\xi = \frac{2}{\pi} \int_{\pi/2}^\pi 25 \sin(n\xi) \, d\xi = \frac{50}{n\pi} (\cos(n\pi/2) - (-1)^n).$$

The solution is

$$u(x, t) = \sum_{n=1}^\infty \left(\frac{1}{2} \frac{-4n^2 t + 8n^4 t^2 + 1 - e^{-4n^2 t}}{n^5 \pi (4n^2 - 1)} \right) \sin(nx)$$

$$+ \sum_{n=1}^\infty \frac{50}{n\pi} (\cos(n\pi/2) - (-1)^n)\sin(nx)e^{-4n^2 t}.$$

The second summation in $u(x, t)$ is the solution of the initial-boundary value problem if the source term $t^2 \cos(x/2)$ is omitted. If we denote this solution as $u_h(x, t)$ (the subscript h is for the fact that the heat equation is now homogeneous), then

$$u_h(x, t) = \sum_{n=1}^\infty \frac{50}{n\pi} (\cos(n\pi/2) - (-1)^n)\sin(nx)e^{-4n^2 t}$$

and

$$u(x, t) = u_h(x, t) + \sum_{n=1}^{\infty} \left(\frac{1}{2} \frac{-4n^2t + 8n^4t^2 + 1 - e^{-4n^2t}}{n^5\pi(4n^2 - 1)} \right) \sin(nx).$$

This way of writing the solution clarifies which terms in the solution arise from the $t^2 \cos(x/2)$ term in the partial differential equation.

Figures 5.6(a) through (e) show graphs the initial temperature function and of $u(x, t)$ as a function of x for selected values of t. These graphs exhibit the evolution of a temperature distribution if we think of the initial-boundary value problem as modeling heat flow in a homogeneous bar, with a heat source defined by the nonhomogeneous term $t^2 \cos(x/2)$. Figures 5.7(a) through (e) compare $u(x, t)$ with $u_h(x, t)$ at selected times to display the influence of the source term on the temperature distribution. As the graphs suggest, the homogeneous temperature distribution $u_h(x, t)$ decreases rapidly toward zero as t increases, while $u(x, t)$, which includes the effects of the source term, does not (at least in the range of t shown in the graphs). ■

EXERCISE 221   In each of problems 1 through 5, solve the problem 5.13. In each case, graph the solution as a function of x for different values of t.

1. $F(x, t) = t$, $f(x) = x(L - x)$
2. $F(x, t) = x \sin(t)$, $f(x) = 1$
3. $F(x, t) = \cos(x)$, $f(x) = x^2(L - x)$
4. $F(x, t) = \begin{cases} K \text{ for } 0 \le x \le L/2 \\ 0 \text{ for } L/2 < x \le L, \end{cases}$ and $f(x) = \sin(\pi x/L)$
5. $F(x, t) = xt$, $f(x) = K$

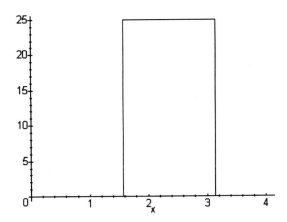

FIGURE 5.6(a)   Initial temperature distribution in Example 40.

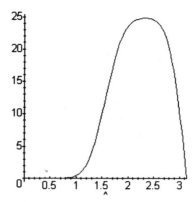

FIGURE 5.6(b)   Temperature at time $t = 0.009$.

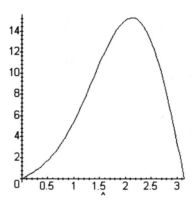

FIGURE 5.6(c) $t = 0.067$.

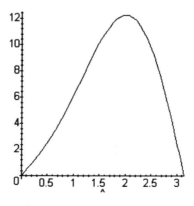

FIGURE 5.6(d) $t = 0.098$.

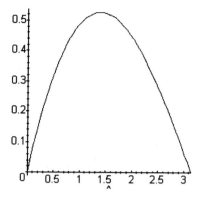

FIGURE 5.6(e) $t = 1.16$.

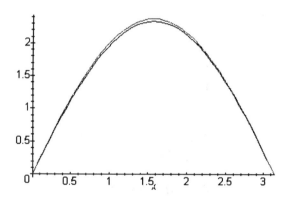

FIGURE 5.7(a)   Comparison of the temperature distributions in Example 40 with and without the heat source, at time $t = 0.48$.

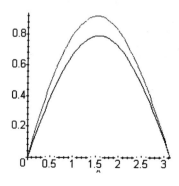

FIGURE 5.7(b)   Comparison at time $t = 0.75$.

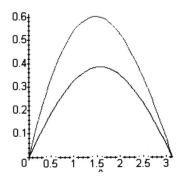

FIGURE 5.7(c) $t = 0.93$.

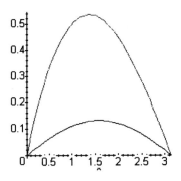

FIGURE 5.7(d) $t = 1.2$.

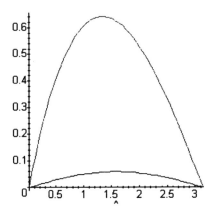

FIGURE 5.7(e) $t = 1.4$.

EXERCISE 222   Adapt the method of solution of the problem 5.13 to obtain a solution of the problem

$$u_t = ku_{xx} + F(x, t) \text{ for } 0 < x < L, \, t > 0$$

$$u_x(0, t) = u_x(L, t) = 0 \text{ for } t > 0$$

$$u(x, 0) = f(x) \text{ for } 0 < x < L.$$

Hint: Begin with

$$u(x, t) = \frac{1}{2} T_0(t) + \sum_{n=1}^{\infty} T_n(t)\cos(n\pi x/L),$$

where

$$T_n(t) = \frac{2}{L} \int_0^L u(\xi, t)\cos(n\pi\xi/L) \, d\xi.$$

Write

$$F(x, t) = \frac{1}{2} A_0(t) + \sum_{n=1}^{\infty} A_n(t)\cos(n\pi x/L)$$

with

$$A_n(t) = \frac{2}{L} \int_0^L F(\xi, t)\cos(n\pi\xi/L) \, d\xi.$$

Now follow the reasoning used in this section in solving problem 5.13.

EXERCISE 223   Solve the initial-boundary value problem of Exercise 222 for the case $F(x, t) = xt$, $f(x) = 1$.

EXERCISE 224   Solve

$$u_t = ku_{xx} \text{ for } 0 < x < L, \, t > 0$$

$$u(0, t) = \alpha(t), \, u_x(L, t) = \beta(t) \text{ for } t > 0$$

$$u(x, 0) = f(x) \text{ for } 0 < x < L.$$

Hint: Attempt a solution

$$u(x, t) = \sum_{n=1}^{\infty} T_n(t)\sin\left(\frac{(2n - 1)\pi x}{2L}\right)$$

(note the hint to Exercise 183). Let

$$\lambda_n = \left(\frac{(2n - 1)\pi}{2L}\right)^2.$$

Show that

$$T_n'(t) + k\lambda_n T_n(t) = b_n(t),$$

where

$$b_n(t) = \frac{2k}{L}\left[\sqrt{\lambda_n}\alpha(t) + (-1)^{n+1}\beta(t)\right].$$

Solve for $T_n(t)$, subject to $T_n(0) = 0$.

EXERCISE 225   Solve the problem of Exercise 224, with $\alpha(t) = 1$, $\beta(t) = t$.

EXERCISE 226   Solve

$$u_t = ku_{xx} + F(x, t) \text{ for } 0 < x < L, t > 0$$

$$u(x, 0) = 0 \text{ for } 0 < x < L$$

$$u_x(0, t) = \alpha(t), u_x(L, t) = 0 \text{ for } t > 0.$$

5.7   THE HEAT EQUATION IN SEVERAL SPACE VARIABLES

We will consider an initial-boundary value problem modeling temperature distribution in a thin flat homogeneous (constant density) rectangular plate having initial temperature $f(x, y)$ at (x, y), if the sides are maintained at temperature zero. This setting is modeled by the problem:

$$u_t = k(u_{xx} + u_{yy}) \text{ for } 0 < x < a, 0 < y < b, t > 0$$

$$u(x, 0, t) = u(x, b, t) = 0 \text{ for } 0 \leq x \leq a, t \geq 0$$

$$u(0, y, t) = u(a, y, t) = 0 \text{ for } 0 \leq y \leq b, t \geq 0$$

$$u(x, y, 0) = f(x, y) \text{ for } 0 \leq x \leq a, 0 \leq y \leq b.$$

Let $u(x, y, t) = X(x)Y(y)T(t)$ to attempt a separation of variables. The heat equation yields

$$XYT' = k(X''YT + XY''T)$$

or

$$\frac{T'}{kT} - \frac{Y''}{Y} = \frac{X''}{X}.$$

Both sides must equal the same constant, which we will call $-\lambda$. Then

$$\frac{X''}{X} = -\lambda \text{ and } \frac{T'}{kT} - \frac{Y''}{Y} = -\lambda.$$

Write the last equation as

$$\frac{T'}{kT} + \lambda = \frac{Y''}{Y}.$$

Both sides of this equation must also be constant (since t and y are independent). Call this constant $-\mu$. Then

$$\frac{T'}{kT} + \lambda = \frac{Y''}{Y} = -\mu.$$

We now have separated the variables, obtaining

$$X'' + \lambda X = 0, \ Y'' + \mu Y = 0 \text{ and } T' + (\lambda + \mu)kT = 0.$$

Now consider the boundary conditions. Since

$$u(0, y, t) = X(0)Y(y)T(t) = 0$$

we conclude that $X(0) = 0$. Using the other boundary conditions, we obtain

$$X(a) = Y(0) = Y(b) = 0.$$

The problems for X and Y are therefore

$$X'' + \lambda X = 0; \ X(0) = X(a) = 0$$

and

$$Y'' + \mu Y = 0; \ Y(0) = Y(b) = 0.$$

Except for notation, these problems for X and Y are the same and have been solved before. The eigenvalues and eigenfunctions are

$$\lambda_n = \frac{n^2\pi^2}{a^2}, \ X_n(x) = \sin\left(\frac{n\pi x}{a}\right)$$

for $n = 1, 2, \ldots$, and

$$\mu_m = \frac{m^2 \pi^2}{b^2}, \quad Y_m(y) = \sin\left(\frac{m\pi y}{b}\right)$$

for $m = 1, 2, \ldots$. These indices are independent—the choice of n does not influence the choice of m.

The equation for T is

$$T' + k\left(\frac{n^2\pi^2}{a^2} + \frac{m^2\pi^2}{b^2}\right) T = 0.$$

Solutions are constant multiples of

$$e^{-k\alpha_{nm}t}$$

in which

$$\alpha_{nm} = \frac{n^2\pi^2}{a^2} + \frac{m^2\pi^2}{b^2}$$

for $n, m = 1, 2, \ldots$.

We can now form functions

$$u_{nm}(x, y, t) = c_{nm} \sin\left(\frac{n\pi x}{a}\right) \sin\left(\frac{m\pi y}{b}\right) e^{-k\alpha_{nm}t}.$$

These functions satisfy the heat equation and the boundary conditions for n and m any positive integers, and for any numbers a_{nm}. In order to satisfy the initial condition, try a superposition

$$u(x, y, t) = \sum_{n=1}^{\infty} \sum_{m=1}^{\infty} c_{nm} \sin\left(\frac{n\pi x}{a}\right) \sin\left(\frac{m\pi y}{b}\right) e^{-k\alpha_{nm}t}.$$

The initial condition requires that

$$u(x, y, 0) = f(x, y) = \sum_{n=1}^{\infty} \sum_{m=1}^{\infty} c_{nm} \sin\left(\frac{n\pi x}{a}\right) \sin\left(\frac{m\pi y}{b}\right).$$

This is the double Fourier sine expansion of $f(x, y)$ on the rectangle $0 \le x \le a$, $0 \le y \le b$. In Section 4.12 we found the coefficients of such an expansion:

$$c_{nm} = \frac{4}{ab} \int_0^b \int_0^a f(\xi, \eta)\sin\left(\frac{n\pi\xi}{a}\right) \sin\left(\frac{m\pi\eta}{b}\right) d\xi \, d\eta$$

For $n = 1, 2, \ldots$ and $m = 1, 2, \ldots$.

For example, suppose

$$f(x, y) = x(a - x)y(b - y).$$

Then

$$c_{nm} = \frac{4}{ab} \int_0^b \int_0^a \xi(a - \xi)\eta(b - \eta)\sin\left(\frac{n\pi\xi}{a}\right) \sin\left(\frac{m\pi\eta}{b}\right) d\xi$$

$$= \frac{16a^2b^2}{n^3 m^3 \pi^6} [1 - (-1)^n][1 - (-1)^m].$$

The solution in this case is

$$u(x, y, t) = \frac{16a^2b^2}{\pi^6} \sum_{n=1}^{\infty} \sum_{m=1}^{\infty} \frac{1 - (-1)^n}{n^3} \frac{1 - (-1)^m}{m^3} \sin\left(\frac{n\pi x}{a}\right) \sin\left(\frac{m\pi y}{b}\right) e^{-k\alpha_{nm}t}.$$

EXERCISE 227 Solve

$$u_t = u_{xx} + u_{yy} \text{ for } 0 < x < \pi, 0 < y < \pi, t > 0$$

$$u(x, 0, t) = u(x, \pi, t) = 0 \text{ for } 0 \le x \le \pi, t \ge 0$$

$$u(0, y, t) = u(\pi, y, t) = 0 \text{ for } 0 \le y \le \pi, t \ge 0$$

$$u(x, y, 0) = \sin(x)y(y - \pi) \text{ for } 0 \le x \le \pi, 0 \le y \le \pi.$$

EXERCISE 228 Solve

$$u_t = u_{xx} + u_{yy} \text{ for } 0 < x < 1, 0 < y < 1, t > 0$$

$$u_x(0, y, t) = u_x(1, y, t) = 0 \text{ for } 0 \le y \le 1, t \ge 0$$

$$u_y(x, 0, t) = u_y(x, 1, t) = 0 \text{ for } 0 \le x \le 1, t \ge 0$$

$$u(x, y, 0) = K \text{ for } 0 \le x \le 1, 0 \le y \le 1.$$

EXERCISE 229 Solve

$$u_t = k(u_{xx} + u_{yy}) \text{ for } 0 < x < \pi, 0 < y < \pi, t > 0$$

$$u(x, 0, t) = u(x, \pi t) = 0 \text{ for } 0 \le x \le \pi, t \ge 0$$

$$u_x(0, y, t) = u_x(\pi, y, t) = 0 \text{ for } 0 \le y \le \pi, t \ge 0$$

$$u(x, y, 0) = x \cos(x/2)y(\pi - y) \text{ for } 0 \le x \le \pi, 0 \le y \le \pi.$$

EXERCISE 230   Let C be a simple closed curve in the x, y-plane, bounding a region D of the plane. Formulate and prove a weak maximum principle for the two-dimensional heat equation

$$u_t = k(u_{xx} + u_{yy})$$

on the three-dimensional region consisting of all points (x, y, t) with $0 \le t \le T$ and (x, y) in D.

EXERCISE 231   Let D be a region of the x, y-plane bounded by a simple closed curve C. Consider the problem

$$u_t = k(u_{xx} + u_{yy}) \text{ for } (x, y) \text{ in } D, t > 0.$$

$$u(x, y, 0) = \varphi(x, y) \text{ for } (x, y) \text{ in } D$$

$$u(x, y, t) = \psi(x, y, t) \text{ for } (x, y) \text{ on } C, t > 0.$$

Let φ and ψ be continuous. Assuming that this problem has a solution, prove that this solution is unique. *Hint:* Use the result of Exercise 230.

6

DIRICHLET AND NEUMANN PROBLEMS

6.1 THE SETTING OF THE PROBLEMS

Having considered the hyperbolic and parabolic equations, we will now take up the elliptic second order partial differential equation with constant coefficients. We will begin with some notation and terminology used to discuss this equation.

The *Laplacian* $\nabla^2 u$ of a function u of $(x_1, x_2, \ldots, x_n)$ is

$$\nabla^2 u = \sum_{j=1}^{n} \frac{\partial^2 u}{\partial x_j^2}.$$

In the case of two independent variables we often write x and y instead of x_1 and x_2, and in 3-space we usually write x, y, and z.

The partial differential equation

$$\nabla^2 u = 0$$

is *Laplace's equation*. A continuous function satisfying Laplace's equation, and having continuous first and second partial derivatives, is called a *harmonic function*. Usually we will discuss harmonic functions for the variables restricted to a specified set of values. We will say more about this shortly.

The Laplacian appears in the heat equation, which can be written

$$u_t = k\nabla^2 u.$$

The steady state case of the heat equation occurs when $u_t = 0$, and then the heat equation becomes Laplace's equation $\nabla^2 u = 0$.

We will need some notation and definitions to analyze solutions of Laplace's equation. Let R^n denote the usual n-dimensional space of ordered n-tuples $(x_1, \ldots, x_n)$, in which each x_j is a real number. R^2 is interpreted as the plane. We may also think of R^n as consisting of vectors with n components. We will denote vectors in bold type.

The distance between $\mathbf{x} = (x_1, \ldots, x_n)$ and $\mathbf{y} = (y_1, \ldots, y_n)$ in R^n is denoted $|\mathbf{x} - \mathbf{y}|$ and is defined by the usual metric

$$|\mathbf{x} - \mathbf{y}| = \sqrt{\sum_{j=1}^{n} (x_j - y_j)^2}.$$

This distance function is symmetric, which means that

$$|\mathbf{x} - \mathbf{y}| = |\mathbf{y} - \mathbf{x}|.$$

It also satisfies the *triangle inequality*:

$$|\mathbf{x} - \mathbf{y}| \leq |\mathbf{x} - \mathbf{z}| + |\mathbf{z} - \mathbf{y}|.$$

We will use the standard notation $\mathbf{x} \in A$ if $\mathbf{x}$ is an element of a set A.

Given a point $\mathbf{x}_0 \in R^n$ and a positive number r, the open ball of radius r about $\mathbf{x}_0$ consists of all points in R^n at distance less than r from $\mathbf{x}_0$. This set is denoted $B(\mathbf{x}_0, r)$. Thus $\mathbf{y} \in B(\mathbf{x}_0, r)$ if and only if $|\mathbf{y} - \mathbf{x}_0| < r$. In the plane an open ball consists of all points inside a circle, and in 3-space an open ball consists of all points inside a sphere.

A *neighborhood* of $\mathbf{x}_0$ is an open ball $B(\mathbf{x}_0, r)$ centered at $\mathbf{x}_0$.

If S is a set of points in R^n, then a point $\mathbf{x}$ in R^n is a *boundary point* of S if every neighborhood of $\mathbf{x}$ contains at least one point in S and one point not in S. We call $\mathbf{x}$ an *interior point* of S if there is some neighborhood of $\mathbf{x}$ which contains only points of S. A point cannot be both a boundary point and an interior point of S. An interior point of S necessarily belongs to S (because it is the center of some neighborhood containing only points of S), while a boundary point of S may or may not belong to S.

The student can check that every point belonging to S must be either a boundary point or an interior point. However, there may be boundary points of S that do not belong to S.

S is an *open set* if all of its points are interior points. A set is *closed* if it contains all of its boundary points. A set may be neither open nor closed. Indeed, if a set contains a boundary point then it cannot be open, but it need not be closed either, because there may be other boundary points the set does not contain.

Example 41   Let S consist of all points (x, y) on the line $x + y = 1$ in the plane. S is shown in Figure 6.1. In the plane, a neighborhood of a point consists of points within a circle about the point.

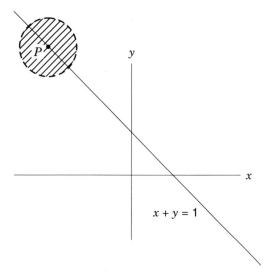

FIGURE 6.1   The set of points on the line $x + y = 1$ in R^2 has no interior point.

S has no interior points. Given any point P of S, no disk about this point can contain only points of the line, so P cannot be an interior point.

Every point of S is a boundary point. About any point of S, every circle contains this point (hence a point of S), as well as infinitely many points not in S (not on the line). Points of S are the only boundary points of S, hence S is closed.   ∎

Example 42   Let S consist of points (x, y) in the plane with $x \geq 0$ and $y > 0$. S is depicted in Figure 6.2, with part of the y-axis drawn solidly to indicate that points $(0, y)$ with $y > 0$ belong to S. No points on the x-axis are in S.

The points (x, y) with $x > 0$ and $y > 0$ are interior points of S. About any such point we can draw an open ball containing only points with positive coordinates, hence lying in S.

The boundary points of S are the points $(0, y)$ with $y \geq 0$ and points $(x, 0)$ with $x \geq 0$. The boundary points $(0, y)$ with $y > 0$ belong to S, while the boundary points $(x, 0)$ with $x \geq 0$ do not belong to S.

This set is not closed because it does not contain all of its boundary points.

This set is not open because it contains some of its boundary points.   ∎

Example 43   An open ball $B(\mathbf{x}_0, r)$ in R^n is an open set. This is most easily visualized in the plane, where an open ball consists of all points in some disk about a point, not counting the points on the boundary circle (Figure 6.3). About any point P in the ball we can construct a smaller disk containing only points of the original disk. This makes every point an interior point, hence the original ball is an open set.   ∎

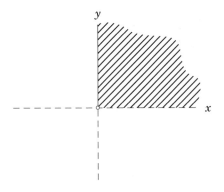

FIGURE 6.2   The right quarter plane $x \geq 0$, $y > 0$.

Given a set S of points in R^n, the *boundary* of S consists of all the boundary points of S, and is denoted ∂S.

Example 44   Let S consist of all points (x, y) in R^2 with $-1 < x < 1$. S may be visualized as the strip shown in Figure 6.4. S is open, every point being an interior point. The boundary points are points $(1, y)$ and points $(-1, y)$, lying on the vertical sides of the strip. Thus ∂S consists of all points (α, y) with $\alpha = \pm 1$.   ■

The boundary of a ball $B(\mathbf{x}_0, r)$ consists of all points $\mathbf{x}$ at distance r from $\mathbf{x}_0$:

$$|\mathbf{x} - \mathbf{x}_0| = r.$$

These boundary points of $B(\mathbf{x}_0, r)$ form the *n-sphere* $S(\mathbf{x}_0, r)$ of radius r about

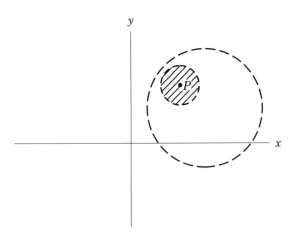

FIGURE 6.3   An open ball is an open set.

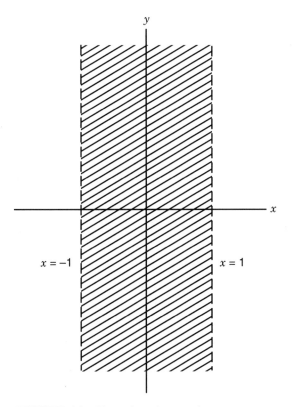

FIGURE 6.4 The strip $-1 < x < 1$, $-\infty < y < \infty$.

$\mathbf{x}_0$. In the plane this consists of all points on a circle about some point (Figure 6.5(a)); in 3-space, of all points on a sphere about a point (Figure 6.5(b)).

The *closure* of a set S consists of all points of S together with all the boundary points of S, and is denoted $\bar{S}$. If S is closed then S already contains all of its boundary points, so $S = \bar{S}$. Conversely, $\bar{S}$ is always a closed set (containing all of its boundary points). Hence $S = \bar{S}$ if and only if S is closed. The closure of the strip S in Example 44 is the set $\bar{S}$ consisting of all points (x, y) with $-1 \le x \le 1$. The closure of $B(\mathbf{x}_0, r)$ consists of all points in this ball, together with all points on the n-sphere $S(\mathbf{x}_0, r)$.

A set of S points in R^n is *connected* if between any two points $\mathbf{x}$ and $\mathbf{y}$ of S there is a polygonal path consisting of a finite number of straight line segments, lying entirely in S, and having $\mathbf{x}$ and $\mathbf{y}$ as end points. (We are using the word *connected* here where a topologist might use the term *polygonally connected*.) For example, the set of points in the plane consisting of all (x, y) with $x < 0$ and $y < 0$, together with all (x, y) with $y > 1 - x$, is not connected. This set is shown in Figure 6.6. It is impossible to connect a point lying above the line $x + y = 1$ with a point in the lower left quarter plane without going outside this set.

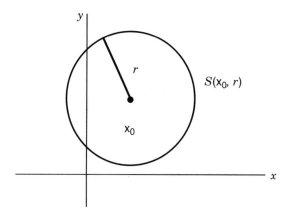

FIGURE 6.5(a)   Boundary of an open ball in the plane.

S is a *bounded set* if there is some number M such that $|\mathbf{x}| \leq M$ for all $\mathbf{x}$ in S. This means that a set is bounded if it can be fit entirely within some open ball about the origin. If a set is not bounded we say that it is *unbounded*.

Boundedness says nothing about whether the set is open, closed, not open, or not closed. Note also the difference between the concepts "boundary of a set" and "a set is bounded." A set may be unbounded (contain points arbitrarily far from the origin), and still have a nonempty boundary (consisting of boundary points). For example, the upper half plane in R^2 consists of all points

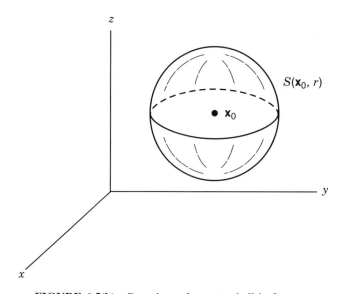

FIGURE 6.5(b)   Boundary of an open ball in 3-space.

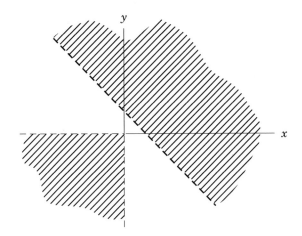

FIGURE 6.6 A disconnected set of points in the plane.

(x, y) with $y > 0$. This set is unbounded. Its boundary ∂S consists of all points $(x, 0)$ on the horizontal axis in the plane.

A closed, bounded subset of R^n is called a *compact set*. Compactness is characterized by the Heine-Borel property, which we will now discuss. Let K be a subset of R^n. A set $\mathfrak{O}$ of subsets of R^n is called an *open cover* of K if each set in $\mathfrak{O}$ is open and each point of K is in at least one of the open sets in $\mathfrak{O}$—that is,

$$K \subset \bigcup_{A \in \mathfrak{O}} A.$$

We say that the open cover $\mathfrak{O}$ is reducible to a finite subcover if there is a finite subset $\mathfrak{A} \subset \mathfrak{O}$ which also covers K—that is,

$$K \subset \bigcup_{A \in \mathfrak{A}} A.$$

This means that we can pick a finite number of the open sets of $\mathfrak{O}$ so that every point of K is in one of this finite subcollection of open sets.

The Heine-Borel Theorem states that a subset $K \subset R^n$ is compact if and only if every open cover is reducible to a finite subcover.

For example, the set ω of all nonnegative integers is not a compact subset of R^1. The open intervals $(j - 1/4, j + 1/4)$ for $j \in \omega$ cover ω, and so form an open cover. But each nonnegative integer is in exactly one set of the cover, so if we attempt to reduce the cover we must fail to include some elements of ω. This cover cannot be reduced to a finite subcover, and ω is not compact. Of course, ω is not bounded (although it is closed in the real line R^1).

In this treatment of partial differential equations, the Heine-Borel Theorem plays a role in proving Harnack's Second Theorem for harmonic functions

(Exercise 277, Section 6.10). This result can in turn be used in proving an existence theorem for solutions of the Dirichlet problem, which we will define shortly and which is the primary object of interest in this chapter.

An open, connected set of points in R^n is called a *domain*. For example, the right half plane in R^2, consisting of (x, y) with $x > 0$, is a domain. Any open ball is a domain.

Now return to Laplace's equation and harmonic functions. Given a domain Ω, a *Dirichlet problem* for Ω consists of finding a function that is harmonic in Ω and assumes given values on the boundary of Ω. That is, given Ω and a function f that is defined on $\partial\Omega$, we seek a function u such that

$$\nabla^2 u(\mathbf{x}) = 0 \text{ for } \mathbf{x} \in \Omega$$

and

$$u(\mathbf{x}) = f(\mathbf{x}) \text{ for } \mathbf{x} \in \partial\Omega.$$

In the plane the boundary of a domain will often be a piecewise smooth curve, having a continuous tangent except possibly at infinitely many points. In R^3 we often consider domains whose boundaries are piecewise smooth surfaces, composed of finitely many smooth pieces, each of which has a continuous normal vector. For example, a sphere in R^3 is smooth, while a cube is piecewise smooth, consisting of six smooth faces.

A *Neumann problem* for Ω consists of finding a function that is harmonic in Ω, and whose normal derivative assumes given values on the boundary. The normal derivative of a function at a boundary point is the directional derivative of the function in the direction of the outer normal to the boundary at that point. If $\mathbf{n}(\mathbf{x})$ is the unit outer, or exterior, normal to $\partial\Omega$ (Figure 6.7(a) for a domain in R^3, Figure 6.7(b) in R^2), then the normal derivative of $u(\mathbf{x})$ is denoted

$$\frac{\partial u}{\partial n}$$

and is the dot product of the gradient of u, with $\mathbf{n}$:

$$\frac{\partial u(\mathbf{x})}{\partial n} = \nabla u(\mathbf{x}) \cdot \mathbf{n}(\mathbf{x}).$$

A Neumann problem therefore consists of determining a function u satisfying:

$$\nabla^2 u = 0 \text{ in } \Omega$$

$$\frac{\partial u}{\partial n} = f \text{ on } \partial\Omega$$

in which f is a given function.

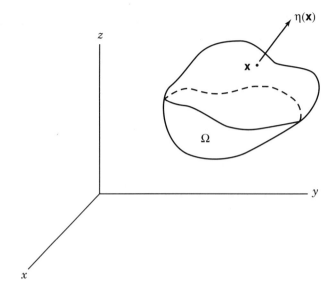

FIGURE 6.7(a)   Outer normal to a surface bounding a domain in R^3.

Notice that both the Dirichlet and Neumann problems are boundary value problems. Initial conditions are not relevant for Dirichlet and Neumann problems.

Sometimes the Dirichlet problem is called the *first boundary value problem* for a domain, and the Neumann problem, the *second boundary value problem*. There is also a *third*, or *mixed, boundary value problem*, in which we seek a function that is harmonic in a domain, and assumes values on the boundary

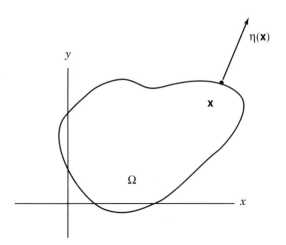

FIGURE 6.7(b)   Outer normal to a curve bounding a domain in R^2.

that are given as a combination of Dirichlet and Neumann type boundary data. This problem, also known as *Robin's problem*, takes the form of seeking u satisfying

$$\nabla^2 u = 0 \text{ in } \Omega$$

$$a(\mathbf{x})u(\mathbf{x}) + b(\mathbf{x})\frac{\partial u(\mathbf{x})}{\partial n} = f(\mathbf{x}) \text{ for } \mathbf{x} \in \partial\Omega$$

in which a, b, and f are given functions.

In this chapter we will concern ourselves primarily with Dirichlet and Neumann problems. However, to maintain consistency with the treatments of the wave and heat equations, we will begin in the next section with an example involving Cauchy data for the Dirichlet problem.

EXERCISE 232   Let S consist of all points (x, y) in R^2 with $0 \le x < 1$, $0 < y \le 1$. Determine all interior points of S, all boundary points, the closure of S, and whether S is open, closed, or neither open nor closed. Is S connected?

EXERCISE 233   Let S consist of all points (x, y) in R^2 with $x < 0$ and $1 \le y \le 4$. Determine all interior points of S, all boundary points, the closure of S, and whether S is open, closed, or neither open nor closed. Is S connected?

EXERCISE 234   Let S consist of all points (x, y) in R^2 with $1 \le x^2 + y^2 < 4$. Determine if S is a domain.

EXERCISE 235   Let S consist of all points (x, y) in R^2 with $-1 < x < 6$. Determine if S is a domain.

EXERCISE 236   Let S consist of all points in R^2 with rational coordinates. Determine all interior points of S, all boundary points, the closure of S, and whether S is open, closed, or neither open nor closed. Is S connected?

Consider the same questions for the set T consisting of all points in R^2 with irrational coordinates.

EXERCISE 237   Does the set of all points on a plane in R^3 constitute a domain?

6.2   THE CAUCHY PROBLEM FOR LAPLACE'S EQUATION IS ILL POSED

We have defined the Dirichlet problem as one of solving Laplace's equation consistent with given boundary information. We do not consider Cauchy data for Laplace's equation because this Cauchy problem is not well posed. The following example, due to the French mathematician Jacques Hadamard, demonstrates this. Consider the problem

$$\nabla^2 u(x, y) = 0 \text{ for all real } x \text{ and } y$$

$$u(x, 0) = 0 \text{ for all } x$$

$$u_y(x, 0) = \frac{1}{n} \sin(nx) \text{ for all } x$$

in which n can be any positive integer. This problem has a solution (in fact a unique one) given by

$$u(x, y) = \frac{1}{n^2} \sinh(ny)\sin(nx).$$

EXERCISE 238   Show that, as n is chosen larger, the data given for u_y along the horizontal axis tends to zero, while the solution can be made to assume arbitrarily large values for certain points (x, y). Explain why this means that the solution does not depend continuously on the boundary data, hence that the Cauchy problem is ill posed for Laplace's equation.

EXERCISE 239   Consider the Dirichlet problem

$$\nabla^2 u(x, y) = 0 \text{ for all } x \text{ and } y > 0$$

$$u(x, 0) = 0, \ u_y(x, 0) = ne^{-\sqrt{n}} \sin(nx)$$

for each positive integer n. Show that $u_n(x, y) = e^{-\sqrt{n}} \sin(nx)e^{ny}$ is a solution. Prove that, for any positive y_0,

$$u_n(x, y_0) \to \infty \text{ as } n \to \infty.$$

Hence prove again that the Cauchy problem is ill posed for Laplace's equation.

6.3 SOME HARMONIC FUNCTIONS

Before considering Dirichlet or Neumann problems, we will develop some properties and examples of harmonic functions.

It is obvious that constant multiples of harmonic functions are harmonic, and finite sums of harmonic functions are harmonic.

6.3.1 Harmonic Functions in the Plane

Rectangular Coordinates   In rectangular coordinates in R^2, one way of generating examples of harmonic functions is to recall that the real and imaginary parts of an analytic function of a complex variable are both harmonic functions. This means that, if we take any analytic complex function and separate the real and imaginary parts, we obtain a pair of harmonic functions.

To illustrate, consider $f(z) = z^2$, where $z = x + iy$ and x and y are real. Then

$$f(z) = (x + iy)^2 = x^2 - y^2 + 2ixy = u(x, y) + iv(x, y)$$

where

$$u(x, y) = x^2 - y^2 \text{ and } v(x, y) = 2xy.$$

It is routine to check that $\nabla^2 u = 0$ and $\nabla^2 v = 0$ for all (x, y). These functions are harmonic in the entire plane R^2.

As another example, consider $f(z) = \cos(z)$. Write

$$f(z) = \frac{1}{2}(e^{iz} + e^{-iz}) = \frac{1}{2}(e^{i(x+iy)} + e^{-i(x+iy)}) = \frac{1}{2}(e^{-y}e^{ix} + e^{y}e^{-ix})$$

$$= \frac{1}{2}[e^{-y}(\cos(x) + i\sin(x)) + e^{y}(\cos(x) - i\sin(x))]$$

$$= \cos(x)\frac{1}{2}(e^{y} + e^{-y}) - i\frac{1}{2}\sin(x)(e^{y} - e^{-y})$$

$$= \cos(x)\cosh(y) - i\sin(x)\sinh(y) = u(x, y) + iv(x, y).$$

This yields the harmonic functions $\cos(x)\cosh(y)$ and $\sin(x)\sinh(y)$.

Polar Coordinates   In rectangular coordinates in the plane, Laplace's equation is

$$u_{xx} + u_{yy} = 0.$$

It will be useful to have the polar coordinate form of this equation, which we will derive. Let

$$x = r\cos(\theta), \ y = r\sin(\theta)$$

and

$$U(r, \theta) = u(r\cos(\theta), r\sin(\theta)).$$

Calculate

$$U_r = u_x \cos(\theta) + u_y \sin(\theta),$$

$$U_{rr} = u_{xx} \cos^2(\theta) + 2u_{xy} \cos(\theta)\sin(\theta) + u_{yy} \sin^2(\theta),$$

$$U_\theta = -u_x r \sin(\theta) + u_y r \cos(\theta), \quad U_{\theta\theta} = -u_x r \cos(\theta) - u_y r \sin(\theta)$$

$$- r \sin(\theta)[u_{xx}(-r \sin(\theta)) + u_{xy} r \cos(\theta)]$$

$$+ r \cos(\theta)[u_{yx}(-r \sin(\theta)) + u_{yy}(r \cos(\theta))].$$

Then

$$U_{\theta\theta} = -rU_r + r^2[u_{xx} \sin^2(\theta) - 2u_{xy} \cos(\theta)\sin(\theta) + u_{yy} \sin^2(\theta)].$$

Now observe that

$$U_{rr} + \frac{1}{r^2} U_{\theta\theta} = u_{xx} + u_{yy} - \frac{1}{r} U_r.$$

Therefore $u_{xx} + u_{yy} = 0$ implies that

$$U_{rr} + \frac{1}{r} U_r + \frac{1}{r^2} U_{\theta\theta} = 0. \tag{6.1}$$

This is Laplace's equation in polar coordinates. Using equation 6.1 we can generate additional harmonic functions in the plane as follows.

To begin with a simple case, there are important harmonic functions which depend only on r, the distance from the origin. Such θ-independent functions can be found as follows. If $U_\theta = 0$, equation 6.1 becomes just $U_{rr} + (1/r)U_r = 0$, or $rU_{rr} + U_r = 0$. This can be written

$$(rU_r)_r = 0,$$

hence $rU_r = c$, a constant. But then $U_r = c/r$ with solutions

$$U(r) = c \ln(r) + k.$$

This function is harmonic for $r > 0$, and any choices of the constants c and k. In particular, 1 is harmonic in the entire plane, and $\ln(r)$ is harmonic in the plane with the origin removed.

There are of course harmonic functions in the plane that have a θ-dependence. We can find examples of such functions by separating variables in equation 6.1. Substitute $U(r, \theta) = R(r)\Theta(\theta)$ into Laplace's equation to obtain

$$R''\Theta + \frac{1}{r} R'\Theta + \frac{1}{r^2} R\Theta'' = 0$$

or

$$\frac{r^2 R'' + rR'}{R} = -\frac{\Theta''}{\Theta}.$$

Since the right side depends only on θ, and the left, only on r, and r and θ are independent, both sides must equal the same constant, which we will call λ. Then

$$r^2 R'' + rR' - \lambda R = 0 \text{ and } \Theta'' + \lambda\Theta = 0.$$

Now (r, π) and $(r, -\pi)$ are polar coordinates of the same point, so we will require that

$$\Theta(\pi) = \Theta(-\pi). \tag{6.2}$$

This is a *periodicity condition*.

It is routine to solve for Θ by considering possibilities for λ. If $\lambda = 0$ then $\Theta = a\theta + b$ and we must choose $a = 0$ to satisfy condition 6.2, obtaining $\Theta(\theta)$ = constant. If $\lambda = -k^2$ with $k \neq 0$, then

$$\Theta = ae^{k\theta} + be^{-k\theta}$$

and this function cannot satisfy condition 6.2 with nonzero real a or b for real θ. If $\lambda = k^2$ then

$$\Theta(\theta) = a \cos(k\theta) + b \sin(k\theta)$$

and this function will satisfy equation 6.2 if and only if k is chosen to be an integer.

Now consider the equation for R. This is an Euler differential equation and we attempt solutions $R = r^\alpha$. Substitute this into the differential equation for R, with $\lambda = k^2$, to obtain

$$\alpha(\alpha - 1) + \alpha - k^2 = 0.$$

Hence $\alpha = \pm k$ and R must have the general form $R(r) = cr^k + dr^{-k}$.

This means that $U(r, \theta)$ has the form

$$U(r, \theta) = R(r)\Theta(\theta) = (cr^k + dr^{-k})(a \cos(k\theta) + b \sin(k\theta))$$

in which k is any integer. This includes the case of the constant function obtained when $k = 0$.

In summary, we have obtained the harmonic functions

$$1, \ln(r)$$

which are independent of θ, with $\ln(r)$ defined for $r > 0$; as well as θ-dependent functions

$$r^k \cos(k\theta), \ r^k \sin(k\theta)$$

defined for $r \geq 0$ and all θ, and

$$r^{-k} \cos(k\theta), \ r^{-k} \sin(k\theta)$$

defined for $r > 0$ and all θ, with $k = 1, 2, \ldots$.

These harmonic functions will be of use later in solving Dirichlet and Neumann problems.

EXERCISE 240   Derive the elliptic coordinate form of Laplace's equation. Elliptic coordinates (λ, σ) are defined by

$$x = \cosh(\lambda)\cos(\sigma), \ y = \sinh(\lambda)\sin(\sigma).$$

Use the resulting equation to obtain harmonic functions of λ and σ.

6.3.2   Harmonic Functions in R^3

In R^3 we find that Laplace's equation in spherical coordinates is:

$$u_{rr} + \frac{2}{r} u_r + \frac{1}{r^2 \sin^2(\varphi)} u_{\theta\theta} + \frac{1}{r^2} u_{\varphi\varphi} + \frac{\cot(\varphi)}{r^2} u_\varphi = 0$$

in which θ is the polar angle and φ the azimuthal angle.

EXERCISE 241   Derive the spherical coordinate form of Laplace's equation.

The function

$$u(r) = \frac{1}{r} = \frac{1}{\sqrt{x^2 + y^2 + z^2}}$$

is harmonic in R^3 with the origin removed and is independent of θ and φ. It is possible to use separation of variables to find other harmonic functions in spherical coordinates, but we will not pursue this project.

EXERCISE 242   In 1 through 4, produce harmonic functions u and v by writing the given complex function in the form $f(z) = u(x, y) + iv(x, y)$.

1. $f(z) = z^3 - 2z$
2. $f(z) = \sin(z)$
   Hint: Use Euler's formula and the fact that

$$\sin(z) = \frac{1}{2i} (e^{iz} - e^{-iz}).$$

3. $f(z) = z \cos(z)$
4. $f(z) = \sin^2(z)$

EXERCISE 243   Let u be harmonic in a domain Ω in R^2. Let a and b be real numbers and let Ω^* consist of all translations $(x + a, y + b)$ of points (x, y) in Ω. Define

$$v(x, y) = u(x - a, y - b)$$

for (x, y) in Ω^*. Prove that v is harmonic in Ω^*. (That is, translations of harmonic functions are harmonic junctions.)

EXERCISE 244   Let θ be any real number and let

$$x^* = \cos(\theta)x + \sin(\theta)y$$

$$y^* = -\sin(\theta)x + \cos(\theta)y,$$

a rotation about the origin in the plane. Suppose u is harmonic in Ω, and let Ω^* be obtained by applying the rotation to all points of Ω. Let $w(x^*, y^*) = u(x, y)$. Show that w is harmonic in Ω^*. (In brief, rotations take harmonic functions to harmonic functions.)

EXERCISE 245   Let u be harmonic on the open ball $B(\mathbf{0}, a)$ of radius a about the origin in R^3. Define an inversion ι in this ball to be the mapping that sends a point $\mathbf{x} \neq \mathbf{0}$ in this ball to the point $\iota(\mathbf{x})$ exterior to the ball, with the property that $\iota(\mathbf{x})$ is on the line from the origin through $\mathbf{x}$, and the product of the distance from the origin to $\mathbf{x}$ and from the origin to $\iota(\mathbf{x})$ is a^2:

$$|\mathbf{x}||\iota(\mathbf{x})| = a^2.$$

(a) Show that

$$\iota(\mathbf{x}) = \frac{a^2}{|\mathbf{x}|^2} \, \mathbf{x}$$

for $\mathbf{x} \in B(\mathbf{0}, a)$ and $\mathbf{x} \neq \mathbf{0}$.

(b) Prove that every point outside the closure of $B(\mathbf{0}, a)$ is the inversion of exactly one point in $B(\mathbf{0}, a)$. Conversely, show that every point except the origin in $B(\mathbf{0}, a)$ is the inversion of exactly one point outside of the closure of this ball.

(c) Let $w(\mathbf{x}) = u(\iota(\mathbf{x}))$ for $\mathbf{x}$ outside $\overline{B(\mathbf{0}, a)}$. Prove that w is harmonic in $\overline{B(\mathbf{0}, a)}$ with the origin deleted, and that $w(\mathbf{x}) = u(\mathbf{x})$ for $\mathbf{x}$ in $S(\mathbf{0}, a)$.

6.4 REPRESENTATION THEOREMS

In this section we will develop integral representations of functions which we will use to derive properties of harmonic functions and to seek solutions of Dirichlet and Neumann problems. We will begin with a general representation theorem for (not necessarily harmonic) functions defined on a domain in R^3 and then discuss its analogue for functions defined on a domain in the plane.

If Ω is a bounded domain in R^n, let $C^2(\bar{\Omega})$ denote the set of functions which are continuous, with continuous first and second partial derivatives throughout $\bar{\Omega}$. The statement that u is in $C^2(\bar{\Omega})$, or $u \in C^2(\bar{\Omega})$, is just shorthand for making these assumptions about u and its partial derivatives.

A surface in R^3 is called a closed surface if it bounds a volume. For example, a sphere is a closed surface, as is a cube, while a hemisphere is not. A hemisphere capped by a disk (Figure 6.8) is a closed surface.

6.4.1 A Representation Theorem in R^3

We will use the following result of Green.

Lemma 5 (Green's Second Identity ($n = 3$)) Let Ω be a bounded domain in R^3 and assume that $\partial\Omega$ is a piecewise smooth closed surface. Let u, $v \in C^2(\bar{\Omega})$. Then

$$\iiint\limits_{\Omega} (u\nabla^2 v - v\nabla^2 u) \, dV = \iint\limits_{\partial\Omega} \left(u \frac{\partial v}{\partial n} - v \frac{\partial u}{\partial n} \right) d\sigma. \qquad \blacksquare$$

Notice that u and v are *not* assumed to be harmonic.

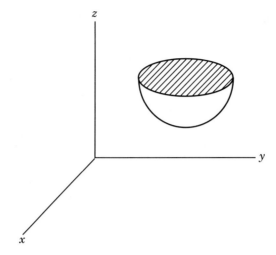

FIGURE 6.8   Hemisphere with cap—a closed surface in R^3.

Proof: Let **n** be the unit outer normal to $\partial\Omega$. Since this surface is piecewise smooth, **n** is piecewise continuous. Apply Gauss's divergence theorem to the following surface integrals:

$$\iint_{\partial\Omega} u \frac{\partial v}{\partial n}\, d\sigma \;-\; \iint_{\partial\Omega} v \frac{\partial u}{\partial n}\, d\sigma \;=\; \iint_{\partial\Omega} u\boldsymbol{\nabla} v\cdot\mathbf{n}\, d\sigma \;-\; \iint_{\partial\Omega} v\boldsymbol{\nabla} u\cdot\mathbf{n}\, d\sigma$$

$$=\; \iiint_{\Omega} div(u\boldsymbol{\nabla} v)\, dV \;-\; \iiint_{\Omega} div(v\boldsymbol{\nabla} u)\, dV.$$

Now it is a routine calculation to show that

$$div(u\boldsymbol{\nabla} v) \;-\; div(v\boldsymbol{\nabla} u) \;=\; u\nabla^2 v \;-\; v\nabla^2 u,$$

completing the proof.   ■

Using this lemma, we can derive an integral representation of $C^2(\bar{\Omega})$ functions, which we will use to write an integral representation for harmonic functions (equation 6.5 below).

Theorem 17 (Representation Theorem for n = 3)   Let Ω be a bounded domain in R^3 with piecewise smooth closed boundary surface $\partial\Omega$. Let $u \in C^2(\bar{\Omega})$. Then, at any **x** in Ω,

$$u(\mathbf{x}) = \frac{1}{4\pi} \iint\limits_{\partial\Omega} \left[\frac{1}{|\mathbf{y} - \mathbf{x}|} \frac{\partial u(\mathbf{y})}{\partial n} - u(\mathbf{y}) \frac{\partial}{\partial n} \frac{1}{|\mathbf{y} - \mathbf{x}|} \right] d\sigma_y$$

$$- \frac{1}{4\pi} \iiint\limits_{\Omega} \frac{\nabla^2 u(\mathbf{y})}{|\mathbf{y} - \mathbf{x}|} \, dV_y. \quad \blacksquare \quad (6.3)$$

The notations $d\sigma_y$ and dV_y are simply reminders that the variable of integration in these surface and volume integrals is $\mathbf{y}$, while $\mathbf{x}$ is any point of Ω. As with Green's second lemma, no assumption is made here that u is harmonic. The theorem gives the value of u at an arbitrary point of Ω in terms of a surface integral, which uses only information about u on the boundary of Ω, and an integral involving the Laplacian of u, over the closure of entire domain (domain together with its boundary surface). This suggests a connection with harmonic functions, which we will explore after proving the theorem.

Proof: Let $v(\mathbf{y}) = 1/|\mathbf{y} - \mathbf{x}|$ for $\mathbf{y}$ in Ω and $\mathbf{y} \neq \mathbf{x}$. We would like to apply Green's second identity to u and v. This cannot be done over Ω because v is not defined at $\mathbf{x}$. However, since Ω is open, $\mathbf{x}$ is an interior point and there is an open ball $B(\mathbf{x}, \epsilon)$ about $\mathbf{x}$ containing only points of Ω (Figure 6.9). Denote this ball as B. Let Ω_ϵ be the set formed by removing all points of $\bar{B}$ from Ω. Note that we remove not only points of B, but points on the sphere bounding

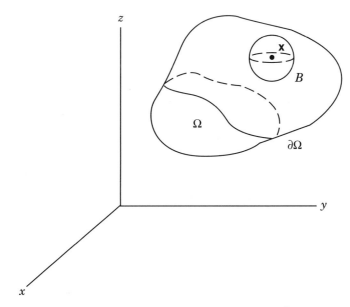

FIGURE 6.9 Open ball B about $\mathbf{x}$ in Ω, with $\bar{B} \subset \Omega$.

B. The student can check that Ω_ϵ is also a domain (open and connected) and that v is harmonic in Ω_ϵ.

Although u is not assumed to be harmonic, v is harmonic for $\mathbf{y} \neq \mathbf{x}$. By Green's second lemma, since $\nabla^2 v = 0$ in Ω_ϵ,

$$-\iiint_{\Omega_\epsilon} \frac{\nabla^2 u(\mathbf{y})}{|\mathbf{y} - \mathbf{x}|} \, dV_y = \iint_{\partial\Omega_\epsilon} \left(u(\mathbf{y}) \frac{\partial}{\partial n} \left(\frac{1}{|\mathbf{y} - \mathbf{x}|} \right) - \frac{1}{|\mathbf{y} - \mathbf{x}|} \frac{\partial u(\mathbf{y})}{\partial n} \right) d\sigma_y.$$

Now (Figure 6.9) $\partial\Omega_\epsilon$ consists of two disjoint pieces, namely $\partial\Omega$ and $S(\mathbf{x}, \epsilon)$, the latter being the sphere bounding B. Denote this sphere as S. The surface integral over $\partial\Omega_\epsilon$ is the sum of the surface integrals over $\partial\Omega$ and S and the last equation can be written

$$-\iiint_{\Omega_\epsilon} \frac{\nabla^2 u(\mathbf{y})}{|\mathbf{y} - \mathbf{x}|} \, dV_y = \iint_{\partial\Omega} \left(u(\mathbf{y}) \frac{\partial}{\partial n} \left(\frac{1}{|\mathbf{y} - \mathbf{x}|} \right) - \frac{1}{|\mathbf{y} - \mathbf{x}|} \frac{\partial u(\mathbf{y})}{\partial n} \right) d\sigma_y$$

$$+ \iint_{S} \left(u(\mathbf{y}) \frac{\partial}{\partial n} \left(\frac{1}{|\mathbf{y} - \mathbf{x}|} \right) - \frac{1}{|\mathbf{y} - \mathbf{x}|} \frac{\partial u(\mathbf{y})}{\partial n} \right) d\sigma_y. \quad (6.4)$$

We want to determine what happens to the terms in this equation as $\epsilon \to 0$.

First we will ask the student to check that the condition that $u \in C^2(\bar{\Omega})$ is sufficient for convergence of the improper integral

$$\iiint_{\Omega} \frac{\nabla^2 u(\mathbf{y})}{|\mathbf{y} - \mathbf{x}|} \, dV_y.$$

Therefore, as $\epsilon \to 0$,

$$\iiint_{\Omega_\epsilon} \frac{\nabla^2(\mathbf{y})}{|\mathbf{y} - \mathbf{x}|} \, dV_y \to \iiint_{\Omega} \frac{\nabla^2 u(\mathbf{y})}{|\mathbf{y} - \mathbf{x}|} \, dV_y.$$

Next, the surface integral over $\partial\Omega$ on the right side of equation 6.4 is independent of ϵ and so remains unchanged as $\epsilon \to 0$.

Finally we will show that the surface integral over S in equation 6.4 has limit $4\pi u(\mathbf{x})$ as $\epsilon \to 0$. For $\mathbf{y}$ on S,

$$\frac{1}{|\mathbf{y} - \mathbf{x}|} = \frac{1}{\epsilon}.$$

To compute the normal derivative of v on S, note that, for any $\mathbf{y}$ on S the vector $\mathbf{y} - \mathbf{x}$ is normal to S and points out of the ball B (Figure 6.10). We want the normal vector oriented away from Ω_ϵ, so we must choose a vector along $-(\mathbf{y} - \mathbf{x})$, toward $\mathbf{x}$ and into the ball (hence exterior to Ω_ϵ). To obtain a unit vector with this orientation, let

$$\mathbf{y} = y_1\mathbf{i} + y_2\mathbf{j} + y_3\mathbf{k} \text{ and } \mathbf{x} = x_1\mathbf{i} + x_2\mathbf{j} + x_3\mathbf{k}$$

and choose

$$\mathbf{n}(\mathbf{y}) = -\frac{\mathbf{y} - \mathbf{x}}{|\mathbf{y} - \mathbf{x}|} = -\frac{(y_1 - x_1)\mathbf{i} + (y_2 - x_2)\mathbf{j} + (y_3 - x_3)\mathbf{k}}{\sqrt{(y_1 - x_1)^2 + (y_2 - x_2)^2 + (y_3 - x_3)^2}}.$$

Now compute

$$\frac{\partial}{\partial n}\left(\frac{1}{|\mathbf{y} - \mathbf{x}|}\right) = \nabla\left(\frac{1}{|\mathbf{y} - \mathbf{x}|}\right) \cdot \mathbf{n}(\mathbf{y})$$

$$= -\frac{(y_1 - x_1)\mathbf{i} + (y_2 - x_2)\mathbf{j} + (y_3 - x_3)\mathbf{k}}{((y_1 - x_1)^2 + (y_2 - x_2)^2 + (y_3 - x_3)^2)^{3/2}} \cdot \mathbf{n}(\mathbf{y})$$

$$= \frac{(y_1 - x_1)^2 + (y_2 - x_2)^2 + (y_3 - x_3)^2}{((y_1 - x_1)^2 + (y_2 - x_2)^2 + (y_3 - x_3)^2)^2}$$

$$= \frac{1}{(y_1 - x_1)^2 + (y_2 - x_2)^2 + (y_3 - x_3)^2} = \frac{1}{\epsilon^2}.$$

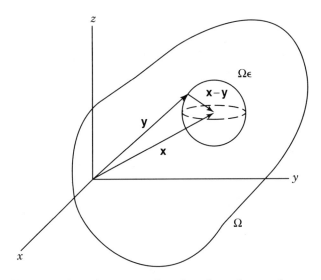

FIGURE 6.10 $\mathbf{x} - \mathbf{y}$ is an inner normal to the sphere about $\mathbf{x}$, for $\mathbf{y}$ on the sphere.

Then

$$\iint_S \left(u(\mathbf{y}) \frac{\partial}{\partial n} \left(\frac{1}{|\mathbf{y} - \mathbf{x}|} \right) - \frac{1}{|\mathbf{y} - \mathbf{x}|} \frac{\partial u(\mathbf{y})}{\partial n} \right) d\sigma_y$$

$$= \iint_S \left(\frac{1}{\epsilon^2} u(\mathbf{y}) - \frac{1}{\epsilon} \frac{\partial u(\mathbf{y})}{\partial n} \right) d\sigma_y = \frac{1}{\epsilon^2} \iint_S u(\mathbf{x}) \, d\sigma_y$$

$$+ \iint_S \left(\frac{1}{\epsilon^2} [u(\mathbf{y}) - u(\mathbf{x})] - \frac{1}{\epsilon} \frac{\partial u(\mathbf{y})}{\partial n} \right) d\sigma_y.$$

Because the integration is with respect to $\mathbf{y}$, the constant $u(\mathbf{x})$ factors through the integral and

$$\frac{1}{\epsilon^2} \iint_S u(\mathbf{x}) \, d\sigma_y = \frac{1}{\epsilon^2} u(\mathbf{x})(\text{area of } S) = 4\pi u(\mathbf{x}).$$

Finally,

$$\left| \iint_S \left(\frac{1}{\epsilon^2} [u(\mathbf{y}) - u(\mathbf{x})] - \frac{1}{\epsilon} \frac{\partial u(\mathbf{y})}{\partial n} \right) d\sigma_y \right| \leq \frac{1}{\epsilon^2} (4\pi\epsilon^2) \max_{\mathbf{y} \in S} |u(\mathbf{y}) - u(\mathbf{x})|$$

$$+ \frac{1}{\epsilon} (4\pi^2) \max_{\mathbf{y} \in S} \left| \frac{\partial u(\mathbf{y})}{\partial n} \right| \to 0$$

as $\epsilon \to 0$ because $u(\mathbf{y}) \to u(\mathbf{x})$ as $\mathbf{y} \to \mathbf{x}$ and, for the last term, because

$$\left| \frac{\partial u(\mathbf{y})}{\partial n} \right|$$

is bounded for $\mathbf{y}$ in S.

Therefore, in the limit as $\epsilon \to 0$, equation 6.4 becomes

$$-\iiint_\Omega \frac{\nabla^2 u(\mathbf{y})}{|\mathbf{y} - \mathbf{x}|} \, dV_y = \iint_{\partial\Omega} \left(u(\mathbf{y}) \frac{\partial}{\partial n} \left(\frac{1}{|\mathbf{y} - \mathbf{x}|} \right) - \frac{1}{|\mathbf{y} - \mathbf{x}|} \frac{\partial u(\mathbf{y})}{\partial n} \right) d\sigma_y + 4\pi u(\mathbf{x}).$$

This is equivalent to the conclusion of the theorem. ■

This representation theorem does not assume that u is harmonic. If, however, $\nabla^2 u = 0$ in Ω, then the triple integral in equation 6.3 vanishes and

$$u(\mathbf{x}) = \frac{1}{4\pi} \iint_{\partial\Omega} \left(\frac{1}{|\mathbf{y} - \mathbf{x}|} \frac{\partial u(\mathbf{y})}{\partial n} - u(\mathbf{y}) \frac{\partial}{\partial n} \frac{1}{|\mathbf{y} - \mathbf{x}|} \right) d\sigma_y \qquad (6.5)$$

for $\mathbf{x} \in \Omega$. This is the integral representation we said we would obtain for a $C^2(\bar{\Omega})$ function u that is harmonic in a bounded domain Ω in R^3.

The importance of this result for the Dirichlet problem is that it relates the value of a harmonic function u at an interior point $\mathbf{x}$ of Ω, to an integral over the boundary, and hence to a quantity that depends only on values of u on the boundary. This is exactly what we want to have, because in a Dirichlet problem we are given values the function is to assume on the boundary.

6.4.2 A Representation Theorem in R^2

The above discussion in R^3 can be adapted to derive an analogous result for functions defined on a domain in the plane. In this adaptation surface integrals are replaced by line integrals, and triple integrals by double integrals, and v is chosen as a logarithm function. We will outline the details, beginning with the R^2 version of Green's second identity.

Lemma 6 (Green's Second Identity ($n = 2$)) Let Ω be a bounded domain in R^2 and suppose $\partial\Omega$ is a piecewise smooth closed curve. Let $u, v \in C^2(\bar{\Omega})$. Then

$$\iint_{\Omega} (u\nabla^2 v - v\nabla^2 u)\, dA = \oint_{\partial\Omega} \left(u\frac{\partial v}{\partial n} - v\frac{\partial u}{\partial n} \right) ds. \qquad \blacksquare$$

The proof is similar to that for the R^3 version, with Green's theorem playing the role of Gauss's divergence theorem.

Now apply Lemma 6 with

$$v(\mathbf{y}) = \ln \left(\frac{1}{|\mathbf{y} - \mathbf{x}|} \right).$$

This function is harmonic in the domain formed by removing $\mathbf{x}$ from Ω. By an argument similar to that just done in R^3 we obtain the following.

Theorem 18 (Representation Theorem, n = 2)   Let Ω be a bounded domain in R^2 whose boundary $\partial\Omega$ is a piecewise smooth closed curve. Let $u \in C^2(\bar{\Omega})$. Then, for any $\mathbf{x}$ in Ω,

$$u(\mathbf{x}) = \frac{1}{2\pi} \oint_{\partial\Omega} \left(\ln\left(\frac{1}{|\mathbf{y} - \mathbf{x}|}\right) \frac{\partial u(\mathbf{y})}{\partial n} - u(\mathbf{y}) \frac{\partial}{\partial n} \ln\left(\frac{1}{|\mathbf{y} - \mathbf{x}|}\right) \right) ds$$

$$- \frac{1}{2\pi} \iint_{\Omega} \nabla^2 u(y) \ln\left(\frac{1}{|y - x|}\right) dA. \quad \blacksquare$$

In R^3 the integrals have a factor $1/4\pi$ which arises from the fact that the area of a sphere of radius ϵ is $4\pi\epsilon^2$. In R^2 we enclose $\mathbf{x}$ by a circle of radius ϵ and length $2\pi\epsilon$ and in computing the limit of the integrals obtain a factor $1/2\pi$.

As in the three-dimensional case, the representation theorem does not assume that u is harmonic. If, however, $\nabla^2 u = 0$ in Ω then the double integral term vanishes and the theorem gives the value of $u(\mathbf{x})$ at any interior point, in terms of information about $u(\mathbf{y})$ for $\mathbf{y}$ on $\partial\Omega$:

$$u(\mathbf{x}) = \frac{1}{2\pi} \oint_{\partial\Omega} \left(\ln\left(\frac{1}{|\mathbf{y} - \mathbf{x}|}\right) \frac{\partial u(\mathbf{y})}{\partial n} - u(\mathbf{y}) \frac{\partial}{\partial n} \ln\left(\frac{1}{|\mathbf{y} - \mathbf{x}|}\right) \right) ds \quad (6.6)$$

for $\mathbf{x} \in \Omega$. This is the two-dimensional analogue of the integral representation 6.5 in R^3.

These representation theorems have versions in R^n, using an n-dimensional analogue of Green's second identity. In this case the integrals have a factor of the reciprocal of the area of the unit sphere in R^n.

It is interesting that for $n \geq 3$ the integral representations all have the same general appearance, allowing for adjustment in the dimension of the integral. Only in the plane is there an intrinsic difference in the integrand itself, in the appearance of the logarithm term $\ln(1/|\mathbf{y} - \mathbf{x}|)$ replacing the simpler term $1/|\mathbf{y} - \mathbf{x}|$ that appears in higher dimensions. This is just one of many examples one can find in mathematics in which the behavior of some object of interest is significantly different in the plane than in higher dimensions.

EXERCISE 246   In the proof of the representation theorem for R^3 it is claimed that

$$\iiint_{\Omega} \frac{\nabla^2 u(\mathbf{y})}{|\mathbf{y} - \mathbf{x}|} dV_y$$

converges for u in $C^2(\bar{\Omega})$. Prove this assertion.

EXERCISE 247   Prove Green's second identity for a domain in the plane.

EXERCISE 248   Prove the representation theorem for R^2.

6.5   THE MEAN VALUE PROPERTY

We will now begin to reap some benefits of the considerable background we have developed. To begin, we will use representation theorems to show that, if u is harmonic in a bounded domain Ω, then at any $\mathbf{x}$ in Ω, $u(\mathbf{x})$ is the average of function values on any sphere centered at $\mathbf{x}$ and lying entirely in Ω. This is the *mean value property for harmonic functions* and it is valid in R^n. We will prove it for $n = 3$.

Theorem 19 (Mean Value Property, n = 3)   Let u be harmonic in a bounded domain Ω in R^3 and let $\mathbf{x} \in \Omega$. Then

$$u(\mathbf{x}) = \frac{1}{4\pi\epsilon^2} \iint\limits_{S(\mathbf{x},\epsilon)} u(\mathbf{y}) \, d\sigma_y,$$

provided that ϵ is sufficiently small that all points in $\overline{B(\mathbf{x}, \epsilon)}$ are in Ω. ∎

Proof: Since u is harmonic, the representation theorem (Theorem 17), applied to the ball of radius ϵ about $\mathbf{x}$, enables us to write

$$u(\mathbf{x}) = \frac{1}{4\pi} \iint\limits_{S} \left(\frac{1}{|\mathbf{y} - \mathbf{x}|} \frac{\partial u(\mathbf{y})}{\partial n} - u(\mathbf{y}) \frac{\partial}{\partial n} \frac{1}{|\mathbf{y} - \mathbf{x}|} \right) d\sigma_y$$

in which $S = S(\mathbf{x}, \epsilon)$. Repeating a calculation done in proving Theorem 17, we obtain

$$\frac{1}{|\mathbf{y} - \mathbf{x}|} = \frac{1}{\epsilon} \text{ and } \frac{\partial}{\partial n} \frac{1}{|\mathbf{y} - \mathbf{x}|} = -\frac{1}{\epsilon^2}$$

for $\mathbf{y}$ on S. This normal derivative is negative because the exterior normal points out of the ball away from $\mathbf{x}$, opposite the orientation in the representation theorem. Now

$$u(\mathbf{x}) = \frac{1}{4\pi} \iint\limits_{S} \frac{1}{\epsilon} \frac{\partial u(\mathbf{y})}{\partial n} \, d\sigma_y + \frac{1}{4\pi} \iint\limits_{S} \frac{1}{\epsilon^2} u(\mathbf{y}) \, d\sigma_y.$$

The first integral is zero by Gauss's divergence theorem:

$$\iint\limits_{S} \frac{\partial u(\mathbf{y})}{\partial n} \, d\sigma_y = \iint\limits_{S} \boldsymbol{\nabla} u(\mathbf{y}) \cdot \mathbf{n} \, d\sigma_y = \iiint\limits_{B} div(\boldsymbol{\nabla} u) \, dV_y = \iiint\limits_{B} \nabla^2 u = 0.$$

Therefore

$$u(\mathbf{x}) = \frac{1}{4\pi\epsilon^2} \iint\limits_S u(\mathbf{y}) \, d\sigma_y. \qquad \blacksquare$$

Since $4\pi\epsilon^2$ is the area of S, this is the average of u on S.

Here is the mean value property for dimension 2.

Theorem 20   Let u be harmonic in a bounded domain Ω in R^2 and let $\mathbf{x}$ be in Ω. Then

$$u(\mathbf{x}) = \frac{1}{2\pi\epsilon} \oint_C u(\mathbf{y}) \, ds_y$$

in which C is a circle of radius ϵ about $\mathbf{x}$ and ϵ is sufficiently small that this circle and all points interior to it are in Ω. $\blacksquare$

Note the appearance of $2\pi\epsilon$, which is the length of C. The theorem is proved by using the representation theorem in R^2 (Theorem 18).

EXERCISE 249   Write a complete proof of the mean value property for a function harmonic in a bounded domain in R^2.

EXERCISE 250   Let Ω be a domain in R^2 bounded by the piecewise smooth closed curve $\partial\Omega$. Let $u, v \in C^2(\bar{\Omega})$. Prove that

$$\oint_{\partial\Omega} u \frac{\partial v}{\partial n} \, ds = \iint\limits_{\Omega} (u\nabla^2 v + \nabla u \cdot \nabla v) \, dA.$$

This is Green's first identity (in the plane).

EXERCISE 251   Use Green's first identity to prove Green's second identity for a bounded domain in the plane.

EXERCISE 252   Let Ω be a bounded domain in R^2 having a piecewise smooth curve as boundary. Let $u \in C(\bar{\Omega})$ and suppose

$$\nabla^2 u = 0 \text{ in } \Omega$$

$$u = 0 \text{ on } \partial\Omega.$$

Use Green's first identity to show that $u \equiv 0$ on $\bar{\Omega}$. *Hint:* Choose $v = u$ in Green's identity.

EXERCISE 253   Fill in the details of this alternate proof of the mean value theorem for functions harmonic in a bounded domain in the plane. Let (x, y) be in Ω and consider an open ball B of radius ϵ about (x, y) and lying entirely in Ω. If (ξ, η) is in B we can write

$$\xi = x + r \cos(\theta), \ \eta = y + r \sin(\theta)$$

with $0 \leq r < \epsilon$ and $0 \leq \theta \leq 2\pi$. Apply Green's first identity (Exercise 250) to show that

$$\int_0^{2\pi} u_r(x + r \cos(\theta), y + r \sin(\theta))r \ d\theta = \iint_B r\nabla^2 u \ dr \ d\theta$$

and conclude that the integral on the left is zero. Therefore

$$\int_0^{2\pi} \frac{\partial}{\partial r} u(x + r \cos(\theta), y + r \sin(\theta)) \ d\theta = 0.$$

Conclude that

$$\int_0^{2\pi} u(x + r \cos(\theta), y + r \sin(\theta)) \ d\theta$$

is independent of r, hence is equal to the particular value of this integral at $r = 0$. Hence conclude that

$$\int_0^{2\pi} u(x, y) \ d\theta = \int_0^{2\pi} u(x + r \cos(\theta), y + r \sin(\theta)) \ d\theta.$$

From this, show that

$$u(x, y) = \frac{1}{2\pi} \int_0^{2\pi} u(x + r \cos(\theta), y + r \sin(\theta)) \ d\theta.$$

Finally, explain how this shows that $u(x, y)$ is the average of function values on a circle of radius ϵ about (x, y).

6.6   THE MAXIMUM PRINCIPLE

Another important consequence of the representation theorems is known as the maximum principle.

Suppose u is harmonic in a bounded domain Ω and continuous on $\bar{\Omega}$. Since $\bar{\Omega}$ is a compact set, $u(\mathbf{x})$ must achieve a maximum and a minimum at points of $\bar{\Omega}$. We will prove that these extreme values must occur on the boundary of Ω, and cannot occur at an interior point. This is the *maximum principle for harmonic functions*.

The proof makes use of the following fact from topology. Suppose Γ is a polygonal path in a domain Ω, consisting of a finite number of line segments of finite length. Then there is a number ρ such that every point P of Γ is at distance $\geq \rho$ from each point of the boundary Ω (Figure 6.11). That is, Γ cannot come arbitrarily close to the boundary of Ω, but instead must have a positive minimum distance from $\partial\Omega$.

Theorem 21 (Maximum Principle)   Let u be harmonic and nonconstant in a bounded domain Ω in R^n and continuous on $\bar{\Omega}$. Then u achieves its maximum and minimum values on $\bar{\Omega}$ only at points of $\partial\Omega$.   ∎

Proof: Let $M = \max_{\mathbf{x} \in \bar{\Omega}} u(\mathbf{x})$. We want to show that this maximum can be achieved only at points on the boundary of Ω.

Suppose instead that $\mathbf{x}_0$ is in Ω and $u(\mathbf{x}_0) = M$. Let $B = B(\mathbf{x}_0, \delta)$, with δ chosen so that $\bar{B}$ is entirely within Ω. This can be done because $\mathbf{x}_0$ is an interior point of Ω. We will prove that $u(\mathbf{x}) = M$ for all $\mathbf{x}$ in $\bar{B}$. To prove this, choose any ϵ with $0 < \epsilon < \delta$ and consider the open ball B_ϵ of radius ϵ about $\mathbf{x}_0$ (see Figure 6.12—diagrams are made in the plane for this proof for ease in visu-

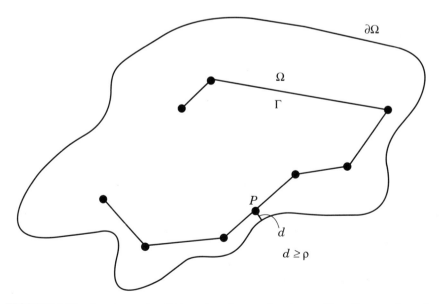

FIGURE 6.11   A polygonal path in a domain remains at a positive distance from the boundary of the domain.

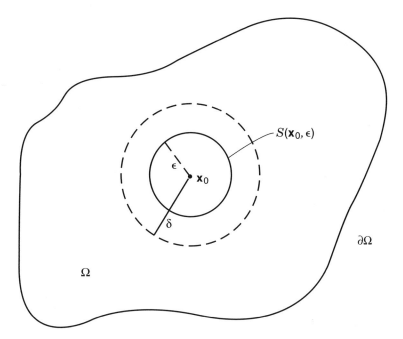

FIGURE 6.12   Sphere of radius ϵ within the open ball of radius δ about $\mathbf{x}_0$.

alization). If $u(\mathbf{x}) < M$ for any $\mathbf{x}$ on $S(\mathbf{x}_0, \epsilon)$, then by continuity of u we would have $u(\mathbf{x}) < M$ at all points on some set of positive area about $\mathbf{x}$ on $S(\mathbf{x}_0, \epsilon)$. Then the average of u over $S(\mathbf{x}_0, \epsilon)$ would be less than M. But by the mean value property, $u(\mathbf{x}_0) = M$ is the average of u over $S(\mathbf{x}_0, \epsilon)$. This is a contradiction. We conclude that $u(\mathbf{x}) = M$ for every point on $S(\mathbf{x}_0, \epsilon)$. Since ϵ is any number with $0 < \epsilon < \delta$, we must have $u(\mathbf{x}) = M$ at every point of B. By continuity, $u(\mathbf{x}) = M$ at every point of $\bar{B}$.

We now claim that $u(\mathbf{x}) = M$ at every point of Ω. For, suppose $\mathbf{y}$ is any point of Ω with $\mathbf{y} \neq \mathbf{x}_0$. We will show that $u(\mathbf{y}) = M$. Because Ω is connected, there is a polygonal path Γ from $\mathbf{x}_0$ to $\mathbf{y}$ in Ω and consisting of a finite number of straight line segments. Suppose every point of Γ has distance at least ρ from any point of $\partial\Omega$. Since ρ is a positive number and Γ has finite length, we can move along Γ from $\mathbf{x}_0$ to $\mathbf{y}$, choosing a finite number of intermediary points $\mathbf{x}_1, \ldots, \mathbf{x}_n$ along Γ, and positive numbers $\epsilon_0, \epsilon_1, \ldots, \epsilon_n$ such that each $\bar{B}(\mathbf{x}_j, \epsilon_j)$ contains only points of Ω, and

$$\mathbf{x}_1 \text{ is in } B(\mathbf{x}_0, \epsilon_0), \quad \mathbf{x}_j \text{ is in } B(\mathbf{x}_{j-1}, \epsilon_{j-1}) \text{ for } j = 2, \ldots, n$$

and

$$\mathbf{y} \text{ is in } B(\mathbf{x}_n, \epsilon_n).$$

This idea is illustrated in Figure 6.13.

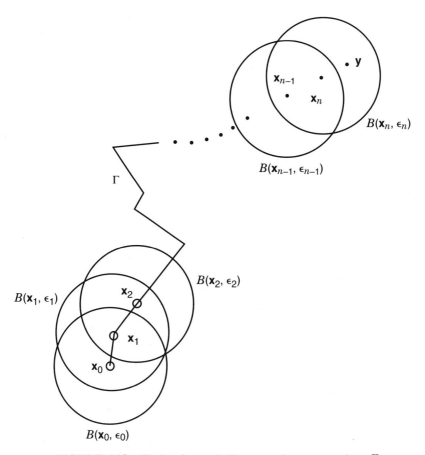

FIGURE 6.13   Chain of open balls connecting $\mathbf{y}$ to $\mathbf{x}_0$ along Γ.

Now u has the constant value M in $B(\mathbf{x}_0, \epsilon_0)$, and $\mathbf{x}_1$ is in this ball, so $u(\mathbf{x}_1)$ $= M$. But then u has the constant value M on $B(\mathbf{x}_1, \epsilon_1)$. Since $\mathbf{x}_2$ is in this ball, $u(\mathbf{x}_2) = M$. Proceeding along Γ, we obtain

$$u(\mathbf{x}_2) = \cdots = u(\mathbf{x}_n) = M$$

and, at the last step, $u(\mathbf{y}) = M$. Therefore u is constant on Ω, hence by continuity, on $\bar{\Omega}$. This contradiction proves that u cannot assume a maximum value at a point of Ω, and must therefore assume it at a point on $\partial\Omega$.

By applying this argument to $-u$ we find that u can achieve its minimum only at a boundary point.   ■

EXERCISE 254   We will outline another proof of the maximum principle for functions in the plane. Fill in the details.

Let u be harmonic in a bounded domain in R^2 having piecewise continuous closed boundary $\partial\Omega$. We want to show that the maximum of $u(x, y)$ on $\bar{\Omega}$ is achieved only on $\partial\Omega$.

First prove that, if v is continuous on $\bar{\Omega}$ and $v_{xx} + v_{yy} > 0$ on Ω, then $v(x, y)$ achieves its maximum only on $\partial\Omega$. (Suppose v achieved its maximum at an interior point (x_0, y_0) of Ω. Show that $v_{xx}(x_0, y_0) \leq 0$ and $v_{yy}(x_0, y_0) \leq 0$, obtaining a contradiction.)

Now let $M_{\partial\Omega} = \max_{(x,y)\in\partial\Omega} u(x, y)$. Let

$$v(x, y) = u(x, y) + \epsilon(x^2 + y^2)$$

for any positive ϵ. Show that v must achieve its maximum M_v on $\bar{\Omega}$ at a point of $\partial\Omega$. Then

$$v(x, y) \leq M_v \leq M_{\partial\Omega} + \epsilon D,$$

where D is the largest value of $x^2 + y^2$ for (x, y) in $\bar{\Omega}$. Thus show that

$$u(x, y) \leq M_{\partial\Omega} + \epsilon D.$$

Finally, use the fact that ϵ can be as close to zero as we like.

EXERCISE 255 Let Ω be a bounded domain in R^n. Let $\{u_n\}$ be a sequence of functions continuous on $\bar{\Omega}$ and harmonic in Ω. Suppose $\{u_n\}$ converges uniformly on $\partial\Omega$. Prove that $\{u_n\}$ converges uniformly on $\bar{\Omega}$.

EXERCISE 256 Prove the Normal Derivative Lemma. Let Ω be a domain in R^2 and let u be continuous on $\bar{\Omega}$ and harmonic in Ω. Let the maximum of u on $\bar{\Omega}$ be achieved at a boundary point $\mathbf{x}_0$, and suppose $\partial\Omega$ has a tangent at $\mathbf{x}_0$. Suppose also that the normal derivative

$$\left.\frac{\partial u}{\partial n}\right]_{\mathbf{x}=\mathbf{x}_0}$$

exists and that u is not the constant function. Then

$$\left.\frac{\partial u}{\partial n}\right]_{\mathbf{x}=\mathbf{x}_0} > 0.$$

Hint: Recall that

$$\left.\frac{\partial u}{\partial n}\right]_{\mathbf{x}=\mathbf{x}_0} = \lim_{h\to 0+} \frac{1}{h}\left(u(\mathbf{x}_0) - u(\mathbf{x}_0 - h\mathbf{n})\right),$$

where **n** is the unit outer normal to $\partial\Omega$ at $\mathbf{x}_0$. It is enough to prove the lemma when $\Omega = B(\mathbf{x}_0, r)$. Define

$$v(\mathbf{x}) = \ln(r/|\mathbf{x}|)$$

for $\mathbf{x} \neq \mathbf{0}$. Then $w(\mathbf{x}) = u(\mathbf{x}) + hv(\mathbf{x})$ is harmonic for $r/2 < |\mathbf{x}| < r$ and continuous on the closure of this domain. Show that $\max_{|\mathbf{x}|=r/2} u(\mathbf{x}) < u(\mathbf{x}_0)$. Hence show that $w(\mathbf{x}) < u(\mathbf{x}_0)$ for h sufficiently small. Conclude that w achieves a maximum at $\mathbf{x} = \mathbf{x}_0$; hence conclude that

$$\left.\frac{\partial w}{\partial n}\right]_{\mathbf{x}=\mathbf{x}_0} \geq 0.$$

From this complete the proof.

The lemma is valid for domains in R^n, with the concept of tangent line at $\mathbf{x}_0$ appropriately generalized (tangent plane in R^3, and so on). For $n > 2$ adapt the above proof by letting

$$v(\mathbf{x}) = |\mathbf{x}|^{2-n} - r^{2-n}.$$

6.7   UNIQUENESS AND CONTINUOUS DEPENDENCE OF SOLUTIONS ON BOUNDARY DATA

We will show that the Dirichlet problem for a bounded domain can have at most one continuous solution, and that this solution depends continuously on the boundary data. First, here is the uniqueness result.

Theorem 22 (Uniqueness)   Let Ω be a bounded domain in R^n and let f be continuous on $\partial\Omega$. Then the Dirichlet problem

$$\nabla^2 u = 0 \text{ in } \Omega$$

$$u = f \text{ on } \partial\Omega$$

has at most one solution which is continuous on $\bar{\Omega}$.   ■

Proof: Suppose w and v are solutions which are continuous on $\bar{\Omega}$. Let $h = w - v$. Then h is harmonic in Ω and $h = 0$ on $\partial\Omega$. The maximum and minimum values of h on $\bar{\Omega}$ are achieved on $\partial\Omega$ and are therefore zero. But then

$$|h(\mathbf{x})| = |w(\mathbf{x}) - v(\mathbf{x})| = 0$$

for all $\mathbf{x}$ in Ω, hence $u = v$.   ■

This theorem does not assert the existence of a solution, only that there can be at most one solution which is continuous on the closure of the domain.

Existence of a solution of a Dirichlet problem is a subtle issue which we will not fully address. At the end of this chapter we will discuss examples showing how a Dirichlet problem can fail to have a solution, and then state an existence theorem. A proof based on subharmonic functions can be found in John [9], and a proof using superharmonic functions is developed in Petrovsky [12]. There are also proofs based on a maximum principle (see, for example, Rauch [13]).

We will next show that the solution of a Dirichlet problem is perturbed by less than ϵ if the boundary data is perturbed by less than ϵ. This implies continuous dependence of the solution on the boundary data.

Theorem 23 (Continuous Dependence on Data)   Let Ω be a bounded domain in R^n. Let f and g be continuous on $\partial\Omega$. Let w be the continuous solution of

$$\nabla^2 u = 0 \text{ in } \Omega; \ u = f \text{ on } \partial\Omega.$$

Let v be the continuous solution of

$$\nabla^2 u = 0 \text{ in } \Omega; \ u = g \text{ on } \partial\Omega.$$

Suppose ϵ is a positive number and

$$|f(\mathbf{x}) - g(\mathbf{x})| < \epsilon \text{ for } \mathbf{x} \in \partial\Omega.$$

Then

$$|w(\mathbf{x}) - v(\mathbf{x})| < \epsilon \text{ for } \mathbf{x} \in \bar{\Omega}. \qquad \blacksquare$$

EXERCISE 257   Prove Theorem 23. *Hint:* Let $h = w - v$ and apply the Maximum Principle to h.

We now have some information about solutions of Dirichlet problems. Next we will turn to techniques for writing solutions of specific Dirichlet problems. As we might expect, the feasibility of writing an explicit solution is very much dependent on how "complicated" the domain is. We will write solutions for some regular domains in the plane, such as rectangles and disks. This is not as restrictive as it might at first appear, however, since, for problems in the plane, if there is a conformal mapping between one domain and another, a solution for one domain can be mapped into a solution for the other. This suggests some lines of investigation we will pursue in Section 6.15. For now, we will solve our first Dirichlet problem, for the case that Ω is bounded by a rectangle in the plane.

6.8   DIRICHLET PROBLEM FOR A RECTANGLE

Consider the Dirichlet problem

$$\nabla^2 u(x, y) = 0 \text{ for } 0 < x < a, \, 0 < y < b$$

$$u(x, 0) = 0 \text{ for } 0 \le x \le a$$

$$u(0, y) = u(a, y) = 0 \text{ for } 0 \le y \le b$$

$$u(x, b) = f(x) \text{ for } 0 \le x \le a.$$

this problem models the steady state temperature distribution in a thin flat plate, with temperatures on the lower and vertical sides kept at zero, and temperature $f(x)$ along the top side.

We can solve this problem by separation of variables. Let $u(x, y) = X(x)Y(y)$ and substitute into Laplace's equation to obtain

$$X''Y + XY'' = 0$$

or

$$\frac{X''}{X} = -\frac{Y''}{Y} = -\lambda,$$

in which both X''/X and Y''/Y must be constant because x and y are independent. Then

$$X'' + \lambda X = 0, \, Y'' - \lambda Y = 0.$$

From the boundary conditions,

$$u(x, 0) = X(x)Y(0) = 0,$$

hence $Y(0) = 0$. Similarly, $X(0) = X(a) = 0$.

The problem for X is the familiar boundary value problem

$$X'' + \lambda X = 0; \, X(0) = X(a) = 0$$

with eigenvalues

$$\lambda = \frac{n^2 \pi^2}{a^2} \text{ for } n = 1, 2, \ldots .$$

and eigenfunctions

$$\sin\left(\frac{n\pi x}{a}\right).$$

For any eigenvalue, the corresponding equation for y is

$$Y'' - \frac{n^2\pi^2}{a^2} Y = 0$$

with general solution

$$Y = ce^{n\pi y/a} + de^{-n\pi y/a}.$$

But $Y(0) = 0$ so $d = -c$ and Y is of the form

$$Y = 2c \sinh \left(\frac{n\pi y}{a} \right).$$

For $n = 1, 2, \ldots$, we now have functions

$$u_n(x, y) = b_n \sin \left(\frac{n\pi x}{a} \right) \sinh \left(\frac{n\pi y}{a} \right)$$

which are harmonic in the rectangle and satisfy the homogeneous boundary conditions on the lower and vertical sides. To find a solution satisfying the condition on the side $y = b$, try a superposition

$$u(x, y) = \sum_{n=1}^{\infty} b_n \sin \left(\frac{n\pi x}{a} \right) \sinh \left(\frac{n\pi y}{a} \right). \tag{6.7}$$

We must choose the b_n's so that

$$u(x, b) = \sum_{n=1}^{\infty} b_n \sin \left(\frac{n\pi x}{a} \right) \sinh \left(\frac{n\pi b}{a} \right) = f(x).$$

This is a Fourier sine expansion of f on $[0, a]$ and we choose the (entire) coefficient of $\sin(n\pi x/a)$ to be the Fourier sine coefficient:

$$b_n \sinh \left(\frac{n\pi b}{a} \right) = \frac{2}{a} \int_0^a f(\xi)\sin \left(\frac{n\pi\xi}{a} \right) d\xi.$$

Therefore

$$b_n = \frac{2}{a \sinh \left(\dfrac{n\pi b}{a} \right)} \int_0^a f(\xi)\sin \left(\frac{n\pi\xi}{a} \right) d\xi.$$

With this choice of the b_n's, equation 6.7 gives the solution of this Dirichlet problem for a rectangle (with nonzero data prescribed on the top side only).

For example, suppose $f(x) = x^2(a - x)$. Compute

$$\int_0^a \xi^2(a - \xi)\sin\left(\frac{n\pi\xi}{a}\right) d\xi = -\frac{a^4}{n^3\pi^3}[1 + 2(-1)^n].$$

In this case the solution is

$$u(x, y) = -2\frac{a^3}{\pi^3}\sum_{n=1}^{\infty}\frac{1 + 2(-1)^n}{n^3\sinh\left(\frac{n\pi b}{a}\right)}\sin\left(\frac{n\pi x}{a}\right)\sinh\left(\frac{n\pi y}{a}\right).$$

In this example the boundary conditions were homogeneous (identically zero) on the vertical sides and bottom side of the rectangle, and a given function on the top side. This problem was solved in a straightforward fashion by separation of variables. We may have nonzero boundary conditions specified on each side of the rectangle. In this event, write four Dirichlet problems, in each of which there is nonzero data on only one side. The solution of the original problem is the sum of the solutions of these four problems.

EXERCISE 258   In each of problems 1 through 5, solve the Dirichlet problem for the rectangle, with the given boundary conditions.

1. $u(0, y) = u(1, y) = 0$ for $0 \leq y \leq \pi$; $u(x, 0) = \sin(\pi x)$, $u(x, \pi) = 0$ for $0 \leq x \leq 1$

2. $u(0, y) = y(2 - y)$, $u(3, y) = 0$ for $0 \leq y \leq 2$; $u(x, 0) = u(x, 2) = 0$ for $0 \leq x \leq 3$

3. $u(0, y) = u(1, y) = 0$ for $0 \leq y \leq 4$; $u(x, 0) = 0$, $u(x, 4) = x\cos(\pi x/2)$ for $0 \leq x \leq 1$

4. $u(0, y) = \sin(y)$, $u(\pi, y) = 0$ for $0 \leq y \leq \pi$; $u(x, 0) = x(\pi - x)$, $u(x, \pi) = 0$ for $0 \leq x \leq \pi$

5. $u(0, y) = 0$, $u(2, y) = \sin(y)$ for $0 \leq y \leq \pi$; $u(x, 0) = 0$, $u(x, \pi) = x\sin(\pi x)$ for $0 \leq x \leq 2$

EXERCISE 259   Apply separation of variables to solve the following mixed boundary value problems

1. $\nabla^2 u = 0$ for $0 < x < a$, $0 < y < b$
    $u(x, 0) = u_y(x, b) = 0$ for $0 \leq x \leq a$
    $u(0, y) = 0$, $u(a, y) = g(y)$ for $0 \leq y \leq b$.

2. $\nabla^2 u = 0$ for $0 < x < a$, $0 < y < b$
    $u(x, 0) = 0$, $u(x, b) = f(x)$ for $0 \leq x \leq a$
    $u(0, y) = u_x(a, y) = 0$ for $0 \leq y \leq b$.

6.9 DIRICHLET PROBLEM FOR A DISK

We will solve the Dirichlet problem for a disk of radius ρ about the origin in the plane. In polar coordinates, the problem is:

$$\nabla^2 u(r, \theta) = 0 \text{ for } 0 \le r < \rho, \ -\pi \le \theta \le \pi$$

$$u(\rho, \theta) = f(\theta) \text{ for } -\pi \le \theta \le \pi.$$

We have derived the polar coordinate form of Laplace's equation:

$$\nabla^2 u(r, \theta) = u_{rr} + \frac{1}{r} u_r + \frac{1}{r^2} u_{\theta\theta} = 0.$$

In Section 6.3 we applied separation of variables to this problem and found functions

$$1, \ r^n \cos(n\theta), \ r^n \sin(n\theta)$$

which are harmonic on the disk, for $n = 1, 2, \ldots$. These are the eigenfunctions for this problem, just as $\sin(n\pi x/L)$ and/or $\cos(n\pi x/L)$ are the eigenfunctions for various problems involving wave motion and heat conduction. This suggests that we attempt a solution which is a superposition of these functions:

$$u(r, \theta) = \frac{1}{2} a_0 + \sum_{n=1}^{\infty} a_n r^n \cos(n\theta) + b_n r^n \sin(n\theta). \tag{6.8}$$

The harmonic functions $r^{-n} \cos(n\theta)$ and $r^{-n} \sin(n\theta)$ also found in Section 6.3 are not used in this superposition because these are undefined at $r = 0$, the center of the disk.

To satisfy the boundary condition, we must choose the coefficients so that

$$u(\rho, \theta) = \frac{1}{2} a_0 + \sum_{n=1}^{\infty} a_n \rho^n \cos(n\theta) + b_n \rho^n \sin(n\theta) = f(\theta).$$

This is a Fourier expansion of f on $[-\pi, \pi]$, leading us to choose:

$$a_0 = \frac{1}{\pi} \int_{-\pi}^{\pi} f(\xi) \, d\xi$$

$$a_n \rho^n = \frac{1}{\pi} \int_{-\pi}^{\pi} f(\xi) \cos(n\xi) \, d\xi$$

and

$$b_n\rho^n = \frac{1}{\pi} \int_{-\pi}^{\pi} f(\xi)\sin(n\xi) \; d\xi.$$

Notice that the factor ρ^n is included in these expressions, since the coefficient of $\cos(n\theta)$ in the Fourier expansion of f is $a_n\rho^n$, and the coefficient of $\sin(n\theta)$ is $b_n\rho^n$. The coefficients are therefore a_0 as just written, and for $n = 1, 2, \ldots,$

$$a_n = \frac{1}{\rho^n\pi} \int_{-\pi}^{\pi} f(\xi)\cos(n\xi) \; d\xi$$

and

$$b_n = \frac{1}{\rho^n\pi} \int_{-\pi}^{\pi} f(\xi)\sin(n\xi) \; d\xi.$$

As a specific example, suppose we want to solve

$$\nabla^2 u = 0 \text{ for } 0 \le r < 3, \; -\pi \le \theta \le \pi$$

$$u(3, \theta) = |\cos(\theta/2)| \text{ for } -\pi \le \theta \le \pi.$$

The solution is given by equation 6.8 and we need only compute the coefficients:

$$a_0 = \frac{1}{\pi} \int_{-\pi}^{\pi} |\cos(\xi/2)| \; d\xi = \frac{4}{\pi};$$

$$a_n = \frac{1}{3^n\pi} \int_{-\pi}^{\pi} |\cos(\xi/2)| \cos(n\xi) \; d\xi = \frac{4(-1)^{n+1}}{3^n\pi(4n^2 - 1)};$$

and

$$b_n = \frac{1}{3^n\pi} \int_{-\pi}^{\pi} |\cos(\xi/2)|\sin(n\xi) \; d\xi = 0.$$

The solution is

$$u(r, \theta) = \frac{2}{\pi} + \frac{4}{\pi} \sum_{n=1}^{\infty} \frac{(-1)^{n+1}}{3^n(4n^2 - 1)} r^n \cos(n\theta)$$

for $0 \le r \le 3$ and $-\pi \le \theta \le \pi$. Notice that $|\cos(\theta/2)|$ has period 2π, and is continuous with a continuous derivative. Further, $f(\pi) = f(-\pi)$. The Fourier series of f therefore converges to f absolutely and uniformly on $[-\pi, \pi]$.

EXERCISE 260   Solve the Dirichlet problem for the disk $r < \rho$ if $u(\rho, \theta) = \cos^2(\theta)$.

EXERCISE 261   Solve the Dirichlet problem for the disk $r < \rho$ if $u(\rho, \theta) = \sin^3(\theta) + \cos^2(\theta)$.

EXERCISE 262   Solve the Dirichlet problem for the disk $r < \rho$ if $u(\rho, \theta) = \sin(\theta)$.

EXERCISE 263   Solve the Dirichlet problem

$$\nabla^2 u(x, y) = 0 \text{ for } x^2 + y^2 < 9$$

$$u(x, y) = x^2 \text{ for } x^2 + y^2 = 9.$$

Hint: Convert the problem to polar coordinates.

EXERCISE 264   Solve the Dirichlet problem

$$\nabla^2 u(x, y) = 0 \text{ for } x^2 + y^2 < 16$$

$$u(x, y) = x^2 y^2 \text{ for } x^2 + y^2 = 16.$$

EXERCISE 265   Consider the Dirichlet problem for the disk $r < \rho$, with $u(\rho, \theta) = f(\theta)$. Suppose f is periodic of period 2π and is an odd function of θ on $[-\pi, \pi]$. Show that the solution $u(r, \theta)$ is also an odd function of θ.
   If f is an even function of θ, is the solution even in θ?

EXERCISE 266   Solve the Dirichlet problem for an annulus (the domain between the concentric circles). The problem is:

$$\nabla^2 u(r, \theta) = 0 \text{ for } \rho_1 < r < \rho_2, 0 \leq \theta \leq 2\pi$$

$$u(\rho_1, \theta) = g(\theta), u(\rho_2, \theta) = f(\theta) \text{ for } 0 \leq \theta \leq 2\pi.$$

Hint: Attempt a series similar to that of equation 6.8, except now include terms $\ln(r)$, $r^{-n} \sin(n\theta)$ and $r^{-n} \cos(n\theta)$ in the expansion, since the origin is not in the annulus. Use the boundary conditions on both of the boundary circles to find the coefficients.

EXERCISE 267   In each of the following, data is given on the bounding circles of an annulus. Using the solution of Exercise 266, write a series solution of the Dirichlet problem for the annulus.

1. $u(1, \theta) = 1$, $u(2, \theta) = 2$
2. $u(1, \theta) = 1$, $u(2, \theta) = \cos(\theta)$
3. $u(1, \theta) = \sin(\theta)$, $u(2, \theta) = \cos(\theta)$
4. $u(1, \theta) = 1$, $u(2, \theta) = \cos^2(\theta)$
5. $u(2, \theta) = \sin(2\theta)$, $u(4, \theta) = \sin(4\theta)$
6. $u(2, \theta) = \sin^2(\theta)$, $u(4, \theta) = \cos^2(\theta)$

6.10   POISSON'S INTEGRAL REPRESENTATION FOR A DISK

We have just obtained an infinite series solution for the Dirichlet problem for a disk about the origin in the plane. We will write an integral formula for this solution. Initially take the radius to be $\rho = 1$.

Begin with the solution 6.8 with the integral formulas for the coefficients inserted:

$$u(r, \theta) = \frac{1}{2\pi} \int_{-\pi}^{\pi} f(\xi)\, d\xi$$

$$+ \sum_{n=1}^{\infty} \frac{1}{\pi} r^n \left[\int_{-\pi}^{\pi} f(\xi)\cos(n\xi)\, d\xi \, \cos(nx) + \int_{-\pi}^{\pi} f(\xi)\sin(n\xi)\, d\xi \, \sin(nx) \right]$$

$$= \frac{1}{2\pi} \int_{-\pi}^{\pi} \left[1 + 2 \sum_{n=1}^{\infty} r^n(\cos(n\xi)\cos(nx) + \sin(n\xi)\sin(nx)) \right] f(\xi)\, d\xi$$

$$= \frac{1}{2\pi} \int_{-\pi}^{\pi} \left[1 + 2 \sum_{n=1}^{\infty} r^n \cos(n(\theta - \xi)) \right] f(\xi)\, d\xi.$$

The interchange of the summation and integral is justified if the series converges uniformly, and we know conditions on f sufficient to guarantee this. The quantity

$$P(r; \zeta) = \frac{1}{2\pi} \left[1 + 2 \sum_{n=1}^{\infty} r^n \cos(n\zeta) \right]$$

is called the *Poisson kernel*, and in terms of it the solution is

$$u(r, \theta) = \int_{-\pi}^{\pi} P(r; \theta - \xi) f(\xi)\, d\xi \tag{6.9}$$

for $0 \leq r < 1$ and $-\pi \leq \theta \leq \pi$. This is another example of a general theme in which kernel functions are used to write a quantity of interest as the integral of the product of the kernel function with a given function. We saw this idea previously when we wrote the partial sum of a Fourier series as the integral of the product of the function and the Dirichlet kernel, and we will see it again in Section 6.14 when we write an integral solution of the Dirichlet problem in terms of the Green's function.

Thus far we have just manipulated the solution of the preceding section and given part of it a name. We will now write the Poisson kernel in closed form (that is, without summation, a tactic we also pursued with the Dirichlet kernel). Recall Euler's formula

$$e^{i\zeta} = \cos(\zeta) + i \sin(\zeta).$$

Let $z = re^{i\zeta}$. Then

$$z^n = r^n e^{in\zeta} = r^n \cos(n\zeta) + ir^n \sin(n\zeta)$$

and $r^n \cos(\zeta)$ may be thought of as the real part of z^n:

$$r^n \cos(n\zeta) = \text{Re}(z^n).$$

Now

$$1 + 2 \sum_{n=1}^{\infty} r^n \cos(n\zeta) = \text{Re}\left(1 + 2 \sum_{n=1}^{\infty} z^n\right).$$

The geometric series $\sum_{n=1}^{\infty} z^n$ converges because $|z| = r < 1$. We will use the familiar result that

$$\sum_{n=1}^{\infty} z^n = \frac{z}{1 - z},$$

yielding

$$1 + 2 \sum_{n=1}^{\infty} r^n \cos(n\zeta) = \text{Re}\left(1 + 2\frac{z}{1 - z}\right) = \text{Re}\left(\frac{1 + z}{1 - z}\right) = \text{Re}\left(\frac{1 + re^{i\zeta}}{1 - re^{i\zeta}}\right).$$

One way to compute the real part of the last quantity is to multiply numerator and denominator by $1 - re^{-i\zeta}$ (the complex conjugate of the denominator):

$$\frac{1 + re^{i\zeta}}{1 - re^{i\zeta}} = \left(\frac{1 + re^{i\zeta}}{1 - re^{i\zeta}}\right)\left(\frac{1 - re^{-i\zeta}}{1 - re^{-i\zeta}}\right) = \frac{1 - r^2 + r(e^{i\zeta} - e^{-i\zeta})}{1 + r^2 - r(e^{i\zeta} + e^{-i\zeta})}$$

$$= \frac{1 - r^2 + 2ir \sin(\zeta)}{1 + r^2 - 2r \cos(\zeta)}.$$

The real part of this is $(1 - r^2)/(1 + r^2 - 2r \cos(\zeta))$, hence

$$1 + 2 \sum_{n=1}^{\infty} r^n \cos(n\zeta) = \mathrm{Re} \left(\frac{1 + re^{i\zeta}}{1 - re^{i\zeta}} \right) = \frac{1 - r^2}{1 + r^2 - 2r \cos(\zeta)}.$$

This gives the closed form of the Poisson kernel:

$$P(r; \zeta) = \frac{1}{2\pi} \frac{1 - r^2}{1 + r^2 - 2r \cos(\zeta)}.$$

Upon substituting this into equation 6.9 we obtain

$$u(r, \theta) = \frac{1}{2\pi} \int_{-\pi}^{\pi} \frac{1 - r^2}{1 - 2r \cos(\theta - \xi) + r^2} f(\xi) \, d\xi. \tag{6.10}$$

This is *Poisson's formula* for the solution of the Dirichlet problem for the unit disk, and it is valid for $0 \le r < 1$ and $-\pi \le \theta \le \pi$.

By a change of variables we can write the integral solution for a disk of any positive radius ρ:

$$u(r, \theta) = \frac{1}{2\pi} \int_{-\pi}^{\pi} \frac{\rho^2 - r^2}{r^2 - 2\rho r \cos(\theta - \xi) + \rho^2} f(\xi) \, d\xi \tag{6.11}$$

for $0 \le r < \rho$ and $-\pi \le \theta \le \pi$.

Assuming that f has period 2π, this integral can be evaluated over any interval of length 2π and we sometimes see the equivalent expression

$$u(r, \theta) = \frac{1}{2\pi} \int_{0}^{2\pi} \frac{\rho^2 - r^2}{\rho^2 + r^2 - 2\rho r \cos(\theta - \xi)} f(\xi) \, d\xi,$$

which is also called Poisson's formula.

As one consequence of equation 6.11,

$$u(0, 0) = \frac{1}{2\pi} \int_{-\pi}^{\pi} f(\xi) \, d\xi = \frac{1}{2\pi} \int_{-\pi}^{\pi} u(\rho, \xi) \, d\xi = \frac{1}{2\pi\rho} \int_{-\pi}^{\pi} u(\rho, \xi) \, ds \tag{6.12}$$

since the differential element of arc length on the circle of radius ρ about the origin is $ds = \rho \, d\xi$. This equation states that the value of u at the center of the disk is the average of its values on the bounding circle, consistent with the mean value property for harmonic functions.

We will use Poisson's integral solution 6.11 to derive Harnack's inequality, and from this another important property of harmonic functions. Since

$$-1 \le \cos(\theta - \xi) \le 1,$$

we can replace the cosine term in equation 6.11 by -1 and then by 1 to obtain:

$$\frac{1}{2\pi} \int_{-\pi}^{\pi} \frac{\rho^2 - r^2}{\rho^2 + r^2 + 2\rho r} f(\xi) \, d\xi \le u(r, \theta) \le \frac{1}{2\pi} \int_{-\pi}^{\pi} \frac{\rho^2 - r^2}{\rho^2 + r^2 - 2\rho r} f(\xi) \, d\xi.$$

Then

$$\frac{(\rho - r)(\rho + r)}{(\rho + r)^2} \frac{1}{2\pi} \int_{-\pi}^{\pi} f(\xi) \, d\xi \le u(r, \theta) \le \frac{(\rho - r)(\rho + r)}{(\rho - r)^2} \frac{1}{2\pi} \int_{-\pi}^{\pi} f(\xi) \, d\xi$$

which, in view of equation 6.12, we can write as

$$\frac{\rho - r}{\rho + r} u(0, 0) \le u(r, \theta) \le \frac{\rho + r}{\rho - r} u(0, 0) \tag{6.13}$$

for $0 \le \rho < r$ and $-\pi \le \theta \le \pi$. This is *Harnack's inequality*.

Suppose now that u is harmonic and nonnegative in the entire plane. Then Harnack's inequality must hold for every positive ρ and upon taking the limit as $\rho \to \infty$ we obtain

$$u(0, 0) \le u(r, \theta) \le u(0, 0),$$

hence $u(r, \theta) = u(0, 0)$. We conclude that a nonnegative function that is harmonic in R^2 must be constant. The student who is familiar with complex function theory can relate this to the fact that a bounded entire function must be constant.

EXERCISE 268 Derive equation 6.11 from equation 6.10.

EXERCISE 269 By differentiating under the integral sign, show that the function defined by equation 6.11 is a solution of Laplace's equation in polar coordinates for the disk $r < \rho$.

EXERCISE 270 Show that, for $0 \le r < \rho$ and $-\pi \le \theta \le \pi$,

$$\int_{-\pi}^{\pi} \frac{\rho^2 - r^2}{\rho^2 + r^2 - 2r\rho \cos(\theta - \xi)} \, d\xi = 2\pi.$$

EXERCISE 271 Solve the Dirichlet problem for the exterior of a disk:

$$\nabla^2 u(r, \theta) = 0 \quad \text{for } r > \rho, \quad -\pi \le \theta \le \pi$$

$$u(\rho, \theta) = f(\theta).$$

Hint: Seek a solution that is bounded as $r \to \infty$. Obtain the Poisson integral formula

$$u(r, \theta) = -\frac{1}{2\pi} \int_{-\pi}^{\pi} \frac{\rho^2 - r^2}{\rho^2 + r^2 - 2r\rho \cos(\theta - \xi)} f(\xi) \, d\xi.$$

EXERCISE 272

(a) Derive an integral expression for the solution of the Dirichlet problem for an annulus:

$$\nabla^2 u(r, \theta) = 0 \text{ for } \rho_1 < r < \rho_2, \, 0 \le \theta \le 2\pi$$

$$u(\rho_1, \theta) = g(\theta), \, u(\rho_2, \theta) = f(\theta) \text{ for } 0 \le \theta \le 2\pi.$$

Hint: One way to proceed is to begin with the series solution requested in Exercise 266 and mimic the derivation of the Poisson integral formula for the disk. A second approach is to add the solution of a Dirichlet problem for the disk $r < \rho_2$ to the solution of a Dirichlet problem for the exterior of a disk $r > \rho_1$ (Exercise 271).

(b) Use the solution derived in part (a) to write an integral solution for each of the Dirichlet problems in (1) through (6) of Exercise 267.

EXERCISE 273

Consider the integral solution of the Dirichlet problem for an annulus (Exercise 272(a)). As $\rho_1 \to 0$, does this yield the Poisson integral solution of the Dirichlet problem for a disk?

EXERCISE 274

Complete the details of the following alternate derivation of Poisson's integral formula (equation 6.11). This derivation is based on the complex form of a Fourier series (see Exercises 66, 67 in Section 3.3.5). The problem is

$$\nabla^2 u(r, \theta) = 0 \text{ for } 0 \le r < \rho, \, 0 \le \theta \le 2\pi$$

$$u(\rho, \theta) = f(\theta) \text{ for } 0 \le \theta \le 2\pi.$$

(a) Begin by expanding f in a complex Fourier series

$$f(\theta) = \sum_{n=-\infty}^{\infty} a_n e^{in\theta},$$

where

$$a_n = \frac{1}{2\pi} \int_{-\pi}^{\pi} f(\xi) e^{-in\xi} \, d\xi \text{ for } n = \cdots, -2, -1, 0, 1, 2, \ldots.$$

(b) Assume that $u(r, \theta) = \sum_{n=-\infty}^{\infty} R_n(r)e^{in\theta}$. Show that

$$R_n(\rho) = a_n \text{ for } n = \cdots, -2, -1, 0, 1, 2, \ldots.$$

(c) By substituting the series for $u(r, \theta)$ into Laplace's equation in polar coordinates, show that

$$R_n''(r) + \frac{1}{r} R_n'(r) - \frac{n^2}{r^2} R_n(r) = 0$$

for $n = \cdots, -2, -1, 0, 1, 2, \ldots$ and $0 < r < \rho$.

(d) By setting $R_n(r) = r^\lambda$, show that $\lambda = \pm n$. Conclude that a bounded solution for R_n has the form

$$R_n(r) = c_n r^n \text{ for } n = 0, 1, 2, \ldots$$

and

$$R_n(r) = c_n r^{-n} \text{ for } \cdots, -2, -1.$$

(e) Show that the solution for R_n satisfying $R_n(\rho) = a_n$ is

$$R_n(r) = a_n \left(\frac{r}{\rho}\right)^{|n|}.$$

(f) Conclude that the solution of the Dirichlet problem for the disk is

$$u(r, \theta) = \sum_{n=-\infty}^{\infty} a_n \left(\frac{r}{\rho}\right)^{|n|} e^{in\theta}.$$

(g) Substitute the integral formula for a_n into this solution and interchange the sum and the integral to obtain

$$u(r, \theta) = \frac{1}{2\pi} \int_{-\pi}^{\pi} f(\xi) \sum_{n=-\infty}^{\infty} \left(\frac{r}{\rho}\right)^{|n|} e^{in(\theta-\xi)} d\xi.$$

(h) Use the fact that in general

$$\sum_{n=-\infty}^{\infty} G(n) = \sum_{n=0}^{\infty} G(-n) + \sum_{n=0}^{\infty} G(n) - G(0),$$

choose G appropriately from part (g), and use the fact that we know the sum of a geometric series, to show that

$$\sum_{n=-\infty}^{\infty} \left(\frac{r}{\rho}\right)^{|n|} e^{in\theta} = \frac{1 - (r/\rho)^2}{1 - 2(r/\rho)\cos(\theta) + (r/\rho)^2}.$$

(i)   From parts (g) and (h), show that

$$u(r, \theta) = \frac{1}{2\pi} \int_{-\pi}^{\pi} f(\xi) \frac{1 - (r/\rho)^2}{1 - 2(r/\rho)\cos(\theta - \xi) + (r/\rho)^2} \, d\xi.$$

(j)   Use the result of part (i) to derive the Poisson integral formula.

EXERCISE 275   Suppose u is harmonic in R^2, but is not nonnegative. Does it follow that u must be constant?

EXERCISE 276   Prove Harnack's First Theorem. Let $\{u_n\}$ be a sequence of functions harmonic in a bounded domain Ω of the plane, and continuous on $\bar{\Omega}$. Suppose $\{u_n\}$ converges uniformly on $\partial\Omega$. Then $\{u_n\}$ converges uniformly on Ω to a harmonic function. *Hint:* First note Exercise 255 in Section 6.6. Let u_n converge uniformly to u on $\bar{\Omega}$. If $(x_0, y_0) \in \Omega$, choose ρ so that $\bar{B}((x_0, y_0), \rho) \subset \Omega$. Denote $B = B((x_0, y_0), \rho)$. By equation 6.11 write

$$u_n(x, y) = \frac{1}{2\pi} \int_0^{2\pi} f_n(\xi) \frac{\rho^2 - r^2}{\rho^2 + r^2 - 2r\rho \cos(\theta - \xi)} \, d\xi,$$

where $u_n(x, y) = f_n(\theta)$ on the boundary circle C of B. Let $n \to \infty$ and use the uniform convergence to show that

$$u(x, y) = \frac{1}{2\pi} \int_0^{2\pi} f(\xi) \frac{\rho^2 - r^2}{\rho^2 + r^2 - 2r\rho \cos(\theta - \xi)} \, d\xi,$$

where $u_n(x, y) \to f(\theta)$ on C. Thus show that u is harmonic in B. From this complete the proof.

EXERCISE 277   Prove Harnack's Second Theorem. Let u_n be harmonic on a domain in R^2 for $n = 1, 2, \ldots$. Let P_0 be a point of Ω and suppose the sequence $\{u_n(P_0)\}$ converges. Then $\{u_n(P)\}$ converges at each P in Ω, and the limit function is harmonic in Ω. Further, this convergence is uniform on each compact subset of Ω. *Hint:* First let B be a neighborhood of radius r_1 about P_0, with $B \subset \Omega$. Choose a circle C of radius $r_1 + \epsilon$ about P_0, with ϵ chosen sufficiently small that C is contained in Ω. Now use an argument like that employed to prove Harnack's inequality to show that

$$\frac{r_1 + \epsilon - \rho}{r_1 + \epsilon + \rho} u_n(P_0) \le u_n(\rho, \varphi) \le \frac{r_1 + \epsilon + \rho}{r_1 + \epsilon - \rho} u_n(P_0)$$

for $0 \le \rho < r_1 + \epsilon$. Hence show that $\{u_n(P)\}$ converges at each P in B.

Next prove that $\{u_n(P)\}$ converges at an arbitrary P in Ω. To do this, form a polygonal path L from P_0 to P in Ω. Using this path, form a finite sequence of overlapping open disks, the first centered at P_0 and the last at P, and use the conclusion of the first part of the proof. (This argument is reminiscent of the proof of the Maximum Principle for harmonic functions.)

There remains to show that $\{u_n\}$ converges uniformly on any closed subset M of Ω. To do this, cover M by forming an open disk about each point of M, with the closure of each disk a subset of Ω. Now use the Heine-Borel Theorem to obtain a finite cover of M by open disks, and use the fact that $\{u_n\}$ converges uniformly on the closure of each such disk.

EXERCISE 278 Prove Liouville's Theorem. Let u be harmonic in the plane R^2, and suppose that u is not a constant function. Then u can have neither an upper bound nor a lower bound. (Liouville's Theorem is stated here for the plane because that is where we have just developed some machinery, but the theorem extends to nonconstant harmonic functions in R^n.)

EXERCISE 279 Sometimes an opportunistic use of a formula can yield additional results. Here are some examples.

1. Choose $u(r, \theta) = r^n \sin(n\theta)$ in equation 6.11 and evaluate $u(\rho/2, \pi/2)$ to derive the integral

$$\int_0^{2\pi} \frac{\sin(n\xi)}{5 - 4 \sin(\xi)} d\xi = \frac{\pi}{3 \cdot 2^{n-1}} \sin(n\pi/2),$$

   in which n is any positive integer.

2. What integral formula do you obtain by evaluating $u(\rho/2, \pi)$ in (1)?

3. Derive integral formulas by choosing $u(r, \theta) = r^n \cos(n\theta)$ in equation 6.11 and evaluating $u(\rho/2, \pi/2)$ and $u(\rho/2, \pi)$.

4. Derive an integral formula by using $u(r, \theta) \equiv 1$ in Poisson's formula.

6.11   DIRICHLET PROBLEM FOR THE UPPER HALF PLANE

Dirichlet problems for unbounded domains can sometimes be solved using Fourier integrals and transforms. We will illustrate with the problem for the upper half plane, whose boundary is the horizontal axis:

$$\nabla^2 u(x, y) = 0 \text{ for } -\infty < x < \infty, y > 0$$

$$u(x, 0) = f(x) \text{ for } -\infty < x < \infty. \tag{6.14}$$

Here Ω is not a bounded domain, and we impose the condition that the solution must be a bounded function.

6.11.1   Solution by Fourier Integral

We can solve this problem by separation of variables. Let $u(x, y) = X(x)Y(y)$ and substitute into the differential equation to obtain

$$\frac{X''}{X} = -\frac{Y''}{Y} = -\lambda$$

or

$$X'' + \lambda X = 0, \quad Y'' - \lambda Y = 0.$$

Consider cases on λ.

If $\lambda = 0$ then $X'' = 0$ so $X = ax + b$. We must choose $a = 0$ to have a bounded solution.

If $\lambda = -\omega^2$ with $\omega > 0$ then $X = ae^{\omega x} + be^{-\omega x}$. Now $e^{\omega x} \to \infty$ as $x \to \infty$ so we must choose $a = 0$. And $e^{-\omega x} \to \infty$ as $x \to -\infty$ so we must also choose $b = 0$. This case yields no nontrivial bounded solution for X.

If $\lambda = \omega^2$ with $\omega > 0$ then $X = a \cos(\omega x) + b \sin(\omega x)$, a bounded solution. This includes the constant solution in the case $\omega = 0$.

Now consider the y-dependence. With $\lambda = \omega^2$ we have $Y'' - \omega^2 Y = 0$ so $Y = ae^{\omega y} + be^{-\omega y}$. Since $e^{\omega y} \to \infty$ as $y \to \infty$, choose $a = 0$. Since $y > 0$, $be^{-\omega y}$ is a bounded solution for Y.

For each $\omega \geq 0$ we now have a function

$$u_\omega(x, y) = [a_\omega \cos(\omega x) + b_\omega \sin(\omega x)]e^{-\omega y}$$

that is harmonic on the half plane $y > 0$. To satisfy the boundary condition we must generally superimpose these solutions over all $\omega \geq 0$ and this is done by integrating. Let

$$u(x, y) = \int_0^\infty [a_\omega \cos(\omega x) + b_\omega \sin(\omega x)]e^{-\omega y} \, d\omega. \qquad (6.15)$$

The boundary condition requires that

$$u(x, 0) = \int_0^\infty [a_\omega \cos(\omega x) + b_\omega \sin(\omega x)] \, d\omega = f(x).$$

This is the Fourier integral expansion of f; hence choose

$$a_\omega = \frac{1}{\pi} \int_{-\infty}^\infty f(\xi)\cos(\omega\xi) \, d\xi$$

and

$$b_\omega = \frac{1}{\pi} \int_{-\infty}^\infty f(\xi)\sin(\omega\xi) \, d\xi.$$

Here we are assuming that f has a Fourier integral representation on the entire real line.

The solution 6.15 can be written in a more compact form by inserting these coefficients, changing the order of integration and performing some routine manipulations:

$$u(x, y) = \frac{1}{\pi} \int_0^\infty \int_{-\infty}^\infty [\cos(\omega\xi)\cos(\omega x) + \sin(\omega\xi)\sin(\omega x)]f(\xi)e^{-\omega y} \, d\xi \, d\omega$$

$$= \frac{1}{\pi} \int_{-\infty}^\infty \left(\int_0^\infty \cos(\omega(\xi - x))e^{-\omega y} \, d\omega \right) f(\xi) \, d\xi.$$

The inner integral is

$$\int_0^\infty e^{-\omega y} \cos(\omega(\xi - x)) \, d\omega = \frac{e^{-\omega y}}{y^2 + (\xi - x)^2} (-y \cos(\omega(\xi - x))$$

$$+ (\xi - x)\sin(\omega(\xi - x))) \Bigg]_0^\infty = \frac{y}{y^2 + (\xi - x)^2}.$$

This yields a relatively simple expression for the solution of the Dirichlet problem for the upper half plane:

$$u(x, y) = \frac{y}{\pi} \int_{-\infty}^\infty \frac{f(\xi)}{y^2 + (\xi - x)^2} \, d\xi. \qquad (6.16)$$

EXERCISE 280   Use of the Fourier integral to solve the Dirichlet problem for the upper half plane imposes conditions on the boundary data function f, since we must be able to write a Fourier integral representation for f. This method therefore fails for certain choices of f. In particular, consider the Dirichlet problem

$$\nabla^2 u = 0 \text{ for } -\infty < x < \infty, \ y > 0$$

$$u(x, 0) = k \text{ for } -\infty < x < \infty,$$

in which k is a nonzero constant. Find the bounded solution of this problem, and then produce infinitely many solutions which are unbounded. What does equation 6.16 give for $f(x) = c$, constant?

6.11.2   Solution by Fourier Transform

We will solve the problem for the upper half plane again by Fourier transform to illustrate the technique. Apply the Fourier transform in x to Laplace's equation, letting $\mathfrak{F}[u(x, y)](\omega) = \hat{u}(\omega, y)$. Since y is independent of x,

$$\mathfrak{F}[u_{yy}(x, y)](\omega) = \int_{-\infty}^{\infty} \frac{\partial^2 u(x, y)}{\partial y^2} e^{-i\omega x} \, dx = \frac{\partial^2}{\partial y^2} \left(\int_{-\infty}^{\infty} u(x, y) e^{-i\omega x} \, dx \right) = \frac{\partial^2}{\partial y^2} \hat{u}(\omega, y).$$

Next, by the operational formula 3.28 for the Fourier transform (Section 3.7),

$$\mathfrak{F}[u_{xx}(x, y)](\omega) = -\omega^2 \hat{u}(\omega, y).$$

The result of applying the transform to Laplace's equation is

$$\frac{\partial^2 \hat{u}}{\partial y^2} - \omega^2 \hat{u} = 0.$$

This has general solution

$$\hat{u}(\omega, y) = a(\omega)e^{\omega y} + b(\omega)e^{-\omega y},$$

in which the coefficients may be functions of ω. Now $y > 0$ in the upper half plane. Since $e^{\omega y} \to \infty$ as $y \to \infty$ if $\omega > 0$, we must have $a(\omega) = 0$ if $\omega > 0$ in order to have a bounded solution. But $e^{-\omega y} \to \infty$ as $y \to \infty$ if $\omega < 0$, so we must have $b(\omega) = 0$ if $\omega < 0$. Thus,

$$\hat{u}(\omega, y) = \begin{cases} b(\omega)e^{-\omega y} \text{ for } \omega \geq 0 \\ a(\omega)e^{\omega y} \text{ for } \omega \leq 0. \end{cases}$$

We can consolidate this notation by writing

$$\hat{u}(\omega, y) = c(\omega)e^{-|\omega|y}.$$

To solve for $c(\omega)$ recall that $u(x, 0) = f(x)$. Therefore

$$\hat{u}(\omega, 0) = \hat{f}(\omega) = c(\omega)$$

and so

$$\hat{u}(\omega, y) = \hat{f}(\omega)e^{-|\omega|y}.$$

This is the Fourier transform of the solution. Apply the inverse Fourier transform:

$$u(x, y) = \mathfrak{F}^{-1}[\hat{u}(\omega, y)](x) = \frac{1}{2\pi} \int_{-\infty}^{\infty} \hat{u}(\omega, y)e^{i\omega x} \, d\omega = \frac{1}{2\pi} \int_{-\infty}^{\infty} \hat{f}(\omega)e^{-|\omega|y}e^{i\omega x} \, d\omega$$

$$= \frac{1}{2\pi} \int_{-\infty}^{\infty} \left(\int_{-\infty}^{\infty} f(\xi)e^{-i\omega\xi} \, d\xi \right) e^{-|\omega|y}e^{i\omega x} \, d\omega$$

$$= \frac{1}{2\pi} \int_{-\infty}^{\infty} \left(\int_{-\infty}^{\infty} e^{-|\omega|y}e^{-i\omega(\xi-x)} \, d\omega \right) f(\xi) \, d\xi.$$

Now

$$e^{-i\omega(\xi-x)} = \cos(\omega(\xi - x)) - i\,\sin(\omega(\xi - x))$$

and a straightforward integration gives

$$\int_{-\infty}^{\infty} e^{-|\omega|y}e^{-i\omega(\xi-x)} \, d\omega = \frac{2y}{y^2 + (\xi - x)^2}.$$

The solution by Fourier transform is

$$u(x, y) = \frac{y}{\pi} \int_{-\infty}^{\infty} \frac{f(\xi)}{y^2 + (\xi - x)^2} \, d\xi$$

in agreement with the solution obtained by separation of variables.

6.12 DIRICHLET PROBLEM FOR THE RIGHT QUARTER PLANE

Sometimes it is possible to exploit the solution of a Dirichlet problem on one domain to solve a Dirichlet problem on another domain. We will illustrate this idea with an example.

Consider a Dirichlet problem for the right quarter plane:

$$\nabla^2 u = 0 \text{ for } x > 0, \ y > 0$$

$$u(x, 0) = f(x) \text{ for } x \geq 0$$

$$u(0, y) = 0 \text{ for } y \geq 0. \tag{6.17}$$

Here the boundary consists of the nonnegative x- and y-axes, and we are pre-scribing nonzero boundary data only on the horizontal boundary.

This problem can be solved by separation of variables, as we did for the upper half plane. We can also obtain the solution by using the Fourier sine transform. We leave these calculations to the student.

Here is another approach. We know a bounded solution of the Dirichlet problem for the upper half plane. If we fold the upper half plane across the vertical axis, we obtain the right half plane. This suggests that we explore the possibility of using the solution 6.16 to produce the solution for the right quarter plane.

Define

$$g(x) = \begin{cases} f(x) \text{ for } x \geq 0 \\ \text{any for } x < 0 \end{cases}$$

where by "any" we mean, for the moment, give $g(x)$ any continuous definition for negative x.

We know that the Dirichlet problem

$$\nabla^2 u = 0 \text{ for } -\infty < x < \infty, \ y > 0$$

$$u(x, 0) = g(x) \text{ for } -\infty < x < \infty$$

for the upper half plane has the solution

$$u_{hp}(x, y) = \frac{y}{\pi} \int_{-\infty}^{\infty} \frac{g(\xi)}{y^2 + (\xi - x)^2} \, d\xi.$$

Write this as

$$u_{hp}(x, y) = \frac{y}{\pi} \left[\int_{-\infty}^{0} \frac{g(\xi)}{y^2 + (\xi - x)^2} \, d\xi + \int_{0}^{\infty} \frac{g(\xi)}{y^2 + (\xi - x)^2} \, d\xi \right].$$

If we put $w = -\xi$ in the first integral in this equation for $u_{hp}(x, y)$, we obtain

$$\int_{-\infty}^{0} \frac{g(\xi)}{y^2 + (\xi - x)^2} \, d\xi = \int_{\infty}^{0} \frac{g(-w)}{y^2 + (w + x)^2} (-1) \, dw.$$

Using ξ again for the variable of integration, this integral is

$$\int_{0}^{\infty} \frac{g(-\xi)}{y^2 + (\xi + x)^2} \, d\xi.$$

Therefore

$$u_{hp}(x, y) = \frac{y}{\pi} \int_{0}^{\infty} \left(\frac{g(-\xi)}{y^2 + (\xi + x)^2} + \frac{g(\xi)}{y^2 + (\xi - x)^2} \right) d\xi$$

$$= \frac{y}{\pi} \int_{0}^{\infty} \left(\frac{g(-\xi)}{y^2 + (\xi + x)^2} + \frac{f(\xi)}{y^2 + (\xi - x)^2} \right) d\xi.$$

Now we will fill in the "any" in the definition of g. The function $u_{hp}(x, y)$ will vanish on the positive y-axis, where $x = 0$, if $g(-\xi) = -f(\xi)$ for $\xi \geq 0$. This occurs if we set

$$g(x) = -f(-x) \text{ for } x < 0.$$

That is, make g the odd extension of f to the entire real line. Now

$$u_{hp}(x, y) = \frac{y}{\pi} \int_{0}^{\infty} \left(\frac{1}{y^2 + (\xi - x)^2} - \frac{1}{y^2 + (\xi + x)^2} \right) f(\xi) \, d\xi \qquad (6.18)$$

is the solution of the Dirichlet problem 6.14 for the upper half plane. But u_{hp} is also harmonic on the right quarter plane, vanishes when $x = 0$, and equals $f(x)$ for $x \geq 0$ if $y = 0$. Therefore equation 6.18 also gives the solution of the problem 6.17 for the right quarter plane.

As a specific example, consider the problem

$$\nabla^2 u = 0 \text{ for } x > 0, y > 0$$

$$u(0, y) = 0 \text{ for } y \geq 0$$

$$u(x, 0) = xe^{-x} \text{ for } x \geq 0.$$

The solution of this problem is

$$u(x, y) = \frac{y}{\pi} \int_{0}^{\infty} \left(\frac{1}{y^2 + (\xi - x)^2} - \frac{1}{y^2 + (\xi + x)^2} \right) \xi e^{-\xi} \, d\xi.$$

As another example, consider the problem 6.17 when $f(x) \equiv 1$. Now the solution 6.18 is

$$u(x, y) = \frac{y}{\pi} \int_0^\infty \frac{1}{y^2 + (\xi - x)^2} \, d\xi - \frac{y}{\pi} \int_0^\infty \frac{1}{y^2 + (\xi + x)^2} \, d\xi.$$

In this simple case these integrals can be evaluated in closed form. First,

$$\int_0^\infty \frac{1}{y^2 + (\xi - x)^2} \, d\xi = \int_0^\infty \frac{1}{y^2 \left[1 + \left(\dfrac{\xi - x}{y} \right)^2 \right]} \, d\xi$$

$$= \frac{1}{y} \left(\frac{\pi}{2} - \arctan \left(-\frac{x}{y} \right) \right) = \frac{\pi}{2y} + \frac{1}{y} \arctan \left(\frac{x}{y} \right).$$

And, by a similar calculation,

$$\int_0^\infty \frac{1}{y^2 + (\xi + x)^2} \, d\xi = \frac{\pi}{2y} - \frac{1}{y} \arctan \left(\frac{x}{y} \right).$$

Therefore

$$u(x, y) = \frac{y}{\pi} \left[\frac{\pi}{2y} + \frac{1}{y} \arctan \left(\frac{x}{y} \right) - \frac{\pi}{2y} + \frac{1}{y} \arctan \left(\frac{x}{y} \right) \right] = \frac{2}{\pi} \arctan \left(\frac{x}{y} \right).$$

It is routine to check that this function is harmonic in the right quarter plane, that $u(0, y) = 0$ and that for $x > 0$,

$$\lim_{y \to 0+} \frac{2}{\pi} \arctan \left(\frac{x}{y} \right) = \frac{2}{\pi} \frac{\pi}{2} = 1.$$

EXERCISE 281   Find a bounded solution of the Dirichlet problem for the left quarter plane:

$$\nabla^2 u = 0 \text{ for } x < 0, y > 0$$

$$u(0, y) = 0 \text{ for } y > 0$$

$$u(x, 0) = f(x) \text{ for } x < 0.$$

EXERCISE 282   Find a bounded solution of the Dirichlet problem for the quarter plane $x < 0, y < 0$:

$$\nabla^2 u = 0 \text{ for } x < 0, y < 0$$

$$u(0, y) = 0 \text{ for } y < 0$$

$$u(x, 0) = f(x) \text{ for } x < 0.$$

6.13   DIRICHLET PROBLEM FOR A CUBE

We will illustrate a Dirichlet problem in three independent variables. Consider

$$\nabla^2 u(x, y, z) = 0 \text{ for } 0 < x < a, \, 0 < y < b, \, 0 < z < c$$

$$u(x, y, 0) = u(x, y, c) = 0$$

$$u(0, y, z) = u(a, y, z) = 0$$

$$u(x, 0, z) = 0, \, u(x, b, z) = f(x, z).$$

Homogeneous boundary data are prescribed on five faces of the rectangular cube (not all sides need have equal length), and boundary data f is given on the side $y = b$. If nonhomogeneous data were given on more than one side, we would independently consider the Dirichlet problem for only one nonzero side at a time and add these solutions.

Let $u(x, y, z) = X(x)Y(y)Z(z)$, substitute into Laplace's equation and divide by XYZ to obtain

$$\frac{X''}{X} + \frac{Y''}{Y} + \frac{Z''}{Z} = 0$$

or

$$\frac{X''}{X} = -\frac{Y''}{Y} - \frac{Z''}{Z} = -\lambda$$

for some constant λ because x, y, and z are independent. Then

$$X'' + \lambda X = 0.$$

Further, from the conditions $u(0, y, z) = u(a, y, z) = 0$ we obtain the boundary conditions

$$X(0) = X(a) = 0.$$

This problem for X has eigenvalues $\lambda_n = n^2\pi^2/a^2$ and eigenfunctions $X_n(x) = \sin(n\pi x/a)$ for $n = 1, 2, \ldots$.

Now write

$$\frac{Z''}{Z} = \lambda - \frac{Y''}{Y} = -\mu,$$

with μ constant, because y and z are independent. Now

$$Z'' + \mu Z = 0,$$

and the conditions $u(x, y, 0) = u(x, y, c) = 0$ imply that

$$Z(0) = Z(c) = 0.$$

This problem for Z has solutions $\mu_m = m^2\pi^2/c^2$ for $m = 1, 2, \ldots$ and $Z_m(z) = \sin(m\pi z/c)$ for $m = 1, 2, \ldots$.
   The equation for Y is

$$Y'' - \left(\frac{n^2\pi^2}{a^2} + \frac{m^2\pi^2}{c^2}\right) Y = 0$$

for $n, m = 1, 2, \ldots$. This has general solution of the form

$$Y = pe^{\beta_{nm}y} + qe^{-\beta_{nm}y}$$

in which $\beta_{nm} = \sqrt{n^2\pi^2/a^2 + m^2\pi^2/c^2}$. Since $u(x, 0, z) = 0$ then $Y(0) = 0$ and $p = -q$. Thus Y must be a constant multiple of $\sinh(\beta_{nm}y)$.
   For each positive integer n and m we now have harmonic functions

$$u_{nm}(x, y, z) = b_{nm} \sin\left(\frac{n\pi x}{a}\right) \sin\left(\frac{m\pi z}{c}\right) \sinh(\beta_{nm}y).$$

In order to satisfy the initial condition, we generally need a superposition

$$u(x, y, z) = \sum_{n=1}^{\infty} \sum_{m=1}^{\infty} b_{nm} \sin\left(\frac{n\pi x}{a}\right) \sin\left(\frac{m\pi z}{c}\right) \sinh(\beta_{nm}y).$$

We must choose the coefficients so that

$$u(x, b, z) = f(x, z) = \sum_{n=1}^{\infty} \sum_{m=1}^{\infty} b_{nm} \sin\left(\frac{n\pi x}{a}\right) \sin\left(\frac{m\pi z}{c}\right) \sinh(\beta_{nm}b).$$

This is the double Fourier sine expansion of f on $0 \le x \le a$, $0 \le z \le c$. We have encountered such expansions before (Sections 4.12 and 5.7). The coefficients are

$$b_{nm} = \frac{4}{ac \ \sinh(\beta_{nm}b)} \int_0^a \int_0^c f(\xi, \zeta)\sin\left(\frac{n\pi\xi}{a}\right) \sin\left(\frac{m\pi\zeta}{c}\right) d\zeta \ d\xi.$$

EXERCISE 283   Solve

$$\nabla^2 u(x, y, z) = 0 \text{ for } 0 < x < 1, \ 0 < y < 1, \ 0 < z < 1$$

$$u(x, y, 0) = u(x, y, 1) = 0$$

$$u(0, y, z) = u(1, y, z) = 0$$

$$u(x, 0, z) = 0, \ u(x, 1, z) = x \ \cos(\pi x/2)z(1 - z).$$

EXERCISE 284   Solve

$$\nabla^2 u(x, y, z) = 0 \text{ for } 0 < x < 1, \ 0 < y < 1, \ 0 < z < \pi$$

$$u(x, y, 0) = 0, \ u(x, y, \pi) = \sin(\pi x)\sin(\pi y)$$

$$u(x, 0, z) = u(x, 1, z) = 0$$

$$u(0, y, z) = u(1, y, z) = 0.$$

EXERCISE 285   Solve

$$\nabla^2 u(x, y, z) = 0 \text{ for } 0 < x < \pi, \ 0 < y < \pi, \ 0 < z < \pi$$

$$u(x, y, 0) = \sin(3x)\sin(y), \ u(x, y, \pi) = x(\pi - x)y(\pi - y)$$

$$u(x, 0, z) = u(x, \pi, z) = 0$$

$$u(0, y, z) = u(\pi, y, z) = 0.$$

6.14   GREEN'S FUNCTION FOR A DIRICHLET PROBLEM

In this section we will develop an integral representation for the solution of a Dirichlet problem, based on a function called the *Green's function*. We will do this for problems in R^3. This discussion can be adapted to problems in the plane.

Let Ω be a bounded domain in R^3 whose boundary $\partial\Omega$ is a piecewise smooth closed surface. If $u \in C^2(\bar\Omega)$ and $\nabla^2 u = 0$ in Ω, then we know that, at any $\mathbf{x} \in \Omega$,

$$u(\mathbf{x}) = \frac{1}{4\pi} \iint_{\partial\Omega} \left(\frac{1}{|\mathbf{y} - \mathbf{x}|} \frac{\partial u(\mathbf{y})}{\partial n} - u(\mathbf{y}) \frac{\partial}{\partial n} \frac{1}{|\mathbf{y} - \mathbf{x}|} \right) d\sigma_y.$$

Now refer to Green's second identity in R^3 (Lemma 5, Section 6.4). If v is any function that is also harmonic in Ω, then the volume integral in Green's second identity is zero and we obtain

$$\iint_{\partial\Omega} \left(v \frac{\partial u}{\partial n} - u \frac{\partial v}{\partial n} \right) d\sigma_y = 0.$$

Since this integral is zero, we can add it to the integral representation of u and obtain, after rearranging terms,

$$u(\mathbf{x}) = \iint_{\partial\Omega} \left(\frac{1}{4\pi} \frac{1}{|\mathbf{y} - \mathbf{x}|} + v(\mathbf{y}) \right) \frac{\partial u(\mathbf{y})}{\partial n} d\sigma_y$$

$$- \iint_{\partial\Omega} u(\mathbf{y}) \frac{\partial}{\partial n} \left(v(\mathbf{y}) + \frac{1}{4\pi} \frac{1}{|\mathbf{y} - \mathbf{x}|} \right) d\sigma_y. \quad (6.19)$$

Equation 6.19 represents $u(\mathbf{x})$ as a sum of two surface integrals. The first involves the normal derivative of u on the boundary of Ω, and no information is given about this quantity. We can cause this integral to be zero if we choose v so that the term in large parentheses in this integral is zero. This will leave the second surface integral in equation 6.19, which involves values of u on the boundary (these are given to us), and the normal derivative of a function that is known once we choose v.

Thus, choose v so that

$$\nabla^2 v(\mathbf{y}) = 0 \text{ in } \Omega$$

$$v(\mathbf{y}) = -\frac{1}{4\pi} \frac{1}{|\mathbf{y} - \mathbf{x}|} \text{ for } \mathbf{y} \text{ in } \partial\Omega.$$

For each $\mathbf{x}$ in Ω, this problem for v is a Dirichlet problem on Ω. Therefore v depends both on $\mathbf{y}$ (the name given to the variable in the problem for v) and on $\mathbf{x}$ (the point at which we are seeking to represent u). For this reason we will denote v as $v(\mathbf{x}, \mathbf{y})$.

Now define

$$G(\mathbf{x}, \mathbf{y}) = v(\mathbf{x}, \mathbf{y}) + \frac{1}{4\pi} \frac{1}{|\mathbf{y} - \mathbf{x}|}. \tag{6.20}$$

G is the *Green's function* for the Dirichlet problem for u. By equation 6.19 and the way G has been defined, the solution of the Dirichlet problem for u is

$$u(\mathbf{x}) = -\iint_{\partial\Omega} f(\mathbf{y}) \frac{\partial}{\partial n} G(\mathbf{x}, \mathbf{y}) \, d\sigma_y. \tag{6.21}$$

One can show that G is symmetric:

$$G(\mathbf{x}, \mathbf{y}) = G(\mathbf{y}, \mathbf{x}).$$

As a function of $\mathbf{y}$, G is harmonic in $\Omega_x = \Omega - \{\mathbf{x}\}$ and vanishes on $\partial\Omega$. Finally, as $\mathbf{y} \to \mathbf{x}$, $G(\mathbf{x}, \mathbf{y}) \to \infty$ in the same way that $1/|\mathbf{y} - \mathbf{x}|$ does.

EXERCISE 286 Adapt the discussion of this section to write the analogue of equation 6.21 for the Dirichlet problem for a bounded domain in the plane. In doing this it is necessary to write the R^2 analogue of the expression 6.20. *Hint:* Use a logarithm in adapting equation 6.20 to R^2.

EXERCISE 287 Poisson's equation for a domain Ω in R^3 has the form

$$\nabla^2 u(\mathbf{x}) = F(\mathbf{x}) \text{ on } \Omega.$$

This is Laplace's equation if $F(x)$ is identically zero. Consider the problem of solving Poisson's equation subject to the boundary condition

$$u(\mathbf{x}) = f(\mathbf{x}) \text{ for } \mathbf{x} \text{ in } \partial\Omega.$$

Beginning with the representation theorem (Theorem 17, Section 6.4), adapt the line of reasoning of this section to derive the solution

$$u(\mathbf{x}) = -\iiint_{\Omega} F(\mathbf{y}) G(\mathbf{x}, \mathbf{y}) \, dV_y - \iint_{\partial\Omega} f(\mathbf{y}) \frac{\partial}{\partial n} G(\mathbf{x}, \mathbf{y}) \, d\sigma_y$$

in which $G(\mathbf{x}, \mathbf{y})$ is the Green's function for the Dirichlet problem

$$\nabla^2 u = 0 \text{ in } \Omega; \ u = f \text{ on } \partial\Omega.$$

Assume that F and f are continuous and that Ω is a bounded domain.

EXERCISE 288   Use the representation theorem for R^2 to write an integral formula for the solution of Poisson's equation satisfying given boundary data, for a bounded domain in the plane.

6.14.1   Green's Function for a Sphere

Equation 6.21 reduces the problem of solving a Dirichlet problem on a bounded domain in R^3, to one of finding a Green's function for this domain. This is a formidable task in itself for many domains. Sometimes, however, a domain has a particular property, often symmetry, that enables us to construct a Green's function explicitly. We will illustrate this for a ball about the origin.

Let $\Omega = B(\mathbf{0}, a)$, the open ball of radius a about the origin in R^3. The first step in constructing the Green's function for Ω is to solve, for each $\mathbf{x}$ in Ω, the problem

$$\nabla^2 v(\mathbf{y}) = 0 \text{ for } \mathbf{y} \text{ in } \Omega$$

$$v(\mathbf{y}) = -\frac{1}{4\pi} \frac{1}{|\mathbf{y} - \mathbf{x}|} \text{ for } \mathbf{y} \text{ in } S(\mathbf{0}, a).$$

An ideal candidate for v might appear to be $-(1/4\pi)|\mathbf{y} - \mathbf{x}|$ itself, since this is a harmonic function and equals itself on $\partial\Omega$. The problem is that this function is not defined at $\mathbf{x}$ in Ω. We will therefore perform an inversion in the sphere. As Figure 6.14 suggests, an inversion ι maps a point $\mathbf{y}$ inside the ball $B(\mathbf{0}, a)$ to a point $\iota(\mathbf{y})$ outside and on the line from the origin through $\mathbf{y}$ and having the property that

$$|\mathbf{y}||\iota(\mathbf{y})| = a^2.$$

The inversion is defined by

$$\iota(\mathbf{y}) = \frac{a^2}{|\mathbf{y}|^2} \mathbf{y}$$

(see Exercise 245). This function maps the origin to ∞. Points on the sphere $S(\mathbf{0}, a)$ map to themselves.

Now take the function

$$-\frac{1}{4\pi} \frac{1}{|\mathbf{y} - \mathbf{x}|}$$

and replace $\mathbf{y}$ with its image under the inversion to form the function

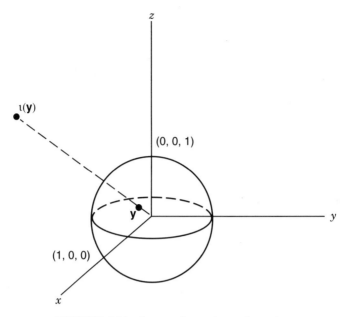

FIGURE 6.14   Image of **y** under an inversion.

$$-\frac{1}{4\pi}\frac{1}{\left|\dfrac{a^2}{|\mathbf{y}|^2}\mathbf{y}-\mathbf{x}\right|}.$$

This can be written

$$-\frac{1}{4\pi}\frac{|\mathbf{y}|}{a}\frac{1}{\left|\dfrac{a}{|\mathbf{y}|}\mathbf{y}-\dfrac{|\mathbf{y}|}{a}\mathbf{x}\right|}$$

and it is easy to check that this function is harmonic in Ω as a function of **y** (keep in mind that $\iota(\mathbf{y})$ is outside the ball if **y** is inside). Further, if we omit the factor $|\mathbf{y}|/a$, the resulting function

$$-\frac{1}{4\pi}\frac{1}{\left|\dfrac{a}{|\mathbf{y}|}\mathbf{y}-\dfrac{|\mathbf{y}|}{a}\mathbf{x}\right|}$$

is also harmonic in Ω. This function is equal to $-(1/4\pi)|\mathbf{y}-\mathbf{x}|$ if **y** is on the sphere, because then $|\mathbf{y}|=a$. Therefore choose

$$v(\mathbf{x}, \mathbf{y}) = -\frac{1}{4\pi} \frac{1}{\left|\frac{a}{\|\mathbf{y}\|}\mathbf{y} - \frac{|\mathbf{y}|}{a}\mathbf{x}\right|}.$$

The Green's function for the ball is

$$G(\mathbf{x}, \mathbf{y}) = \frac{1}{4\pi} \frac{1}{|\mathbf{y} - \mathbf{x}|} - \frac{1}{4\pi} \frac{1}{\left|\frac{a}{\|\mathbf{y}\|}\mathbf{y} - \frac{|\mathbf{y}|}{a}\mathbf{x}\right|},$$

and the solution of the Dirichlet problem for $B(\mathbf{0}, a)$ is

$$u(\mathbf{x}) = -\iint\limits_{S(0,a)} f(\mathbf{y}) \frac{\partial}{\partial n} G(\mathbf{x}, \mathbf{y}) \, d\sigma_y.$$

This defines a function that is harmonic on Ω and equals f on the sphere.
  It is possible to show that

$$\frac{\partial G(\mathbf{x}, \mathbf{y})}{\partial n} = -\frac{1}{4\pi a} \frac{a^2 - r^2}{(a^2 + r^2 - 2ar\cos(\Theta))^{3/2}} \tag{6.22}$$

in which $\Theta(\mathbf{x}, \mathbf{y})$ is the angle between $\mathbf{x}$ and $\mathbf{y}$, and $r = |\mathbf{x}|$. This gives the solution of the Dirichlet problem for $B(\mathbf{0}, a)$ in R^3 as

$$u(\mathbf{x}) = \frac{1}{4\pi a} \iint\limits_{S(0,a)} \frac{a^2 - r^2}{(a^2 + r^2 - 2ar\cos(\Theta))^{3/2}} f(\mathbf{y}) \, d\sigma_y \tag{6.23}$$

and this is the three-dimensional analogue of Poisson's integral formula 6.11 for a disk in R^2. It is possible to derive an analogous integral formula for the solution of the Dirichlet problem for an open ball in R^n. This proves the existence of a solution of the Dirichlet problem for an open ball in R^n, assuming that the function given on the boundary is continuous.

EXERCISE 289   Derive equation 6.22.

EXERCISE 290   Prove that, for $0 \leq r < 1$,

$$\iint\limits_{S(0,1)} \frac{1 - r^2}{(1 + r^2 - 2r\cos(\Theta))^{3/2}} \, d\sigma_y = 4\pi.$$

EXERCISE 291   Show that the solution of the Dirichlet problem for the disk $B(\mathbf{0}, a)$ can be written

$$u(\mathbf{x}) = -\oint_{\partial\Omega} f(\mathbf{y}) \frac{\partial}{\partial n} G(\mathbf{x}, \mathbf{y}) \, ds_y,$$

where

$$G(\mathbf{x}, \mathbf{y}) = \frac{1}{2\pi} \left[\ln\left(\frac{1}{|\mathbf{y} - \mathbf{x}|}\right) - \ln\left(\frac{1}{\left|\frac{a}{|\mathbf{y}|}\mathbf{y} - \frac{|\mathbf{y}|}{a}\mathbf{x}\right|}\right) \right]$$

is the Green's function for a disk in R^2.

EXERCISE 292   Use the Green's function for a disk (Exercise 291) to give another derivation of Poisson's integral formula for the solution of the Dirichlet problem for a disk.

EXERCISE 293   Use the integral expression in equation 6.23 to prove a three-dimensional version of Harnack's inequality (inequality 6.13, Section 6.10).

EXERCISE 294   Use the result of Exercise 293 to prove on R^3-version of Liouville's Theorem.

EXERCISE 295   Use the results of Exercise 287 (Section 6.14) and 291 to write an integral formula for the solution of the problem:

$$\nabla^2 u(x) = F(x) \text{ in } \Omega$$

$$u(x) = f(x) \text{ on } \partial\Omega,$$

where Ω is the disk of radius ρ about the origin in R^2. Convert this integral solution to polar coordinates.

EXERCISE 296   Write an integral expression for the solution of the problem

$$\nabla^2 u(r, \theta) = \cos(\theta) \text{ for } 0 \le r < \rho, 0 \le \theta \le 2\pi$$

$$u(\rho, \theta) = 1 \text{ for } 0 \le \theta \le 2\pi.$$

EXERCISE 297   Write an integral expression for the solution of

$$\nabla^2 u(r, \theta) = 1 \text{ for } 0 \le r < \rho, 0 \le \theta \le 2\pi$$

$$u(\rho, \theta) = 1 \text{ for } 0 \le \theta \le 2\pi.$$

6.14.2   The Method of Electrostatic Images

Sometimes a physical interpretation of a mathematical concept provides insight into the solution of equations or the determination of important functions. This happens with Green's functions for certain domains in R^3.

The Green's function $G(\mathbf{x}, \mathbf{y})$ for a domain Ω in R^3 can be interpreted as the electrostatic potential resulting from a unit charge at $\mathbf{x}$ in Ω, with the bounding surface $\partial\Omega$ a grounded conducting surface (the electrostatic potential is zero on such a surface). Now, recall that

$$G(\mathbf{x}, \mathbf{y}) = v(\mathbf{x}, \mathbf{y}) + \frac{1}{4\pi}\frac{1}{|\mathbf{y} - \mathbf{x}|}.$$

The second term on the right side of this expression is the potential due to a unit charge at $\mathbf{x}$. The term $v(\mathbf{x}, \mathbf{y})$ is the potential due to the induced charge distribution on $\partial\Omega$. We can therefore find $v(\mathbf{x}, \mathbf{y})$, and hence the Green's function for Ω, if we can find the induced charge distribution on $\partial\Omega$.

As any student in a classical electricity and magnetism course can testify, this is usually not an easy task. However, there is a clever way of approaching this difficulty, which sometimes meets with success. We can also think of $v(\mathbf{x}, \mathbf{y})$ as the potential due to "imaginary" charges placed at appropriate points outside of Ω. These imaginary charges are called *electrostatic images* of the unit charges in Ω, and the idea is to locate them outside of Ω so that the potential $v(\mathbf{x}, \mathbf{y})$ caused by these charges satisfies

$$v(\mathbf{x}, \mathbf{y}) = -\frac{1}{4\pi}\frac{1}{|\mathbf{y} - \mathbf{x}|}$$

for $\mathbf{y}$ in $\partial\Omega$. The idea of determining $v(\mathbf{x}, \mathbf{y})$ in this way is called the *method of electrostatic images*.

It would be nice if we could now solve all Dirichlet problems, but of course this is not realistic. However, in cases where Ω has some property which enables us to place the electrostatic images, we can write the Green's function and use equation 6.21 to obtain an integral formula for the solution of the Dirichlet problem for Ω.

Example 45   Let Ω be the half-space consisting of all points above the x,y-plane. Thus Ω consists of all points (x, y, z) with $z > 0$. We will determine the Green's function for Ω.

Here $\partial\Omega$ is the x,y-plane, consisting of points $(x, y, 0)$. Corresponding to a unit charge at $\mathbf{x} = (x, y, z)$ in Ω, place an imaginary unit charge at $\mathbf{x}^* = (x, y, -z)$ placed symmetrically with respect to $\mathbf{x}$ across the x,y-plane. These two charges cancel each other, and the potential due to their joint effect is zero on $\partial\Omega$. Therefore the Green's function for Ω is

$$G(\mathbf{x}, \mathbf{y}) = \frac{1}{4\pi} \frac{1}{|\mathbf{y} - \mathbf{x}|} - \frac{1}{4\pi} \frac{1}{|\mathbf{y} - \mathbf{x}^*|}.$$

Since

$$|\mathbf{y} - \mathbf{x}| = |\mathbf{y} - \mathbf{x}^*|$$

if $\mathbf{y}$ is in the x,y-plane, then indeed $G(\mathbf{x}, \mathbf{y}) = 0$ for $\mathbf{y}$ in $\partial\Omega$, as required. ∎

EXERCISE 298 Use the method of electrostatic images to derive the Green's function for the unit ball in R^3, obtained in Section 6.14.1 using the inversion mapping. Explain how the use of the inversion mapping and the method of electrostatic images for the sphere are really different ways of using the same idea.

EXERCISE 299 Let Ω consist of all (x, y, z) with $y > 0$ and $z > 0$. Determine the Green's function for Ω. *Hint:* Corresponding to a positive unit charge at (x, y, z) in Ω, place a negative unit charge at $(x, -y, z)$, then balance this with a positive unit charge at $(x, -y, -z)$, and finally balance the entire system with a negative unit charge at $(x, y, -z)$.

EXERCISE 300 Find the Green's function for the domain bounded by two parallel planes $z = 0$ and $z = 1$. *Hint:* Place an imaginary charge so that each plane $z = 0, \pm 1, \pm 2, \ldots$ has zero potential.

EXERCISE 301 Consider the Dirichlet problem for the upper half-space:

$$\nabla^2 u(x, y, z) = 0 \text{ for } z > 0; \ u(x, y, 0) = f(x, y).$$

Use the Green's function found in Example 45 for the upper half-space, together with equation 6.21, to derive the solution

$$u(x, y, z) = \frac{z}{2\pi} \int_{-\infty}^{\infty} \int_{-\infty}^{\infty} \frac{f(\xi, \eta)}{[(\xi - x)^2 + (\eta - y)^2 + z^2]^{3/2}} \, d\xi \, d\eta.$$

The motivating notion of an electrostatic charge would seem to restrict its use to domains in R^3. However, once we understand the use of imaginary charges (really an exploitation of symmetry) to construct Green's functions for certain domains in 3-space, we can jettison the physical motivation and use the method for certain domains in the plane. However, now we must replace the potential

$$\frac{1}{4\pi} \frac{1}{|\mathbf{y} - \mathbf{x}|}$$

in R^3 due to a unit charge at $\mathbf{x}$, with the potential

$$\frac{1}{2\pi} \ln \left(\frac{1}{|\mathbf{y} - \mathbf{x}|} \right)$$

which we have found to be appropriate for problems in the plane.

EXERCISE 302   Use the last remark and the method of electrostatic images to find the Green's function for the unit disk. Use this Green's function and the two-dimensional analogue of equation 6.21 (Exercise 286) to derive an integral solution of the Dirichlet problem for the unit disk. How does this expression compare with the Poisson integral formula?

EXERCISE 303   Find the Green's function for the upper half-plane $y > 0$ and use this and the two-dimensional analogue of equation 6.21 to write an integral solution for the Dirichlet problem for the upper half plane. Compare this ex- pression with a previously derived integral solution of this problem (Equation 6.16).

EXERCISE 304   Find the Green's function for the domain

$$\Omega = \{(x, y) | 0 < y < 1\}.$$

This is a horizontal strip in the x,y-plane. *Hint:* Recall finding the Green's function for the domain bounded by two parallel planes in R^3, Exercise 300.

EXERCISE 305   Find the Green's function for the right quarter plane $x > 0$, $y > 0$. Use this to write an integral formula for the solution of the Dirichlet problem for the right quarter plane. Compare this solution with the solution (equation 6.18) obtained in Section 6.12.

6.15   CONFORMAL MAPPING TECHNIQUES

Assuming some background in complex analysis, we will discuss the use of conformal mappings to write integral formulas for solutions of Dirichlet prob- lems. Without loss of continuity the reader can skip over this section and pro- ceed to the Neumann problem.

   One strategy for solving certain kinds of problems is to find the solution for a simple domain, such as an open disk or rectangle, then map this domain conformally onto a domain of interest. This mapping may then carry the so- lution for the simple domain to a solution for the given domain. We will recall some facts about conformal mappings and then show how this strategy can be used to solve Dirichlet problems.

6.15.1 Review of Conformal Mappings

For this section, we consider functions mapping sets of complex numbers to sets of complex numbers—that is, complex valued functions of a complex variable. Let f be such a function. In thinking of f from a geometric point of view, it is convenient to make two copies of the complex plane, one for the complex variable z, the independent variable of f, and the other for $w = f(z)$. As z varies over a given set of points in the z-plane, we can observe how $f(z)$ varies over a corresponding set in the w-plane. In this way f can be thought of as mapping given domains in the z-plane to sets in the w-plane, and we refer to f as a mapping.

Now suppose f is defined on a set D of points in the z-plane, and let K be a set of complex numbers in the w-plane. We say that f *maps D into K*, and write $f:D \to K$, if $f(z) \in K$ for each $z \in D$. This means that every point z of D has its image $f(z)$ in K.

The mapping is *onto* if every point of K is the image of some point in D under this mapping. This occurs if, for each $w \in K$, there is some $z \in D$ with $w = f(z)$.

The mapping is *one-to-one* if f maps distinct points to distinct points. This means that $f(z_1) = f(z_2)$ can occur only if $z_1 = z_2$.

The concepts of onto and one-to-one are independent. A mapping may be onto and one-to-one, or one-to-one and not onto, or onto and not one-to-one, or neither onto nor one-to-one.

Now we come to two special features that distinguish conformal mappings.

We say that $f:D \to K$ *preserves angles* if, for any z in D, any two smooth curves in D intersecting at an angle θ at z have images in K that intersect at $f(z)$ at the same angle. This idea is illustrated in Figure 6.15.

We define f to be *orientation preserving* if a counterclockwise rotation in D is mapped by f to a counterclockwise rotation in K. This is illustrated in Figure 6.16(a), where the counterclockwise sense of orientation from L_1 to L_2 is mapped to a similarly counterclockwise orientation from the images of these lines. By contrast, Figure 6.16(b) illustrates a mapping that is not orientation preserving.

If f is both orientation and angle preserving, then we call f a *conformal mapping*.

As restrictive as these definitions may appear, it is easier to produce examples of conformal mappings than of nonconformal mappings, as the following theorem suggests.

Theorem 24   Let $f:D \to K$ be an analytic function mapping a domain D onto a domain K. Suppose $f'(z) \neq 0$ for every $z \in D$. Then f is a conformal mapping. ∎

Thus, for analytic mappings between domains, nonvanishing of the derivative is sufficient to guarantee that the mapping preserves both angles and orientation.

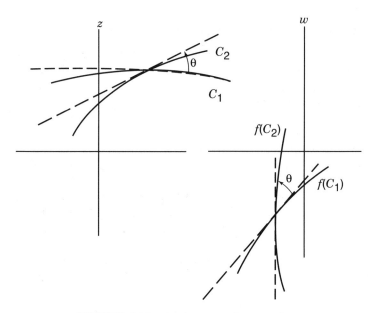

FIGURE 6.15 Angle-preserving mapping.

It is routine to show that a composition of conformal mappings is again conformal. This is geometrically apparent. If f is a conformal mapping of D onto K, and g is a conformal mapping of K onto S (Figure 6.17), then each mapping preserves orientation and angles, so the composition $g \circ f$ will also preserve orientation and angles and also be conformal. It is also possible to show that the inverse of a one-to-one conformal mapping is conformal.

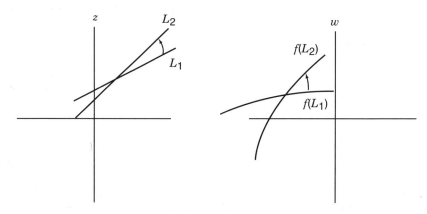

FIGURE 6.16(a) Orientation-preserving mapping.

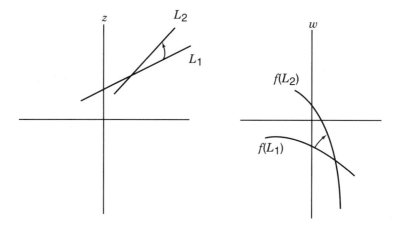

FIGURE 6.16(b)   Orientation-reversing mapping.

Constructing a conformal mapping between two given domains may be a daunting task. However, a classical theorem due to Riemann leads to the existence of such a mapping under very broad conditions.

Theorem 25 (Riemann Mapping Theorem)   Let K be a domain in the w-plane and assume that K is not the entire complex plane. Then there exists a one-to-one conformal mapping of the unit disk $|z| < 1$ onto K.   ∎

Of course, this only gives a conformal mapping of the unit disk onto a given domain. However, suppose we want to map a domain D one-to-one onto a domain K, and K is not the entire plane. Put a copy of the unit disk in another plane between the z- and w-planes, as in Figure 6.18. By the Riemann Mapping Theorem, there is a one-to-one conformal mapping F of this unit disk onto K, and there is also a one-to-one conformal mapping G in the other direction of this unit disk onto D. Then G^{-1} is a one-to-one conformal mapping of D onto

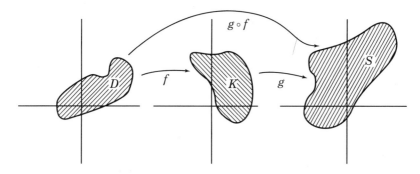

FIGURE 6.17   A composition of conformal mappings is conformal.

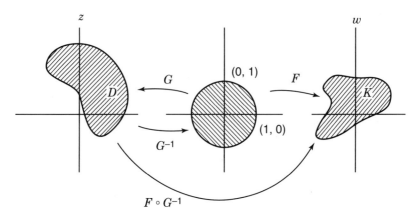

FIGURE 6.18 Constructing a conformal mapping from D to K, using the Riemann Mapping Theorem.

the unit disk, and the composition $F \circ G^{-1}$ is a one-to-one conformal mapping of D onto K.

EXERCISE 306 Let $f(z) = z^2$. Prove that f maps the z-plane onto the w-plane, and is not one-to-one. Show that the image of a vertical line $x = a$ and of a horizontal line $y = b$ are both parabolas in the w-plane. How do these parabolas differ? *Hint:* Write $z = x + iy$ and $w = u + iv$, so

$$f(z) = (x + iy)^2 = x^2 - y^2 + 2ixy = u + iv.$$

Thus in general $u = x^2 - y^2$ and $v = 2xy$. Now for the case $x = a$, determine a relationship between u and v, yielding the image of the vertical line in the w-plane. Similarly, determine a relationship between u and v when $y = b$ to obtain the image of a horizontal line.

EXERCISE 307 Let $f(z) = e^z$. Show that a vertical line $x = a$ maps to a circle of radius e^a about the origin in the w-plane, and a horizontal line $y = b$ maps to a half line from the origin, making an angle of b radians with the positive real axis in the w-plane.

EXERCISE 308 Let $f(z) = \cos(z)$. Show that a vertical line $x = a$ maps to a branch of the hyperbola

$$\frac{u^2}{\cos^2(a)} - \frac{v^2}{\sin^2(a)} = 1$$

provided that $\cos(a)$ and $\sin(a)$ are nonzero. Show that the image of a horizontal line $y = b$ is the ellipse

$$\frac{u^2}{\cosh^2(b)} + \frac{v^2}{\sinh^2(b)} = 1$$

provided that $b \neq 0$.

EXERCISE 309   Show that the mapping

$$w = f(z) = \frac{1}{2}\left(z + \frac{1}{z}\right)$$

maps the circle $|z| = r$ onto an ellipse with foci ± 1.

EXERCISE 310   Show that the mapping $f(z) = \bar{z}$ is not conformal. Here, if $z = x + iy$, then $\bar{z} = x - iy$ is the conjugate of z.

6.15.2   Bilinear Transformations

We have just observed that there always exists a one-to-one conformal mapping of a given domain onto another domain that is not the entire plane. Now suppose we are given two such domains and we want to produce a conformal mapping between them. There is no formula or set procedure for doing this, but there are some techniques that are useful to know. In this section we will develop an important class of conformal mappings.

A *bilinear*, or *linear fractional transformation*, is a function of the form

$$T(z) = \frac{az + b}{cz + d},$$

with a, b, c, and d complex numbers and $ad - bc \neq 0$. This condition assures that the mapping is one-to-one and has an inverse mapping. $T(z)$ is defined for all z except $-d/c$. Further,

$$T'(z) = \frac{ad - bc}{(cz + d)^2}$$

and this is nonzero for $z \neq -d/c$, hence T is a conformal mapping defined on the complex plane with the number $-d/c$ deleted.

There are three special bilinear transformations which are fundamental in a sense we will soon explain.

Translation   A mapping $T(z) = z + b$ is called a *translation*. If $b = \alpha + i\beta$, the effect of applying this mapping to z is to shift z horizontally by α and vertically by β. For example, $T(z) = z + 2 - i$ takes each complex number and moves it two units to the right and one unit down.

Rotation/Magnification   A mapping

$$T(z) = az$$

is called a *rotation/magnification* if a is a nonzero complex number. If we write a and z in polar form as $a = Ae^{i\alpha}$, and $z = re^{i\theta}$, then

$$T(z) = az = Are^{i(\theta + \alpha)}.$$

Thus the image point $T(z)$ has magnitude Ar and argument $\theta + \alpha$. The effect of T on z is to stretch (if $A > 1$) or shrink (if $0 < A < 1$) the distance from the origin to z, and rotate the line from the origin to z through an angle of α radians (counterclockwise if α is positive, clockwise if α is negative). This is shown in Figure 6.19(a) for the case $A > 0$ and $\alpha > 0$, and in Figure 6.19(b) for $0 < A < 1$ and $\alpha < 0$.

Inversion   The mapping $T(z) = 1/z$ is an *inversion*. It maps points inside the unit circle (except the origin) to the exterior, and vice versa (Figure 6.20). If w is on the unit circle, then $1/w$ is on the unit circle at the other intersection of the circle with the vertical line through w.

We will now show that any bilinear transformation can be achieved as a sequence of transformations of these three kinds. Suppose

$$T(z) = \frac{az + b}{cz + d}$$

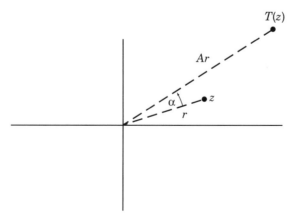

FIGURE 6.19(a)   Magnification (by a factor greater than 1) and rotation (counterclockwise).

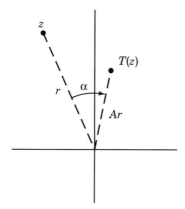

FIGURE 6.19(b)   Magnification (by a factor less than 1) and rotation (clockwise).

and consider two cases. If $c = 0$, then

$$T(z) = \frac{a}{d} z + \frac{b}{d}$$

and we can think of this as the end result of the sequence of mappings

$$z \underset{rot/mag}{\longrightarrow} \frac{a}{d} z \underset{translation}{\longrightarrow} \frac{a}{d} z + \frac{b}{d} = T(z).$$

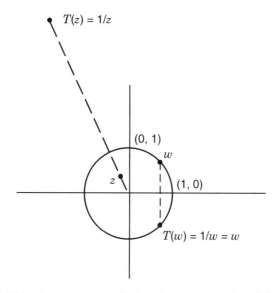

FIGURE 6.20   Inversion maps the interior to the exterior of the unit disk.

If $c \neq 0$, then write

$$z \underset{rot/mag}{\rightarrow} cz \underset{translation}{\rightarrow} cz + d \underset{inversion}{\rightarrow} \frac{1}{cz + d} \underset{rot/mag}{\rightarrow} \frac{bc - ad}{c} \frac{1}{cz + d}$$

$$\underset{translation}{\rightarrow} \frac{bc - ad}{c} \frac{1}{cz + d} + \frac{a}{c} = \frac{az + b}{cz + d} = T(z).$$

We do not recommend memorizing these sequences. The important point is that bilinear transformations can be "factored" into simpler components. This is sometimes useful in analyzing a particular transformation, or constructing a transformation to certain specifications. This way of decomposing bilinear transformations is also important in understanding properties of these mappings.

Theorem 26 Let T be a bilinear transformation. Then T maps a circle to a circle or straight line, and a straight line to a circle or straight line. ■

It is obvious that a translation maps a circle to a circle and a line to a line. Similarly, a rotation/magnification maps a circle to a circle and a line to a line. We need to determine the effect of an inversion on a circle or line. Consider an arbitrary circle or line in the plane, having equation

$$A(x^2 + y^2) + Bx + Cy + R = 0.$$

This is a straight line if $A = 0$ and B and C are not both zero, and a circle if $A \neq 0$. If $z = x + iy$ this equation can be written

$$A|z|^2 + \frac{B}{2} (z + \bar{z}) + \frac{C}{2i} (z - \bar{z}) + R = 0.$$

Now let $w = 1/z$ to see the effect of an inversion on this equation. The image of the locus of this equation in the w-plane is

$$A \frac{1}{|w|^2} + \frac{B}{2} \left(\frac{1}{w} + \frac{1}{\bar{w}} \right) + \frac{C}{2i} \left(\frac{1}{w} - \frac{1}{\bar{w}} \right) + R = 0.$$

Upon multiplying by $w\bar{w}$, which is the same as $|w|^2$, this equation becomes

$$R|w|^2 + \frac{B}{2} (w + \bar{w}) - \frac{C}{2i} (w - \bar{w}) + A = 0,$$

and this is a circle if $R \neq 0$ and a line if $R = 0$ and B and C are not both zero.

With this as background, we now claim that we can always produce a bilinear transformation mapping three given points to three given points.

Theorem 27 Let z_1, z_2, z_3 be distinct complex numbers, and w_1, w_2, w_3 distinct complex numbers. Then there is a bilinear transformation T such that $T(z_j) = w_j$ for $j = 1, 2, 3$. ■

To obtain such a transformation, think of $w = T(z)$ and solve for w in the equation

$$(w_1 - w)(w_3 - w_2)(z_1 - z_2)(z_3 - z) = (z_1 - z)(z_3 - z_2)(w_1 - w_2)(w_3 - w).$$

For example, suppose we want to map $3 \rightarrow i$, $1 - i \rightarrow 4$ and $2 - i \rightarrow 6 + 2i$. Let

$$z_1 = 3, z_2 = 1 - i, z_3 = 2 - i, w_1 = i, w_2 = 4, w_3 = 6 + 2i$$

and consider the equation

$$(i - w)(2 + 2i)(2 + i)(2 - i - z) = (3 - z)(1)(i - 4)(6 + 2i - w).$$

Upon solving for w we obtain

$$w = T(z) = \frac{(20 + 4i)z - (16i + 68)}{(6 + 5i)z - (22 + 7i)}.$$

It is routine to check that this mapping sends the three given points to their respective targets.

EXERCISE 311 Show that the inversion $T(z) = 1/z$ maps the vertical line $x = a \neq 0$ to the circle

$$\left(u - \frac{1}{2a}\right)^2 + v^2 = \frac{1}{4a^2}.$$

EXERCISE 312 Show that the inverse of a bilinear mapping

$$T(z) = \frac{az + b}{cz + d}$$

is also a bilinear mapping.

EXERCISE 313 In each of 1 through 6, find a bilinear mapping that sends th given points to the indicated images.

1. $1 \rightarrow 1, 2 \rightarrow -i, 3 \rightarrow 1 + i$
2. $i \rightarrow i, 1 \rightarrow -i, 2 \rightarrow 0$

3. $1 \rightarrow 1 + i, 2i \rightarrow 3 - i, 4 \rightarrow 4$
4. $6 + i \rightarrow 2 - i, i \rightarrow 3i, 5 \rightarrow -i$
5. $1 \rightarrow 6 - 4i, 1 + i \rightarrow 2, 3 + 4i \rightarrow -2$
6. $2 \rightarrow -3i, 1 \rightarrow 1 - i, 2 + i \rightarrow 0$

EXERCISE 314   In each of 1 through 6, write the bilinear mapping as a sequence of mappings, each of which is a translation, rotation/magnification, or inversion.

1. $$T(z) = \frac{iz - 4}{z}$$

2. $$T(z) = \frac{z - 1}{z + 3 + i}$$

3. $$T(z) = \frac{z - 4}{2z + i}$$

4. $$T(z) = i(z + 6) - 2 + i$$

5. $$T(z) = \frac{(-2 + 3i)z}{z + 4}$$

6. $$T(z) = \frac{6i}{z + 8}$$

6.15.3   Construction of Conformal Mappings Between Domains

From the Riemann Mapping Theorem, "almost" any two domains can be mapped onto one another by a conformal mapping. This does not make it easy to find a mapping between two given domains. In attempting to construct such a mapping, the following observation is crucial: *A conformal mapping of a domain D onto a domain K will map the boundary of D to the boundary of K.*

We use this fact as follows. Suppose D is bounded by a piecewise smooth curve C (not necessarily closed) which separates the z-plane into two domains, D and $\mathcal{D}$. These are called *complementary domains*. Similarly, suppose K is bounded by a piecewise smooth curve P in the w-plane, separating this plane into complementary domains K and $\mathcal{K}$ (Figure 6.21). Try to find a conformal mapping f that sends points of C to points of P. This may be easier than trying

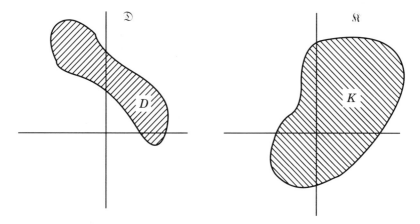

FIGURE 6.21   Pairs of complementary domains.

to find a mapping of the entire domain directly. Such an f will then send D either to K or to $\mathfrak{N}$. To determine which, choose any convenient $z_0 \in D$. If $f(z_0)$ is in K, then $f:D \to K$ and f is a suitable conformal mapping. If $f(z_0)$ is in $\mathfrak{N}$, then $f:D \to \mathfrak{N}$ and f is not the mapping we want. However, it is often possible to produce the conformal mapping we want from f, sometimes by using an inversion.

To sum, first try to produce a mapping between boundaries, then use this to define a conformal mapping between the given domains. We will consider some examples.

Example 46   Map the disk $|z| < 1$ conformally onto the disk $|w| < 3$.

Clearly all we need to do here is expand the unit disk. Use the magnification $f(z) = 3z$ (Figure 6.22). ∎

Notice that this mapping carries the boundary $|z| = 1$ of D onto the boundary $|w| = 3$ of K. Here C is the unit circle $|z| = 1$, and this separates the z-plane into the complementary domains $|z| < 1$ and $|z| > 1$. The complementary domains in the w-plane are $|w| < 3$ and $|w| > 3$, both having the circle $P:|w| = 3$ as boundary. Now pick any point in D, say $z = 0$. Since $f(0) = 0$ is in $|w| < 3$, then f maps $|z| < 1$ onto $|w| < 3$. f also maps $|z| = 1$ onto $|w| = 3$. Although this conclusion is obvious in this simple example, it serves to illustrate the discussion of the preceding paragraph.

Example 47   Map $|z| < 1$ conformally onto $|w| > 4$.

We know that an inversion maps the interior of the unit disk to its exterior, so all we have to do is combine an inversion with a magnification by 4 to expand the radius to 4 (Figure 6.23). This suggests the mapping

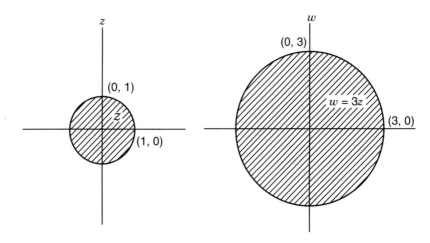

FIGURE 6.22   Mapping $|z| < 1$ onto $|w| < 3$.

$$w = f(z) = \frac{4}{z}.$$ ■

To illustrate again the preceding remarks, notice that the boundary circle $|z| = 1$ maps to the circle $|w| = 4$. This circle bounds two complementary domains in the w-plane, namely the interior of this circle, and its exterior. Since $f(1/2) = 8$ is exterior to $|w| = 4$, then f maps $|z| < 1$ to the exterior of $|w| = 4$, as we want.

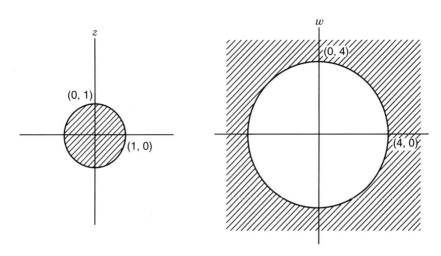

FIGURE 6.23   Mapping $|z| < 1$ onto $|w| > 4$.

Example 48   Map the unit disk $|z| < 1$ onto the disk $|w - i| < 3$ of radius 3 centered at i.

We want to expand the unit disk by a factor of 3 and move it up one unit to have center i. Construct the mapping in two steps, as suggested in Figure 6.24. Place an intermediate ζ-plane between the z- and w-planes and first map

$$\zeta = 3z.$$

This expands the unit disk to the disk of radius 3 about the origin in the ζ-plane. Now translate this circle one unit vertically to center it at i in the w-plane by putting

$$w = \zeta + i.$$

The mapping we want is the composition of these mappings:

$$w = \zeta + i = 3z + i.$$

This takes $|z| = 1$ to $|w - i| = 3$, since

$$|w - i| = |3z| = 3|z| = 3$$

if $|z| = 1$.   ■

Example 49   Map the right half-plane $\mathrm{Re}(z) > 0$ conformally onto the unit disk $|w| < 1$.

The domains are shown in Figure 6.25. The boundary of the half-plane is the imaginary axis $\mathrm{Re}(z) = 0$. We will map this line onto the circle $|w| = 1$, the boundary of the target domain. One way to do this is to pick three points on $\mathrm{Re}(z) = 0$ and map them to three points on $|w| = 1$. There is, however, a subtlety. To maintain positive orientation (counterclockwise on closed curves), choose three points in succession down the imaginary axis, so a person walking along these points sees the right half-plane on the left, and map these to points chosen counterclockwise (positive orientation) around $|w| = 1$ (a person walking around this circle counterclockwise sees the domain $|w| < 1$ on the left). To save on calculation, give a little thought to picking "simple" points.

With this in mind, choose, say, i, 0, and $-i$ moving down the vertical axis, and 1, i, and -1 in the counterclockwise order around $|w| = 1$. Of course, infinitely many other choices are possible. Thus we will seek a bilinear transformation mapping

$$i \rightarrow 1, \; 0 \rightarrow i, \; -i \rightarrow -1.$$

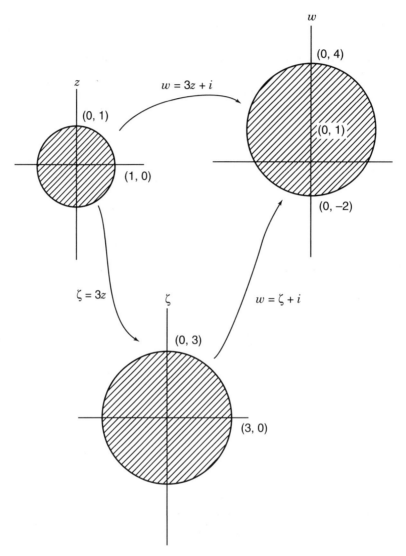

FIGURE 6.24   Constructing a conformal mapping of $|z| < 1$ onto $|w - i| < 3$.

Solve for w in the equation

$$(1 - w)(-1 - i)(i)(-i - z) = (i - z)(-i)(1 - i)(-1 - w)$$

to obtain

$$w = T(z) = -i \left(\frac{z - 1}{z + 1} \right).$$

This mapping must take the right half-plane either to $|w| < 1$ or to its comple-

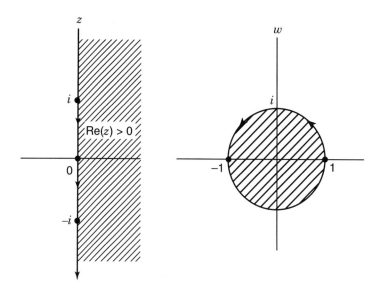

FIGURE 6.25   Constructing a conformal mapping of the right half plane onto the unit disk.

mentary domain $|w| > 1$. To see which, choose a point in the right half-plane, say $z = 1$. Now

$$T(1) = 0$$

is in $|w| < 1$, so T is a conformal mapping sending $\text{Re}(z) > 0$ onto $|w| < 1$. ■

Suppose in the last example we had wanted to map $\text{Re}(z) > 0$ to the exterior of the unit circle, $|w| > 1$. Using the mapping just found onto the interior of the unit circle, we can now perform an inversion to obtain

$$f(z) = \frac{1}{T(z)} = i\left(\frac{z+1}{z-1}\right).$$

For example, $1 + i$ is in the right half-plane, and its image under this mapping is

$$f(1 + i) = i\left(\frac{2+i}{i}\right) = 2 + i,$$

which lies in $|w| > 1$, as expected.

Example 50   Map the right half-plane $\text{Re}(z) > 0$ conformally onto the disk $|w - i| < 3$.

We can do this using a composition of mapping already formed. Put an intermediate ζ-plane between the z- and w-planes, as we did in Example 48. Map $\text{Re}(z) > 0$ onto $|\zeta| < 1$ by using the last example:

$$\zeta = -i\left(\frac{z-1}{z+1}\right).$$

Next map $|\zeta| < 1$ onto $|w - i| < 3$ by

$$w = 3\zeta + i.$$

Finally, form the composition

$$w = 3\zeta + i = -3i\left(\frac{z-1}{z+1}\right) + i = 2i\,\frac{2-z}{z+1}.$$

The sequence of mappings is displayed in Figure 6.26.   ∎

We do not want to suggest that bilinear transformations are all that are needed to construct a conformal mapping between given domains. For domains bounded by polygons, there is a mapping called the Schwarz-Christoffel transformation (an integral) which is often used, and for other domains a full range of complex functions may be required. A full treatment of conformal mappings would take us far away from our main purpose, so we will now return to the objective of solving Dirichlet problems.

EXERCISE 315   Consider the mapping $f(z) = \sin(z)$. Show that the half line $x = -\pi/2$, $y \geq 0$ maps to the interval $(-\infty, -1]$ on the real axis in the w-plane; that the half-line $x = \pi/2$, $y \geq 0$ maps to the interval $[1, \infty)$; and that the interval $[-\pi/2, \pi/2]$ on the real axis in the z-plane maps to $[-1, 1]$. Let S be the strip consisting of all z with $-\pi/2 < \text{Re}(z) < \pi/2$ and $\text{Im}(z) > 0$ in the z-plane. The boundary of this strip consists of the three line segments just mapped. Thus the boundary of S maps to the real axis in the w-plane. Using this fact and the above discussion of boundaries and mappings of domains, explain why f maps S onto the upper half plane $\text{Im}(w) > 0$ in the w-plane.

EXERCISE 316   Determine the image of the rectangle $0 \leq x \leq \pi$, $0 \leq y \leq \pi$ under the mapping $w = e^z$.

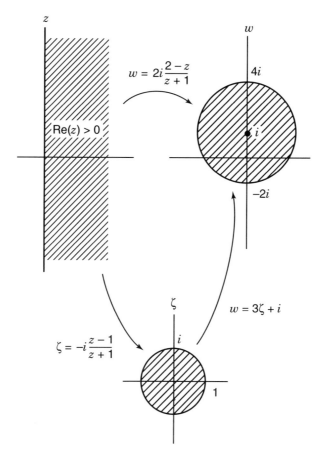

FIGURE 6.26   Constructing a conformal mapping of the right half plane onto $|w - i| < 3$.

EXERCISE 317   Determine the image of the rectangle $\pi/2 \leq x \leq \pi$, $1 \leq y \leq 3$ under the mapping $w = \cos(z)$.

EXERCISE 318   Find a conformal mapping of $|z| < 3$ onto $|w - 1 + i| > 6$. *Hint:* Pick three points in counterclockwise order on $|z| = 3$, and map these by a bilinear transformation to three points chosen in clockwise order on $|w - 1 + i| = 6$. There are infinitely many ways to do this.

EXERCISE 319   Find a conformal mapping of $|z + 2i| < 1$ onto $|w - 3| > 2$.

EXERCISE 320   Find a conformal mapping of $\text{Re}(z) > 1$ onto $\text{Im}(w) > -1$. *Hint:* Choose three points (in the right order) along the boundary of $\text{Re}(z) > 1$

and map them to three chosen points (in the right order) along the boundary of $\text{Im}(w) > -1$.

EXERCISE 321   Find a conformal mapping of $\text{Re}(z) < -4$ onto $|w + 1 - 2i| > 3$.

EXERCISE 322   Find a conformal mapping of $|z - 1 + 3i| > 1$ onto $\text{Re}(w) < -5$.

6.15.4   An Integral Solution of the Dirichlet Problem for a Disk

We have already derived Poisson's integral formula for the solution of the Dirichlet problem for a disk. Here we will use complex analysis to derive a different integral representation of this solution (from which, incidentally, Poisson's formula follows easily). This alternate representation is particularly suited to using conformal mappings to write solutions of Dirichlet problems for other domains.

Begin with a function u that is harmonic in the disk $|u| < 1 + \epsilon$, slightly larger than the unit disk. We will derive an integral representation of $u(x, y)$.

We have already observed that the real and imaginary parts of any analytic complex function are harmonic. It is also true that, given a harmonic function u, there is a harmonic function v (its harmonic conjugate) such that $u + iv$ is an analytic complex function. Thus we can produce a harmonic v so that $f = u + iv$ is analytic on $|z| < 1 + \epsilon$. Further, by adding a constant if necessary, we may choose v so that $v(0, 0) = 0$. It is possible to write a formula for v, but we want v only as a lever to employ complex function methods. Only u appears in the final result.

Expand f in a Maclaurin series

$$f(z) = \sum_{n=0}^{\infty} a_n z^n.$$

Then

$$u(x, y) = \text{Re}[f(x + iy)] = \frac{1}{2}(f(z) + \overline{f(z)})$$

$$= \frac{1}{2} \sum_{n=0}^{\infty} (a_n z^n + \overline{a_n z^n}) = a_0 + \frac{1}{2} \sum_{n=1}^{\infty} (a_n z^n + \overline{a_n z^n}).$$

Now let ζ be on the unit circle γ. Then $|\zeta|^2 = \zeta\overline{\zeta} = 1$ so $\overline{\zeta} = 1/\zeta$ and

$$u(\zeta) = a_0 + \frac{1}{2} \sum_{n=0}^{\infty} (a_n \zeta^n + \overline{a_n} \zeta^{-n}).$$

Choose any integer m, multiply this equation by $\zeta^m/2\pi i$ and integrate over γ. The series and the integral can be interchanged within the open disk of convergence, and we obtain

$$\frac{1}{2\pi i} \int_\gamma u(\zeta)\zeta^m \, d\zeta = \frac{a_0}{2\pi i} \oint_\gamma \zeta^m \, d\zeta$$

$$+ \frac{1}{2} \frac{1}{2\pi i} \sum_{n=1}^\infty \left(a_n \oint_\gamma \zeta^{n+m} \, d\zeta + \bar{a}_n \oint_\gamma \zeta^{-n+m} \, d\zeta \right). \quad (6.24)$$

But

$$\oint_\gamma \zeta^k \, d\zeta = \begin{cases} 0 & \text{if } k \ne -1 \\ 2\pi i & \text{if } k = -1. \end{cases}$$

Thus, if $m = -1$ in equation 6.24, we get

$$\frac{1}{2\pi i} \oint_\gamma u(\zeta) \frac{1}{\zeta} \, d\zeta = a_0.$$

And if $m = -n - 1$ with $n = 1, 2, \ldots$, we get

$$\frac{1}{2\pi i} \oint_\gamma u(\zeta)\zeta^{-n-1} \, d\zeta = \frac{1}{2} a_n.$$

Substitute these coefficients into the Maclaurin series for f, keeping in mind that ζ is the variable of integration on γ, to obtain

$$f(z) = \sum_{n=0}^\infty a_n z^n = \frac{1}{2\pi i} \oint_\gamma u(\zeta)\zeta^{-1} \, d\zeta + \sum_{n=1}^\infty \frac{1}{\pi i} \oint_\gamma u(\zeta)\zeta^{-n-1} \, d\zeta z^n$$

$$= \frac{1}{2\pi i} \oint_\gamma \left[1 + 2 \sum_{n=1}^\infty \left(\frac{z}{\zeta} \right)^n \right] \frac{u(\zeta)}{\zeta} \, d\zeta.$$

Since $|z| < 1$ and $|\zeta| = 1$, $|z/\zeta| < 1$ and the geometric series in the integrand converges to the familiar result:

$$\sum_{n=1}^\infty \left(\frac{z}{\zeta} \right)^n = \frac{z/\zeta}{1 - z/\zeta} = \frac{z}{\zeta - z}.$$

Therefore

$$f(z) = \frac{1}{2\pi i} \oint_\gamma \left(1 + \frac{2z}{\zeta - z}\right) \frac{u(\zeta)}{\zeta} \, d\zeta = \frac{1}{2\pi i} \oint_\gamma u(\zeta) \left(\frac{\zeta + z}{\zeta - z}\right) \frac{1}{\zeta} \, d\zeta.$$

If values of u are prescribed on the boundary circle γ, say $u(\zeta) = g(\zeta)$ for a given function g, then for $|z| < 1$,

$$f(z) = \frac{1}{2\pi i} \oint_\gamma g(\zeta) \left(\frac{\zeta + z}{\zeta - z}\right) \frac{1}{\zeta} \, d\zeta. \tag{6.25}$$

This integral formula determines $f(z)$ at points in the open unit disk, given values of $\mathrm{Re}[f(z)]$ on the boundary unit circle. The solution of the Dirichlet problem for the unit disk, which asks for a harmonic function taking prescribed values on this unit circle, is retrieved from this formula as

$$u(x, y) = \mathrm{Re}[f(x + iy)].$$

EXERCISE 323   Derive Poisson's integral formula by putting $z = re^{i\theta}$ and $\zeta = e^{i\varphi}$ into equation 6.25 and extracting the real part of the resulting expression.

6.15.5   Solution of Dirichlet Problems by Conformal Mapping

We are now poised to solve Dirichlet problems by using conformal mappings. Equation 6.25 was derived with this in mind. The idea is to map a Dirichlet problem for a domain D in the z-plane to a Dirichlet problem for the unit disk in the w-plane. We solve the latter problem for the disk, using equation 6.25, then map this back to the solution of the original problem for D, using the mapping as a change of variables in the integral.

Now for the details. Suppose we know a differentiable, one-to-one conformal mapping $T : D \to K$, with K the unit disk $|w| < 1$. Assume that T maps the boundary curve C of D onto the unit circle γ bounding K, and that T^{-1} is also a differentiable conformal mapping. To help follow the notation, we will use ζ to denote an arbitrary point of γ, ξ for an arbitrary point of C, and $(\tilde{x}, \tilde{y})$ for an arbitrary point of the w-plane (Figure 6.27).

Suppose we want to solve a Dirichlet problem for D:

$$u_{xx} + u_{yy} = 0 \text{ for } (x, y) \in D$$

$$u(x, y) = g(x, y) \text{ for } (x, y) \in C.$$

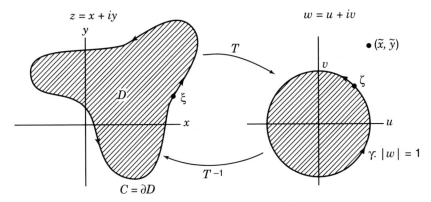

FIGURE 6.27 Constructing an integral solution of the Dirichlet problem using a conformal mapping of a domain onto the unit disk.

If $w = T(z)$, then $z = T^{-1}(w)$. Define

$$\tilde{g}(w) = g(T^{-1}(w)) = g(z)$$

for z on C and consider the Dirichlet problem on the unit disk $|w| < 1$:

$$\tilde{u}_{xx} + \tilde{u}_{yy} = 0 \text{ for } |\tilde{x} + i\tilde{y}| < 1$$

$$\tilde{u}(\tilde{x}, \tilde{y}) = \tilde{g}(\tilde{x}, \tilde{y}) \text{ for } |\tilde{x} + i\tilde{y}| = 1.$$

From equation 6.25 the solution of this problem for the disk in the w-plane is the real part of

$$\tilde{f}(w) = \frac{1}{2\pi i} \oint_{\gamma} \tilde{g}(\zeta) \left(\frac{\zeta + w}{\zeta - w} \right) \frac{1}{\zeta} \, d\zeta.$$

Define

$$f(z) = \tilde{f}(T(z)).$$

Then

$$f(z) = \frac{1}{2\pi i} \oint_{\gamma} \tilde{g}(\zeta) \left(\frac{\zeta + T(z)}{\zeta - T(z)} \right) \frac{1}{\zeta} \, d\zeta \qquad (6.26)$$

and

$$u(x, y) = \text{Re}[f(x + iy)]$$

is the solution of the original Dirichlet problem for the domain D in the x,y-plane.

It is convenient to put the integral expression 6.26 entirely in terms of the boundary curve of D and variables in the original z-plane. Recalling that T maps C onto γ, set $\zeta = T(\xi)$ for ξ on C to change variables in the integral in equation 6.26. This yields

$$f(z) = \frac{1}{2\pi i} \int_C \tilde{g}(T(\xi)) \left(\frac{T(\xi) + T(z)}{T(\xi) - T(z)} \right) \frac{1}{T(\xi)} T'(\xi) \, d\xi.$$

But, by definition of $\tilde{g}$, $\tilde{g}(T(\xi)) = g(T^{-1}(T(\xi))) = g(\xi)$, so

$$f(z) = \frac{1}{2\pi i} \int_C g(\xi) \left(\frac{T(\xi) + T(z)}{T(\xi) - T(z)} \right) \frac{T'(\xi)}{T(\xi)} \, d\xi, \qquad (6.27)$$

and the solution of the Dirichlet problem for D is

$$u(x, y) = \text{Re}[f(x + iy)].$$

We emphasize that the integral in this solution is over C, which need not be a closed curve. This is the reason for the notation $\int_C$ rather than $\oint_C$.

The Dirichlet Problem for the Right Half-Plane   Consider the Dirichlet problem for the right half-plane:

$$u_{xx} + u_{yy} = 0 \text{ for } x > 0, \ -\infty < y < \infty$$

$$u(0, y) = g(0, y) \text{ for } -\infty < y < \infty,$$

with g a given function. We could write $g(y)$ instead of $g(0, y)$, since the first coordinate plays no role in this function defined on the vertical boundary of the right half-plane. However, we denote points on this boundary as $(0, y)$, and hence have adopted this notation.

In Example 49 we constructed a conformal mapping of the right half-plane onto the unit disk:

$$w = T(z) = -i \left(\frac{z - 1}{z + 1} \right).$$

Other conformal mappings between these domains could also be used. Compute

$$T'(z) = \frac{-2i}{(z + 1)^2}.$$

From equation 6.27, the solution of the Dirichlet problem is the real part of

$$f(z) = \frac{1}{2\pi i} \int_C g(\xi) \left(\frac{-i\dfrac{\xi - 1}{\xi + 1} - i\dfrac{z - 1}{z + 1}}{-i\dfrac{\xi - 1}{\xi + 1} + i\dfrac{z - 1}{z + 1}} \right) \frac{\xi + 1}{-i(\xi - 1)} \frac{-2i}{(\xi + 1)^2} \, d\xi$$

in which C is the boundary of the right half-plane (the imaginary axis). Although this integrand may appear formidable, routine algebra yields the more tractable expression

$$f(z) = \frac{1}{\pi i} \int_C g(\xi) \left(\frac{\xi z - 1}{\xi - z} \right) \frac{1}{\xi^2 - 1} \, d\xi.$$

On C, the vertical axis, $\xi = it$, where t varies from ∞ to $-\infty$. Remember that ξ must move down this axis to maintain positive orientation. We obtain

$$f(z) = \frac{1}{\pi i} \int_{\infty}^{-\infty} g(it) \left(\frac{itz - 1}{it - z} \right) \frac{-1}{1 + t^2} \, dt = \frac{1}{\pi} \int_{-\infty}^{\infty} g(it) \left(\frac{itz - 1}{it - z} \right) \frac{1}{1 + t^2} \, dt.$$

We must extract the real part of this integral. Since t is real, $it = (0, t)$ and $g(it) = g(0, t)$ is real. Further, $1/(1 + t^2)$ is real. Therefore the real part of the integral will depend on the real part of the term in large parentheses in the integrand. With $z = x + iy$, calculate

$$\frac{itz - 1}{it - z} = \frac{itx - ty - 1}{i(t - y) - x} = \left(\frac{itx - ty - 1}{i(t - y) - x} \right) \left(\frac{-i(t - y) - x}{-i(t - y) - x} \right)$$

$$= \frac{tx(t - y) - itx^2 + ity(t - y) + txy + i(t - y) + x}{x^2 + (t - y)^2}.$$

The real part of this expression is

$$\frac{x(1 + t^2)}{x^2 + (t - y)^2}.$$

Putting everything together, we finally have the solution of the Dirichlet problem for the right half-plane:

$$u(x, y) = \text{Re}[f(x + iy)] = \frac{1}{\pi} \int_{-\infty}^{\infty} g(it) \frac{x(1 + t^2)}{x^2 + (t - y)^2} \frac{1}{1 + t^2} \, dt$$

$$= \frac{x}{\pi} \int_{-\infty}^{\infty} g(0, t) \frac{1}{x^2 + (t - y)^2} \, dt.$$

This is similar in form to the solution 6.16 for the upper half-plane.

As a specific example, consider the Dirichlet problem for the right half-plane:

$$u_{xx} + u_{yy} = 0 \text{ for } x > 0, \quad -\infty < y < \infty$$

$$u(0, y) = e^{-|y|} \text{ for } -\infty < y < \infty.$$

The solution is

$$u(x, y) = \frac{x}{\pi} \int_{-\infty}^{\infty} \frac{1}{x^2 + (t - y)^2} e^{-|t|} \, dt.$$

EXERCISE 324   Use the conformal mapping method to write an integral solution for the Dirichlet problem for the upper half-plane. Compare this expression with that obtained previously by Fourier integral and transform methods.

EXERCISE 325   Use the conformal mapping method to solve the Dirichlet problem for the right quarter-plane $x > 0$, $y > 0$.

EXERCISE 326   Use a conformal mapping to write the solution of the Dirichlet problem for the disk $|z - z_0| < R$.

EXERCISE 327   Use a conformal mapping to write the solution of the Dirichlet problem for the infinite strip S consisting of all (x, y) with $-\infty < x < \infty$, $-\pi, 2 < y < \pi/2$. Hint: A conformal mapping of S onto the unit disk is given by

$$w = -i \left(\frac{e^z - 1}{e^z + 1} \right).$$

EXERCISE 328   Use a conformal mapping to write an integral solution for the Dirichlet problem for the right half-plane $\text{Re}(z) > 1$. Hint: A translation maps this domain onto $\text{Re}(z) > 0$.

EXERCISE 329   Use a conformal mapping to write an integral solution for the Dirichlet problem for the lower half-plane $\text{Im}(z) < 0$.

EXERCISE 330   Use a conformal mapping to write an integral solution for the Dirichlet problem for the exterior of a disk, $|z - z_0| > R$.

6.16   EXISTENCE OF A SOLUTION OF THE DIRICHLET PROBLEM

Does every Dirichlet problem have a solution? The answer is no. We will give two examples, one in 3-space and one in the plane, of Dirichlet problems having no solution.

6.16.1   An Example in 3-Space

Begin with a physically motivated example in R^3 due to the French mathematician Henri Lebesgue. First define a surface as follows. Imagine a sphere of radius 1 about the origin in R^3, made of stretchable rubber. Take a thin needle and, starting at $(0, 1, 0)$, push the point inward, deforming part of the rubber on the sphere into a thin spike in the interior of the sphere pointing toward the origin (but without puncturing the sphere). The resulting surface is shown in Figure 6.28, on which there is darkened strip to be explained shortly. Let Ω be the domain bounded by this surface (inside the sphere, outside the spike). The deformed surface is $\partial\Omega$.

Next define a function f over $\partial\Omega$. Let $f(x, y, z) = 0$ for (x, y, z) on the spike part of the surface, up to the shaded strip. At the strip, proceeding from the spike outward over the shaded part of the sphere, make $f(x, y, z)$ grow at a

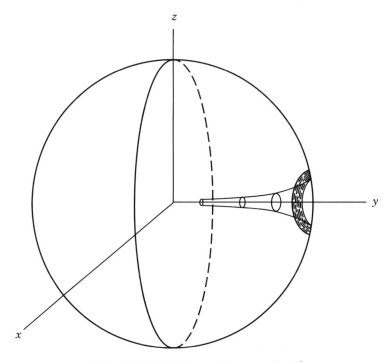

FIGURE 6.28   Sphere with a spike in R^3.

rapid rate until it reaches a value T at the outer edge of the shaded part, beyond which it maintains the constant value T over the rest of the spherical part of the surface. T can be chosen arbitrarily large. We can also make the spike thin enough, and the shaded part of sufficiently small diameter, that the shaded area and spike have arbitrarily small surface area.

Now suppose u is a solution of

$$\nabla^2 u = 0 \text{ in } \Omega, \, u = f \text{ on } \partial\Omega.$$

We may think of $u(x, y, z)$ as the steady state temperature distribution throughout Ω. But the temperature on the surface $\partial\Omega$ is a constant T except over the spike and the darkened layer, which can be constructed to have as small a surface area as we like. Thus we would expect the steady state temperature over Ω to be nearly the constant T, say equal to $T - \epsilon$ for some small positive ϵ:

$$u(x, y, z) = T - \epsilon \text{ for } (x, y, z) \in \Omega.$$

But now u cannot be continuous on $\bar{\Omega}$, since $u(x, y, z) = 0$ at points on the spike and $u(x, y, z) = T - \epsilon$ at points inside the sphere arbitrarily close to the spike.

This example, which can be done rigorously, suggests that an existence theorem for the Dirichlet problem must place some condition on the domain, or perhaps its boundary, as well as on the boundary data function.

One such condition on the boundary can be described as follows. Consider the surface of revolution obtained by rotating the graph of $y = x^n$ about the x-axis for $x \geq 0$, with n any positive integer. Figure 6.29 shows the graphs of several such functions, and Figure 6.30 a typical surface of revolution obtained by rotating one of these graphs about the y-axis. The origin is called the tip of this surface of revolution. The spike at the origin becomes "sharper" as n is chosen larger. We call any such surface of revolution, rotated and/or translated to have its tip at any point in R^3, a *spiked surface*. The following existence theorem uses spiked surfaces to avoid circumstances such as occur in Lebesgue's counterexample.

Theorem 28 (Existence) Let Ω be a domain in R^3 whose boundary is a piecewise smooth closed surface, and let f be continuous on $\partial\Omega$. Then the Dirichlet problem

$$\nabla^2 u = 0 \text{ in } \Omega; \, u = f \text{ on } \partial\Omega$$

has a solution if, for each $\mathbf{x} \in \partial\Omega$, there is a spiked surface $\Sigma_\mathbf{x}$ and a positive number $\rho_\mathbf{x}$ such that all points in $\Sigma_\mathbf{x}$ at distance $\leq \rho_\mathbf{x}$ from $\mathbf{x}$ lie in the exterior of Ω. ∎

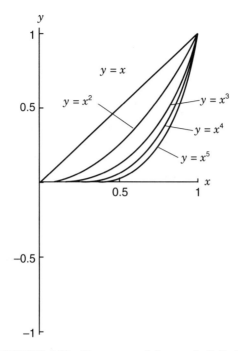

FIGURE 6.29 Curves $y = x^n$ for $n = 1, 2, 3, 4, 5$.

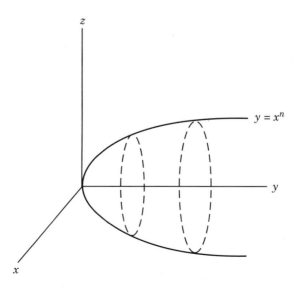

FIGURE 6.30 Surface of revolution generated by the graph of $y = x^n$.

EXERCISE 331   Explain how the bounding surface of the domain in Lebesgue's example fails to satisfy the sufficient condition of Theorem 28.

6.16.2   An Example in the Plane

We will construct an example of a Dirichlet problem for a domain in the plane, for which there is no solution. This example appears in *Three Secrets About Harmonic Functions*, by R.B. Burckel (American Mathematical Monthly, Vol. 104, No. 1, pages 52–56). The example is a modification due to Sadahiro Saeki of an idea of S.J. Gardiner. Throughout the example we identify a complex number $z = x + iy$ with the point (x, y) in R^2.

Using polar coordinates, define the set

$$S_n = \left\{ re^{i\theta} \,\middle|\, 0 < r < n \text{ and } \pi \,\frac{1 + \dfrac{1}{2n}}{n + 1} < \theta < \pi \,\frac{1 + \dfrac{1}{2(n + 1)}}{n} \right\}$$

for each integer $n \geq 3$. Thus

$$S_3 = \left\{ re^{i\theta} \,\middle|\, 0 < r < 3 \text{ and } \frac{7\pi}{24} < \theta < \frac{9\pi}{24} \right\},$$

$$S_4 = \left\{ re^{i\theta} \,\middle|\, 0 < r < 4 \text{ and } \frac{9\pi}{40} < \theta < \frac{11\pi}{40} \right\},$$

and so on. Figure 6.31 shows S_3, S_4, S_5 and S_6. Each S_n is an open subset of R^2.

Notice that

$$\overline{S_n} \cap \overline{S_m} = \{0\}$$

for $m > n \geq 3$. Define

$$\Omega = B(0, 2) \cup \left(\bigcup_{n \geq 3} S_n \right),$$

consisting of all points in $B(0, 2)$ together with each point lying in any S_n. Ω is open, being a union of open sets. Further, Ω is connected. Any two points of a particular S_n can be connected by a straight line path in S_n, and points in different S_n and S_m, or in S_n and $B(0, 2)$, can be connected by drawing a line from one point to 0 and then from 0 to the other point. Thus Ω is a domain in R^2.

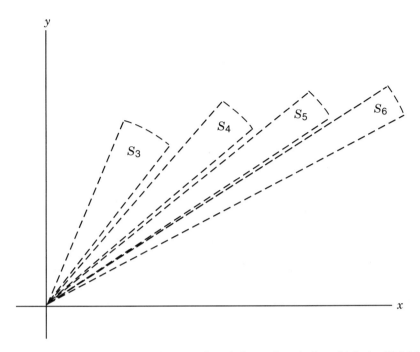

FIGURE 6.31 Subsets of the plane used to define a domain for which the Dirichlet problem has no solution.

EXERCISE 332 Determine $\partial\Omega$. *Hint:* $\partial\Omega$ consists of points in $(2, \infty)$ on the real line, together with certain arcs in the boundary of $B(\mathbf{0}, 2)$ and the sets $\partial S_n - B(\mathbf{0}, 2)$.

Now define a function $f: \partial\Omega \to R^1$ by setting

$$f(z) = \begin{cases} \text{Re}[(ze^{-i\pi/n})^{n(n+1)}] \text{ for } z \in \partial S_n - B(\mathbf{0}, 2) \\ 0 \text{ for all other } z \in \partial\Omega. \end{cases}$$

EXERCISE 333 Prove that f is continuous. *Hint:* Observe that f maps the half-lines from the origin bounding S_n to points on the imaginary axis in the complex plane.

Now consider the Dirichlet problem

$$\nabla^2 u = 0 \text{ in } \Omega$$

$$u = f \text{ on } \partial\Omega.$$

EXERCISE 334 Show that this Dirichlet problem has no solution. *Hint:* Suppose u is a solution. Produce a real number k such that

$$u(x, y) \geq k \text{ for } x + iy \in \overline{B(\mathbf{0}, 2)}.$$

Show that

$$u(x, y) - \operatorname{Re}[(ze^{-i\pi/n})^{n(n+1)}] \geq k$$

for $n \geq 3$ and $z \in \partial S_n$. Now use the Maximum Principle to assert that, for $z = 2e^{i\pi/n} \in \partial S_n$,

$$u(z) - 2^{n(n+1)} \geq k$$

This contradicts the fact that the continuous function u must be bounded on the compact set $\overline{B(\mathbf{0}, 2)}$.

In this example, the function specified on the boundary of the domain is continuous, and is therefore a function we would think of as "well-behaved." The reason there is no solution is that the boundary contains points at which any proposed solution can be made arbitrarily large. As with the preceding example, this suggests that an existence theorem for the Dirichlet problem must place conditions on the boundary of the domain.

6.16.3   The Idea Underlying an Existence Theorem for the Dirichlet Problem

We will discuss an approach to an existence theorem originated by Poincaré and further developed by Perron and others. All functions in this section are real-valued functions defined on sets in R^n.

Let Ω be a bounded domain in R^n. Suppose w is continuous on $\bar{\Omega}$. Let B be any open ball with $B \subset \Omega$. We know that the Dirichlet problem for an open ball has a unique solution (this is the Poisson integral formula for $n = 2$, and for $n > 2$ there are generalizations of this integral formula giving the solution in terms of higher-dimension integrals). Let w^* be the solution of

$$\nabla^2 u = 0 \text{ in } B$$

$$u = w \text{ on } \partial B.$$

We can uniquely define a new function w_B by setting

$$w_B(\mathbf{x}) = \begin{cases} w(\mathbf{x}) \text{ for } \mathbf{x} \in \bar{\Omega} - B \\ w^*(\mathbf{x}) \text{ for } \mathbf{x} \in B. \end{cases}$$

Thus w_B is that function which coincides with w in $\bar{\Omega}$ but outside B, and is harmonic within B. We can find one such function for each open ball contained in Ω.

EXERCISE 335 Under the conditions of the discussion, prove that w is harmonic in Ω if and only if $w = w_B$ for each open ball $B \subset \Omega$.

A continuous function w defined on $\bar{\Omega}$ is called *superharmonic* if, for each open ball $B \subset \Omega$,

$$w_B(\mathbf{x}) \leq w(\mathbf{x}) \text{ for } \mathbf{x} \in \bar{\Omega}.$$

Now suppose f is defined and continuous on $\partial\Omega$. A function w is an *upper function* for f if w is superharmonic (on Ω) and

$$f(\mathbf{x}) \leq w(\mathbf{x}) \text{ for } \mathbf{x} \in \partial\Omega.$$

Here is how these concepts relate to the Dirichlet problem

$$\nabla^2 u = 0 \text{ in } \Omega$$
$$u = f \text{ on } \partial\Omega. \tag{6.28}$$

Define

$$\mathfrak{U} = \{w \,|\, w \text{ is an upper function for } f \text{ on } \Omega\}.$$

$\mathfrak{U}$ is not empty and we can define a function u by setting

$$u(\mathbf{x}) = glb\{w(\mathbf{x}) \,|\, w \in \mathfrak{U}\} \tag{6.29}$$

for each $\mathbf{x} \in \Omega$. Thus, for each $\mathbf{x}$ in Ω, $u(\mathbf{x})$ is the greatest lower bound of the set of real numbers $w(\mathbf{x})$, over all upper functions w for f. The existence proof consists of showing that u is harmonic on Ω, and that, if a certain condition we will state shortly is satisfied on $\partial\Omega$, then $u = f$ on $\partial\Omega$. This means that, subject to a condition on the boundary of Ω, the Dirichlet problem 6.28 has a solution.

EXERCISE 336 Prove that $\mathfrak{U} \neq 0$ and that, for each $\mathbf{x} \in \Omega$, $\{w(\mathbf{x}) \,|\, w \in \mathfrak{U}\}$ is bounded below.

EXERCISE 337 Prove that the sum of a harmonic and superharmonic function is superharmonic.

EXERCISE 338 Prove that the sum of two superharmonic functions is superharmonic. How about the difference of two superharmonic functions?

EXERCISE 339 Define a subharmonic function by reversing the inequalities in the definition of superharmonic function. Prove that, if u is superharmonic and v is subharmonic, then $u - v$ is superharmonic. What can be said about $v - u$?

EXERCISE 340   Let w be superharmonic on Ω. Prove that $\min_{x \in \bar{\Omega}} w$ is assumed on $\partial\Omega$. *Hint:* Suppose w assumes its minimum on $\bar{\Omega}$ at a point $\mathbf{x}_0 \in \Omega$. Let B be the disk of radius ρ about $\mathbf{x}_0$, where

$$\rho = glb_{\mathbf{x} \in \partial\Omega} |\mathbf{x} - \mathbf{x}_0|.$$

Then B intersects $\partial\Omega$ in at least one point. Show that $w(\mathbf{x})$ has the constant value $w(\mathbf{x}_0)$ on the boundary of B. To do this, note that $w_B(\mathbf{x}_0)$ is equal to the average of $w_B(\mathbf{x})$ over the boundary of B. Conclude from this and the assumption that w is superharmonic, that

$$w_B(\mathbf{x}_0) = (\text{average of } w(\mathbf{x}) \text{ for } \mathbf{x} \in \partial B) \leq w(\mathbf{x}_0).$$

From this and the choice of $\mathbf{x}_0$ conclude that $w(\mathbf{x}) = w(\mathbf{x}_0)$ for all $\mathbf{x} \in \partial B$. Hence show that w achieves its minimum on $\bar{\Omega}$ at some point of $\partial\Omega$.

6.16.4   Barrier Functions and an Existence Theorem

We continue the discussion of existence, with Ω and f as defined above.

If $\mathbf{y} \in \partial\Omega$, we say that $\mathbf{y}$ satisfies the *barrier condition* if there exists a superharmonic function ω_y such that

$$\omega_y(\mathbf{y}) = 0$$

and

$$\omega_y(\mathbf{x}) > 0 \text{ for } \mathbf{x} \in \bar{\Omega} - \{\mathbf{y}\}.$$

The function ω_y is called a *barrier function* at $\mathbf{y}$.

We can now state an existence theorem.

Theorem 29 (Existence)   Let Ω be a bounded domain in R^n and let f be a continuous, real-valued function defined on $\partial\Omega$. For each $\mathbf{x} \in \bar{\Omega}$, let

$$u(\mathbf{x}) = glb\{w(\mathbf{x}) | w \text{ is an upper function for } f \text{ on } \Omega\}.$$

Then, u is harmonic on Ω. Further, if each point of $\partial\Omega$ satisfies the barrier condition, then $u = f$ on $\partial\Omega$.   ∎

Thus, for any bounded domain whose boundary points all satisfy the barrier condition, the Dirichlet problem, with continuous information specified on the boundary, has a solution.

For a given domain it may be far from obvious whether or not the barrier condition is satisfied at each point of the boundary. Here is a fairly simple criterion that applies to some simply connected domains in the plane.

EXERCISE 341   Suppose Ω is a bounded domain in R^2 whose boundary is a simple piecewise smooth curve C. Then every point of C satisfies the barrier condition. *Hint:* Let $\mathbf{x}_0 = (x_0, y_0)$ be a point of C. Define

$$\omega_{\mathbf{x}_0} = -\frac{p(x - x_0, y - y_0)}{p^2(x - x_0, y - y_0) + q^2(x - x_0, y - y_0)}$$

where

$$p(x, y) = \mathrm{Re}[\ln(x + iy)] = \frac{1}{2}\ln(x^2 + y^2)$$

and

$$q(x, y) = \mathrm{Im}[\ln(x + iy)] = \arg(x + iy).$$

Verify that $\omega_{\mathbf{x}_0}$ is a barrier function at $\mathbf{x}_0$.

A proof of Theorem 29 can be found in Petrovsky [12]. A "complementary" approach to this existence theorem is discussed in John [9]. In this approach the inequality in the definition of superharmonic function is reversed, leading to the notion of a subharmonic function. With corresponding adjustments in other inequalities, such as occur in the definition of a barrier function, it is possible to arrive at a similar existence theorem, in which the solution value at any point of Ω is the least upper bound (instead of greatest lower bound) of a set of function values. Egorov and Shubin [6] give yet another approach to this idea, giving the solution of the Dirichlet problem as the limit of a constructed sequence. Their discussion exploits the concept of the capacity of a compact subset of R^n, which is a measure theoretic concept.

The idea of basing an existence proof on the construction of a solution as a greatest lower bound (or least upper bound, or limit) of a set or sequence of functions is a development of Poincàre's original approach, which is often referred to as the *method of balayage*.

6.16.5   Dirichlet's Principle

There is another approach to the existence of a solution of the Dirichlet problem which has historical as well as mathematical interest.

Among the first attempts at proving an existence theorem for the Dirichlet problem was an approach taken by Bernhard Riemann. Riemann's observation was that the solution of a Dirichlet problem minimizes a certain integral. That is, if $I(w)$ is the value of this integral when a function w is inserted, then the solution u of the Dirichlet problem has the property that $I(u) \leq I(w)$ for all w in a certain class of functions (the definition of I will be seen to involve the function f specifying the values of u on the boundary of the domain). In the course of his work on the calculus of variations, a branch of which deals with determining functions which extremize integrals, the German mathematician David Hilbert returned to Riemann's approach, which had also been pursued by the self-taught Englishman George Green. We will informally describe the idea behind this approach. Details of this proof can be found in Rauch [13].

Begin with the following simple observation. Instead of directly solving a Dirichlet problem

$$\nabla^2 u = 0 \text{ in } \Omega;\ u = f \text{ on } \partial\Omega, \tag{6.30}$$

we can instead concentrate on the problem

$$\nabla^2 u = F \text{ in } \Omega;\ u = 0 \text{ on } \partial\Omega. \tag{6.31}$$

If we can solve the problem 6.31 for any continuous F, then we can solve the problem 6.30, given f.

EXERCISE 342   Prove the assertion just made about problems 6.30 and 6.31. *Hint:* Assume that we can solve the problem 6.31 for any continuous F. We want to show that we can solve 6.30. Choose any $\tilde{f}$, continuous with continuous partial derivatives of all orders on $\bar{\Omega}$, and such that $\tilde{f}(\mathbf{x}) = f(\mathbf{x})$ for $\mathbf{x} \in \partial\Omega$. Let $F = -\nabla^2\tilde{f}$, be a solution of 6.31. Show that $u + f$ is a solution of 6.30.

Now define the functional

$$I(w) = \int_\Omega (|\nabla w|^2 + 2wF)\, dV,$$

in which $\int_\Omega \ldots dV$ denotes the integral of appropriate dimension over the domain Ω in R^n. We assume that F is defined on Ω, and is restricted to a class of functions for which this integral exists. I maps functions w to real numbers by means of an integral, and hence is traditionally called a functional. *Dirichlet's Principle* states that $\nabla^2 u = F$ on Ω if and only if u minimizes $I(w)$, when this function is evaluated for all "well-behaved" functions vanishing on the boundary of Ω.

Theorem 30 (Dirichlet's Principle)   Suppose $u \in C^1(\bar{\Omega})$ and $u = 0$ on $\partial\Omega$. Then

$$\nabla^2 u = F \text{ on } \Omega$$

if and only if

$$I(u) \leq I(w)$$

for all $w \in C^1(\bar{\Omega})$ such that $w = 0$ on $\partial\Omega$. ∎

The existence proof now consists of showing that there does indeed exist a function u, in the appropriate class of functions, which minimizes I. This function is a solution of problem 6.31, and we can use the existence of solutions of this problem to assert that there exists a solution of 6.30.

6.17 THE NEUMANN PROBLEM

Recall that the Neumann problem for a domain Ω is to find a function that is harmonic on Ω and whose normal derivative takes on given values on the boundary:

$$\nabla^2 u = 0 \text{ in } \Omega$$

$$\frac{\partial u}{\partial n} = f \text{ on } \partial\Omega. \qquad (6.32)$$

Certainly if u were a solution of a Neumann problem, then so would be u plus any constant, since the boundary data is given for the normal derivative, not for the function itself, and the constant will vanish in the differentiation. We will show that continuous solutions of the Neumann problem are unique to within this additive constant.

Theorem 31 Let Ω be a bounded domain in R^n having a closed, piecewise smooth boundary $\partial\Omega$. Let f be continuous on $\partial\Omega$. Let w and v be continuous solutions of the Neumann problem 6.32. Then w and v differ by a constant on $\bar{\Omega}$. ∎

We will prove the theorem for a domain in R^2. The proof makes use of Green's first identity.

Lemma 7 (Green's First Identity) Let Ω be a bounded domain in R^2 having a closed, piecewise smooth curve as boundary. Let g and h be in $C^2(\bar{\Omega})$. Then

$$\oint_{\partial\Omega} g \frac{\partial h}{\partial n} \, ds = \iint_{\Omega} (g\nabla^2 h + \nabla g \cdot \nabla h) \, dA. \qquad \blacksquare$$

Proof: First use Green's theorem to write

$$\oint_{\partial\Omega} g \frac{\partial h}{\partial n} \, ds = \iint_{\Omega} div(g\nabla h) \, dA.$$

Now verify that

$$div(g\nabla h) = g\nabla^2 h + \nabla g \cdot \nabla h, \qquad (6.33)$$

completing the proof of the lemma. $\qquad \blacksquare$

If $g = h$ in the lemma, we obtain

$$\oint_{\partial\Omega} g \frac{\partial g}{\partial n} \, ds = \iint_{\Omega} (g\nabla^2 g + |\nabla g|^2) \, dA. \qquad (6.34)$$

EXERCISE 343   Derive equation 6.33.

Now we can prove Theorem 31. Let $u = v - w$. Then

$$\nabla^2 u = 0 \text{ in } \Omega$$

and

$$\frac{\partial u}{\partial n} = \frac{\partial v}{\partial n} - \frac{\partial w}{\partial n} = 0 \text{ on } \partial\Omega.$$

Apply equation 6.34 with $g = u$ to conclude that

$$\iint_{\Omega} |\nabla u|^2 \, dA = 0.$$

Then

$$u_x^2 + u_y^2 = 0$$

on Ω, hence

$$u_x = u_y = 0$$

on Ω. But then $u(x, y) = $ constant on Ω. By continuity $u(x, y) = $ constant on $\bar{\Omega}$.

A similar argument can be used to prove the theorem for a domain in R^3. In this case Gauss's divergence theorem is used to prove the three-dimensional version of Green's lemma, a surface integral replaces the line integral in Green's lemma, and a volume integral replaces the integral in the plane over the domain Ω.

EXERCISE 344   Prove Theorem 31 for a domain in R^3.

We will not give conditions sufficient for a Neumann problem to have a solution. However, it is easy to derive an important necessary condition. Let u be a solution of a Neumann problem on Ω in R^2. By Lemma 7 with $g = 1$ and $h = u$,

$$\oint_{\partial\Omega} \frac{\partial u}{\partial n}\, ds = 0.$$

Since $\partial u/\partial n = f$ on $\partial\Omega$,

$$\oint_{\partial\Omega} f\, ds = 0.$$

We have proved that a necessary condition for there to exist a solution of the Neumann problem 6.32 in R^2 is that the integral of the data function over the boundary curve must be zero.

For a domain in R^3, the analogous condition is that the surface integral $\iint_{\partial\Omega} f\, d\sigma$ of the data function over the boundary surface must be zero.

This condition has a physical interpretation. If we think of Laplace's equation $\nabla^2 u = 0$ as the steady-state heat equation, this integral condition means that in the steady-state case the net flow of heat energy across the boundary of the domain must be zero. This is a conservation of energy condition, in the absence of sources or sinks of heat energy in the domain.

EXERCISE 345   Prove that $\iint_{\partial\Omega} f\, d\sigma = 0$ is necessary for existence of a solution of the Neumann problem in the case that Ω is a bounded domain in R^3.

EXERCISE 346   Let Ω be a bounded domain in the plane, and suppose $\partial\Omega$ is a piecewise smooth closed curve. Let q be continuous on $\bar\Omega$ and f on $\partial\Omega$. Prove that the problem

$$\nabla^2 u = q \text{ in } \Omega$$

$$\frac{\partial u}{\partial n} = f \text{ on } \partial\Omega$$

can have a solution only if

$$\iint_{\Omega} q(x, y) \, dA = \oint_{\partial\Omega} f \, ds.$$

EXERCISE 347   Consider the boundary value problem

$$\nabla^2 u = 0 \text{ on } \Omega$$

$$\frac{\partial u}{\partial n} + hu = f \text{ on } \partial\Omega.$$

Assume that Ω is a bounded domain in the plane, with piecewise smooth closed boundary curve $\partial\Omega$. Let h and f be continuous on $\partial\Omega$ and assume that h is not identically zero and that $h(\mathbf{x}) \geq 0$ for $\mathbf{x}$ in $\partial\Omega$. Prove that this problem can have only one solution.

EXERCISE 348   Prove the conclusion of Exercise 347 for the case that Ω is a bounded domain in R^3, with piecewise smooth closed boundary surface $\partial\Omega$.

EXERCISE 349   Let f be continuous on the domain Ω in R^2, and let g be continuous on the piecewise smooth curve bounding Ω. Let k be a real number. Prove that the problem

$$\nabla^2 u + ku = f \text{ on } \Omega$$

$$\frac{\partial u}{\partial n} = g \text{ on } \partial\Omega$$

can have at most one solution if $k < 0$.

EXERCISE 350   Let Ω be a domain in R^2, whose boundary consists of two piecewise smooth curves C and K. Let f be continuous on Ω and let g be continuous on C. Let k be a real number. Prove that the mixed Dirichlet/ Neumann type problem

$$\nabla^2 u = f \text{ on } \Omega$$

$$u = g \text{ on } C$$

$$\frac{\partial u}{\partial n} + ku = 0 \text{ on } K$$

can have at most one solution.

6.18 NEUMANN PROBLEM FOR A RECTANGLE

We will solve a Neumann problem for a rectangle:

$$\nabla^2 u(x, y) = 0 \text{ for } 0 < x < a, \, 0 < y < b$$

$$u_y(x, 0) = u_y(x, b) = 0 \text{ for } 0 \le x \le a$$

$$u_x(0, y) = 0 \text{ for } 0 \le y \le b$$

$$u_x(a, y) = g(y) \text{ for } 0 \le y \le b.$$

This is a Neumann problem because the specified partial derivatives are normal to the respective sides of the rectangle.

A necessary condition for the existence of a solution is that

$$\oint_{\partial \Omega} \frac{\partial u}{\partial n} \, ds = 0$$

in which $\partial \Omega$ is the piecewise smooth curve bounding the rectangle. For the given boundary conditions this condition reduces to

$$\int_0^b g(y) \, dy = 0.$$

We assume that g satisfies this condition. We will soon see why there can be no solution if this condition is not met.

In order to use separation of variables, let $u(x, y) = X(x)Y(y)$ and obtain in the usual way

$$X'' + \lambda X = 0, \, Y'' - \lambda Y = 0$$

in which λ is the separation constant. Now

$$u_y(x, 0) = X(x)Y'(0) = 0$$

implies that $Y'(0) = 0$. And $u_y(x, b) = 0$ implies that $Y'(b) = 0$. The problem for Y is

$$Y'' - \lambda Y = 0; \, Y'(0) = Y'(b) = 0.$$

We solved this problem in Section 5.4.2, with X in place of Y, L in place of b, and λ in place of $-\lambda$. The eigenvalues and eigenfunctions are

$$\lambda_n = -\frac{n^2\pi^2}{b^2}, \ Y_n = \cos\left(\frac{n\pi y}{b}\right)$$

for $n = 0, 1, 2, \ldots$.

Now the problem for X is

$$X'' - \frac{n^2\pi^2}{b^2}X = 0.$$

If $n = 0$, $X = cx + d$. But $u_x(0, y) = 0$ implies that $X'(0) = 0$, so $c = 0$ and $X_0 = $ constant for $n = 0$.

If n is a positive integer, the differential equation for X has general solution

$$X = ce^{n\pi x/b} + de^{-n\pi x/b}.$$

$X'(0) = 0$ implies that

$$\frac{n\pi}{b}c - \frac{n\pi}{b}d = 0$$

so $c = d$ and

$$X_n = \cosh\left(\frac{n\pi x}{b}\right).$$

For each nonnegative integer n, we therefore have functions

$$u_0(x, y) = \text{constant}$$

and, for $n = 1, 2, \ldots$,

$$u_n(x, y) = \alpha_n \cosh\left(\frac{n\pi x}{b}\right)\cos\left(\frac{n\pi y}{b}\right).$$

To satisfy the boundary condition $u_x(a, y) = g(y)$, use a superposition of these functions, which we will write as

$$u(x, y) = \alpha_0 + \sum_{n=1}^{\infty} \alpha_n \cosh\left(\frac{n\pi x}{b}\right)\cos\left(\frac{n\pi y}{b}\right).$$

We need

$$u_x(a, y) = \sum_{n=1}^{\infty} \alpha_n \frac{n\pi}{b} \sinh\left(\frac{n\pi a}{b}\right)\cos\left(\frac{n\pi y}{b}\right) = g(y). \qquad (6.35)$$

This is a Fourier cosine expansion of g in which the nth coefficient is given by

$$\alpha_n \frac{n\pi}{b} \sinh\left(\frac{n\pi a}{b}\right) = \frac{2}{b} \int_0^b g(\xi)\cos\left(\frac{n\pi\xi}{b}\right) d\xi,$$

or

$$\alpha_n = \frac{2}{n\pi \sinh\left(\dfrac{n\pi a}{b}\right)} \int_0^b g(\xi)\cos\left(\frac{n\pi\xi}{b}\right) d\xi.$$

The constant term in the Fourier cosine expansion of g on $[0, b]$ is $1/b \int_0^b$ $g(\xi)\, d\xi$. The expansion 6.35 which is needed for a solution requires that this term must be zero. But this is exactly the condition $\oint \partial u/\partial n\; ds = 0$ which is necessary for this Neumann problem to have a solution.

The solution is

$$u(x, y) =$$

$$\alpha_0 + \sum_{n=1}^{\infty} \left(\frac{2}{n\pi \sinh\left(\dfrac{n\pi a}{b}\right)} \int_0^b g(\xi)\cos\left(\frac{n\pi\xi}{b}\right) d\xi \right) \cosh\left(\frac{n\pi x}{b}\right) \cos\left(\frac{n\pi y}{b}\right)$$

in which α_0 is an arbitrary constant. Recall that, for a Neumann problem, we can only expect to obtain a solution to within an additive constant.

EXERCISE 351   Solve

$$\nabla^2 u(x, y) = 0 \text{ or } 0 < x < 1, 0 < y < 1$$

$$u_x(0, y) = u_x(1, y) = 0 \text{ for } 0 \le y \le 1$$

$$u_y(x, 0) = 4\cos(\pi x),\ u_y(x, 1) = 0 \text{ for } 0 \le x \le 1.$$

EXERCISE 352   Solve

$$\nabla^2 u = 0 \text{ for } 0 < x < 1, 0 < y < \pi$$

$$u_x(0, y) = y - \frac{\pi}{2},\ u_x(1, y) = \cos(y) \text{ for } 0 \le y \le \pi$$

$$u_y(x, 0) = u_y(x, \pi) = 0 \text{ for } 0 \le x \le 1.$$

EXERCISE 353 Solve

$$\nabla^2 u = 0 \text{ for } 0 < x < \pi, \ 0 < y < \pi$$

$$u_x(0, y) = u_x(\pi, y) = 0 \text{ for } 0 \le y \le \pi$$

$$u_y(x, 0) = \cos(3x), \ u_y(x, \pi) = 6x - 3\pi \text{ for } 0 \le x \le \pi.$$

EXERCISE 354 Use separation of variables to solve the mixed boundary value problem:

$$\nabla^2 u(x, y) = 0 \text{ for } 0 < x < \pi, \ 0 < y < \pi$$

$$u(x, 0) = f(x), \ u(x, \pi) = 0 \text{ for } 0 \le x \le \pi$$

$$u_x(0, y) = u_x(\pi, y) = 0 \text{ for } 0 \le y \le \pi.$$

Does this problem have a unique solution?

EXERCISE 355 Attempt separation of variables to solve

$$\nabla^2 u(x, y) = 0 \text{ for } 0 < x < 1, \ 0 < y < 1$$

$$u(x, 0) = u(x, 1) = 0 \text{ for } 0 \le x \le 1$$

$$u_x(0, y) = 3y^2 - 2y, \ u_y(1, y) = 0 \text{ for } 0 \le y \le 1.$$

6.19 NEUMANN PROBLEM FOR A DISK

We will solve the Neumann problem for a disk centered about the origin in the plane. In polar coordinates the problem is

$$\nabla^2 u(r, \theta) = 0 \text{ for } 0 \le r < \rho, \ -\pi \le \theta \le \pi$$

$$\frac{\partial u}{\partial n}(\rho, \theta) = f(\theta) \text{ for } -\pi \le \theta \le \pi.$$

At any point on a circle about the origin, the line from the origin through the point is along the normal to the circle at that point, so the normal derivative in polar coordinates is just the radial derivative $\partial/\partial r$. The boundary condition can be written

$$\frac{\partial u}{\partial r}(\rho, \theta) = f(\theta) \text{ for } -\pi \le \theta \le \pi.$$

A necessary condition for a solution to exist is that

$$\int_{-\pi}^{\pi} f(\theta) \, d\theta = 0$$

and we assume that f satisfies this condition.

Attempt a solution

$$u(r, \theta) = \frac{1}{2} a_0 + \sum_{n=1}^{\infty} a_n r^n \cos(n\theta) + b_n r^n \sin(n\theta).$$

We need

$$\frac{\partial u}{\partial r} (\rho, \theta) = \sum_{n=1}^{\infty} a_n n \rho^{n-1} \cos(n\theta) + b_n n \rho^{n-1} \sin(n\theta) = f(\theta).$$

This is a Fourier expansion of f on $[-\pi, \pi]$ and the coefficients are

$$a_n = \frac{1}{\pi n \rho^{n-1}} \int_{-\pi}^{\pi} f(\xi) \cos(n\xi) \, d\xi$$

and

$$b_n = \frac{1}{\pi n \rho^{n-1}} \int_{-\pi}^{\pi} f(\xi) \sin(n\xi) \, d\xi$$

for $n = 1, 2, \ldots$.

Upon inserting these coefficients, the solution is

$$u(r, \theta) = \frac{1}{2} a_0 + \frac{\rho}{\pi} \sum_{n=1}^{\infty} \frac{1}{n} \left(\frac{r}{\rho}\right)^n \int_{-\pi}^{\pi} [\cos(n\xi)\cos(n\theta)$$

$$+ \sin(n\xi)\sin(n\theta)] f(\xi) \, d\xi. \quad (6.36)$$

with a_0 an arbitrary constant. The factor $1/2$ is customary and will prove handy in the ensuing calculation.

It is possible to sum this series to obtain an integral expression for the solution which is analogous to Poisson's formula for the Dirichlet problem for a disk. Interchange the summation and the integral and use a trigonometric identity, as we have done before (Section 6.10), to write equation 6.36 as

$$u(r, \theta) = \frac{1}{2} a_0 + \frac{\rho}{\pi} \int_{-\pi}^{\pi} \left(\sum_{n=1}^{\infty} \frac{1}{n} \left(\frac{r}{\rho}\right)^n \cos(n(\theta - \xi))\right) f(\xi) \, d\xi. \quad (6.37)$$

The interchange of $\sum_{n=1}^{\infty}$ and $\int_{-\pi}^{\pi}$ is justified by the uniform convergence of this series as a function of ξ, for $0 \le r < a$. To sum the series in large parentheses in equation 6.37, let $z = Re^{i\zeta}$. Then

$$z^n = R^n \cos(n\zeta) + iR^n \sin(n\zeta)$$

and

$$\sum_{n=1}^{\infty} R^n \cos(n\zeta) = \sum_{n=1}^{\infty} \text{Re}(z^n) = \text{Re}\left[\sum_{n=1}^{\infty} z^n\right],$$

in which Re denotes the real part of a complex number. We envision letting $R = r/\rho$ after some calculation. Since $0 \le R < \rho$, $|z| < 1$ and

$$\sum_{n=1}^{\infty} z^n = \frac{z}{1-z}.$$

Now compute

$$\frac{1}{2} + \sum_{n=1}^{\infty} z^n = \frac{1}{2} + \frac{z}{1-z} = \frac{1+z}{2(1-z)}.$$

To find the real part of this expression, first multiply numerator and denominator by $1 - \bar{z}$:

$$\frac{1}{2} + \sum_{n=1}^{\infty} z^n = \frac{1}{2}\frac{1+z}{1-z}\frac{1-\bar{z}}{1-\bar{z}} = \frac{1}{2}\frac{1-\bar{z}+z-z\bar{z}}{1-z-\bar{z}+z\bar{z}}.$$

But

$$z + \bar{z} = R\cos(\zeta) + iR\sin(\zeta) + [R\cos(\zeta) - iR\sin(\zeta)] = 2R\cos(\xi),$$

$$z - \bar{z} = 2iR\sin(\xi)$$

and

$$z\bar{z} = R^2.$$

Therefore

$$\frac{1}{2} + \sum_{n=1}^{\infty} z^n = \frac{1}{2}\left(\frac{1 + 2iR\sin(\zeta) - R^2}{1 - 2R\cos(\zeta) + R^2}\right).$$

Then

$$\frac{1}{2} + \sum_{n=1}^{\infty} R^n \cos(n\zeta) = \text{Re}\left[\frac{1}{2} + \sum_{n=1}^{\infty} z^n\right] = \frac{1}{2}\frac{1 - R^2}{1 + R^2 - 2R\cos(\zeta)}$$

so

$$\sum_{n=1}^{\infty} R^n \cos(n\zeta) = \frac{1}{2} \frac{1 - R^2}{1 + R^2 - 2R\cos(\zeta)} - \frac{1}{2} = \frac{R\cos(\zeta) - R^2}{1 + R^2 - 2R\cos(\zeta)}.$$

Upon dividing by R we obtain

$$\sum_{n=1}^{\infty} R^{n-1} \cos(n\zeta) = \frac{\cos(\zeta) - R}{1 + R^2 - 2R\cos(\zeta)}.$$

This series also converges uniformly as a function of R so

$$\int_0^R \sum_{n=1}^{\infty} t^{n-1} \cos(n\zeta)\, dt = \sum_{n=1}^{\infty} \int_0^R t^{n-1} \cos(n\zeta)\, dt = \sum_{n=1}^{\infty} \frac{1}{n} R^n \cos(n\zeta)$$

$$= \int_0^R \frac{\cos(\zeta) - t}{1 + t^2 - 2t\cos(\zeta)}\, dt = -\frac{1}{2} \ln(1 + R^2 - 2R\cos(\zeta)).$$

Now put $R = r/\rho$ and $\zeta = \theta - \xi$ to obtain

$$\sum_{n=1}^{\infty} \frac{1}{n} \left(\frac{r}{\rho}\right)^n \cos(n(\theta - \xi)) = -\frac{1}{2} \ln\left(1 + \frac{r^2}{\rho^2} - 2\frac{r^2}{\rho}\cos(\theta - \xi)\right).$$

Finally, we can write the solution 6.37 as

$$u(r, \theta) = \frac{1}{2} a_0 - \frac{\rho}{2\pi} \int_{-\pi}^{\pi} \ln\left(1 + \frac{r^2}{\rho^2} - 2\frac{r}{\rho}\cos(\theta - \xi)\right) f(\xi)\, d\xi,$$

in which a_0 can be any real number. This is the solution of the Neumann problem for the disk of radius ρ about the origin in R^2.

EXERCISE 356   Use the expression 6.36 to solve:

$$\nabla^2 u(r, \theta) = 0 \text{ for } 0 \le r < \rho, \ -\pi \le \theta \le \pi$$

$$\frac{\partial u}{\partial r}(\rho, \theta) = \sin(3\theta) \text{ for } -\pi \le \theta \le \pi.$$

EXERCISE 357   Solve

$$\nabla^2 u(r, \theta) = 0 \text{ for } 0 \le r < \rho, \ -\pi \le \theta \le \pi$$

$$\frac{\partial u}{\partial r}(\rho, \theta) = \cos(2\theta) \text{ for } -\pi \le \theta \le \pi.$$

EXERCISE 358   Solve

$$\nabla^2 u(x, y) = 0 \text{ for } 0 \le x^2 + y^2 < 9$$

$$\frac{\partial u}{\partial n} = 4xy \text{ for } x^2 + y^2 = 9.$$

Hint: Convert the problem to polar coordinates.

EXERCISE 359   Solve

$$\nabla^2 u(x, y) = 0 \text{ for } x^2 + y^2 < 1$$

$$\frac{\partial u}{\partial n} = x \text{ for } x^2 + y^2 = 1.$$

EXERCISE 360   Solve

$$\nabla^2 u(x, y) = 0 \text{ for } x^2 + y^2 < 1$$

$$\frac{\partial u}{\partial n} = xy^2 \text{ for } x^2 + y^2 = 1.$$

6.20   NEUMANN PROBLEM FOR THE UPPER HALF PLANE

As an illustration of a Neumann problem in an unbounded domain, consider the following problem for the upper half plane:

$$\nabla^2 u(x, y) = 0 \text{ for } -\infty < x < \infty, y > 0$$

$$u_y(x, 0) = f(x) \text{ for } -\infty < x < \infty.$$

u_y is the normal derivative to the horizontal axis.

Although $\partial \Omega$ in this problem is not a closed curve, it is possible to show that a necessary condition for a solution to exist is that the integral of the data function over the boundary is zero. Thus we assume that

$$\int_{-\infty}^{\infty} f(x) \, dx = 0.$$

There is an elegant device for reducing this problem to one we have already solved. Let $v = u_y$. Then

$$\nabla^2 v = v_{xx} + v_{yy} = (u_y)_{xx} + (u_y)_{yy} = (u_{xx})_y + (u_{yy})_y$$

$$= (\nabla^2 u)_y = 0 \text{ for } -\infty < x < \infty, \, y > 0$$

and

$$v(x, 0) = u_y(x, 0) = f(x) \text{ for } -\infty < x < \infty.$$

We conclude that v is the solution of a Dirichlet problem for the upper half plane, if u is a solution of the Neumann problem. But we know the solution of this Dirichlet problem. By equation 6.16 it is

$$v(x, y) = \frac{y}{\pi} \int_{-\infty}^{\infty} \frac{f(\xi)}{y^2 + (\xi - x)^2} \, d\xi.$$

Since $v = u_y$, we can recover u from this solution for v by integrating with respect to y:

$$u(x, y) = \frac{1}{\pi} \int \int_{-\infty}^{\infty} \frac{y}{y^2 + (\xi - x)^2} f(\xi) \, d\xi \, dy = \frac{1}{\pi} \int_{-\infty}^{\infty} \left(\int \frac{y}{y^2 + (\xi - x)^2} \, dy \right) f(\xi) \, d\xi$$

$$= \frac{1}{2\pi} \int_{-\infty}^{\infty} \ln(y^2 + (\xi - x)^2) f(\xi) \, d\xi + c, \tag{6.38}$$

in which c is an arbitrary constant.

EXERCISE 361   Solve the Neumann problem for the lower half plane:

$$\nabla^2 u(x, y) = 0 \text{ for } -\infty < x <, \, y < 0$$

$$u_y(x, 0) = f(x) \text{ for } -\infty < x < \infty.$$

Assume that f is bounded and continuous. Is any other condition on f required for this problem to have a solution?

EXERCISE 362   Solve the Neumann problem for the right quarter plane:

$$\nabla^2 u(x, y) = 0 \text{ for } x > 0, \, y > 0$$

$$u_x(0, y) = 0 \text{ for } y \geq 0$$

$$u_y(x, 0) = f(x) \text{ for } x \geq 0.$$

Assume that f is bounded and continuous. Are any other conditions on f required for this problem to have a solution?

EXERCISE 363   Solve the Neumann problem for the right half plane:

$$\nabla^2 u(x, y) = 0 \text{ for } x > 0, \ -\infty < y < \infty$$

$$u_y(0, y) = g(y) \text{ for } -\infty < y < \infty.$$

Assume that g is bounded and continuous. Is any other condition on g required?

EXERCISE 364   Solve the Neumann problem for the left half plane:

$$\nabla^2 u(x, y) = 0 \text{ for } x < 0, \ -\infty < y < \infty$$

$$u_y(0, y) = g(y) \text{ for } -\infty < y < \infty.$$

EXERCISE 365   Solve the problem

$$\nabla^2 u(x, y) = 0 \text{ for } x > 0, \ y > 0$$

$$u(0, y) = 0 \text{ for } y \geq 0$$

$$u_y(x, 0) = f(x) \text{ for } x \geq 0.$$

Hint: The solution can be obtained using equation 6.38.

7

CONCLUSION

7.1 HISTORICAL PERSPECTIVE

This short section is devoted to some historical remarks on the evolution of the field of partial differential equations. In a broad sense, we may think of the eighteenth century as belonging to the wave equation, and the nineteenth to the heat equation, with Laplace's equation arising from studies of potential theory. Much of the material we touch upon here is discussed in greater detail in Morris Kline's *Mathematical Thought from Ancient to Modern Times*, published by Oxford University Press.

The first systematic attack on a problem involving a partial differential equation was carried out in a sequence of 1746 papers by Jean Le Rond d'Alembert (1717–1783), who sought the fundamental modes of vibration of a vibrating string. In 1727 John Bernoulli had approximated the vibrating string by imagining a finite number of beads connected by a weightless elastic string, and then allowing the number of beads to become infinite. This, however, led to an equation independent of time, and hence not a partial differential equation as we encounter with the wave equation.

D'Alembert's papers, in which he derived his formula for the solution in terms of the initial position and velocity functions, were followed by investigations by Leonhard Euler (1707–1783), the prolific Swiss mathematician, who considered a more general class of functions than had d'Alembert as possible solutions of the wave equation. It was Euler who, in 1749, obtained what we would now recognize as a Fourier series solution, resulting in a protracted and sometimes heated debate over the validity of certain kinds of "functions" as solutions. The argument engaged some of the leading mathematicians of the day, including Euler, d'Alembert, Joseph-Louis Lagrange (1736–1813), Pierre-

Simon de Laplace (1747–1827), and Daniel Bernoulli (1700–1782). Much of the dispute was generated by differing views on the definitions of function and continuity, as well as a lack of the mathematical tools needed to understand convergence of trigonometric series.

Daniel Bernoulli was perhaps the first mathematical physicist, and was in residence at the St. Petersburg academy established by Catherine the Great of Russia when he did fundamental work on hydrodynamics, elasticity, and the motion of vibrating systems. His 1732 paper gave the higher modes of vibration of a vibrating string.

Some work, particularly by Euler and d'Alembert, was done on a general version of the wave equation. Euler obtained a solution when the mass distribution of the string had a special form (hence was allowed to be nonconstant), but was unable to make progress when the mass was an arbitrary function. d'Alembert also attempted solutions when the thickness of the string varied. Euler considered wave motion in two space dimensions, deriving a series solution for a vibrating membrane. This series contained functions we would now recognize as Bessel functions.

Another major impetus for work in partial differential equations came from potential theory, motivated by research on gravitational effects on bodies of different shapes and mass distributions.

First order partial differential equations began to receive attention about 1739, when Alexis-Claude Clairaut (1713–1765) encountered them in studying the shape of the earth. In the 1770s Lagrange gave the first systematic treatment of nonlinear first order equations

$$f(x, y, u, u_x, u_y) = 0.$$

Lagrange had in his possession a form of the method of characteristics, although today Augustin-Louis Cauchy (1789–1857) is credited with this method because he overcame some difficulties that had proved intractable to Lagrange.

Gaspard Monge (1746–1818) was among the first to associate geometry with first order partial differential equations, developing the concepts of characteristic surfaces and characteristic cones (or Monge cones). Monge also did work on the linear, homogeneous, second order partial differential equation.

There was also some work on systems of partial differential equations, primarily by Euler and d'Alembert. This was motivated by the fact that Newton's laws of motion are in vector form, and lead to a system when written in terms of components.

The major innovation of the nineteenth century was the development of Fourier analysis and techniques, and all of the mathematics that arose from studying Fourier series, integrals, and transforms. Joseph Fourier (1768–1830) was a fascinating person. In addition to his mathematical contributions, he accompanied Napoleon on his ill-fated Egyptian military campaign, became known (and is known to this day) as a historian of ancient Egypt, and later, while serving as a government administrator in France, was responsible for a large land reclamation project and the construction of a highway through the Alps. He was also, at the beginning of his career, condemned to the guillotine by the Robsepierre-led Revolution, but was spared at the last minute by the rapidly changing political scene in France at this time.

Fourier's main interest was in phenomena related to the theory of heat, hence the emphasis in the nineteenth century on the heat equation. The paper he submitted to the Academy of Sciences in Paris in 1807 was judged by Lagrange, Laplace, and Legendre and found wanting in rigor, but these eminent mathematicians were impressed by the originality and importance of Fourier's ideas and established a prize for work on the problem of heat conduction. In 1811 Fourier submitted a revised version of his paper in competition for the prize, which he won. Nevertheless, the Academy continued to refuse publication of the work. Finally, in 1822 he published his classic *Theorie analytique de la chaleur*, in which his ideas on separation of variables and the use of Fourier series and integrals were given to the world. Subsequent work devoted to proving convergence theorems about Fourier series opened up whole new areas of mathematics and led to profound developments in analysis.

Potential theory also saw advances in the nineteenth century. Part of the motivation was found in studies of gravitational attraction, but Laplace's equation also arose as the steady state heat equation, and Maxwell's equations gave new initiative to potential theory as well. Major contributors were Simeon-Denis Poisson (1781–1840), Cauchy, Laplace, and Carl Friedrich Gauss (1777–1855), who were all scientists and mathematicians of considerable stature. A surprising figure in potential theory who emerged at this time was the Englishman George Green (1793–1841), who taught himself calculus and, in 1828, produced at his own expense a little book entitled *An Essay on the Application of Mathematical Analysis to the Theories of Electricity and Magnetism*. Because Green was unknown, the booklet went largely unnoticed until Lord Kelvin happened upon a copy, understood the importance of its contents and had it published in a journal. Among other results, Green proved several integral relationships which now bear his name (Green's Theorem, Green's identities). The important Green's Theorem was proved independently in 1828 by the Ukrainian mathematician Michel Ostrogradsky (1801–1861). Green's functions are also named for George Green.

Much of Green's booklet is devoted to applications of his integral theorems to studies of electricity and magnetism. In one section he anticipated the use of Dirichlet's Principle in attempting to prove the existence of a solution of a Dirichlet problem.

As one might expect, early work on partial differential equations focused on attempts to write solutions in terms of series, integrals, or newly developed special functions. Soon it became apparent that subtle existence questions needed to be addressed. Cauchy was among the first to observe that partial differential equations of order two or higher can be exchanged for systems of first order partial differential equations, and he proved an existence theorem which today bears his name and that of Sophie Kowalevski. Kowalevski (1850–

1891) was one of the few women to achieve prominence in mathematics in the nineteenth century. She was a student of the leading German mathematician Karl Weierstrass, and in addition to her results on existence, she did prize-winning work on mathematical treatments of elasticity and the motion of rotating bodies. In the latter stages of her relatively short life she was a professor of mathematics at the University of Stockholm in Sweden.

The theory of characteristics had been developed by Monge and by André-Marie Ampère (1775–1836), and was extended in the early part of this century by Albert Victor Bäcklund (1845–1922). Jacques Hadamard (1865–1963) adapted the theory to partial differential equations of arbitrary order. Important work was also done on the existence of solutions of Dirichlet problems, particularly by Hermann Amandus Schwarz (1843–1921), who, like Kowalevski, was a student of Weierstrass. Important contributions to the understanding of harmonic functions were also made by Henri Poincaré (1854–1912) and David Hilbert (1862–1943), respectively the leading French and German mathematicians of the last part of the nineteenth and early part of this century. It was Hilbert who proved the Dirichlet Principle anticipated by Bernhard Riemann and George Green.

Today partial differential equations cover a vast body of theory and significant areas of applications, particularly to mathematical physics. In applied research requiring the use of partial differential equations, such as studies of turbulence in fluid flow, the advent of modern computing power has made it possible to obtain numerical information about solutions which defy explicit representation. Nevertheless, the need for rigor in establishing results has been dramatically demonstrated by the construction of unanticipated examples. In 1957, Hans Lewy produced a linear partial differential equation having no singular points (that is, the coefficient functions are well behaved everywhere), and for which no solution exists at any point. For a discussion of this result, see Fritz John [9].

7.2 REFERENCES FOR FURTHER READING

[1]   Berg, Paul W., and McGregor, James L., *Elementary Partial Differential Equations*, Holden-Day, 1966.

[2]   Bers, Lipman, John, F., and Schechter, M., *Partial Differential Equations*, Interscience Publishers, 1964.

[3]   Carrier, George F., and Pearson, Carl E., *Partial Differential Equations, Theory and Technique*, second edition, Academic Press, 1988.

[4]   Colton, Davbid, *Partial Differential Equations, An Introduction*. Random House, 1988.

[5]   Churchill, Ruel V., and Brown, James Ward, *Fourier Series and Boundary Value Problems*, third edition, McGraw-Hill, 1978.

[6]   Egorov, Yu. V., and Shubin, M.A., *Partial Differential Equations I*, Springer-Verlag Encyclopaedia of Mathematical Sciences, Volume 30, 1988.

[7]   Garabedian, P.R., *Partial Differential Equations*, John Wiley & Sons, 1964.

[8]   Hörmander, Lars, *Linear Partial Differential Operators*, Springer-Verlag, 1963.

[9]   John, Fritz, *Partial Differential Equations*, fourth edition, Springer-Verlag, 1982.

[10]   Körner, T.W., *Fourier Analysis*, Cambridge University Press, 1988.

[11]   Lax, P.D., *Lectures on Hyperbolic Partial Differential Equations*, Stanford University, 1963.

[12]   Petrovsky, I.G., *Lectures on Partial Differential Equations*, Interscience Publishers, 1954.

[13]   Rauch, Jeffrey, *Partial Differential Equations*, Springer-Verlag, 1991.

[14]   Schechter, M., *Modern Methods in Partial Differential Equations, An Introductions*, McGraw Hill, 1977.

[15]   Treves, F., *Basic Linear Partial Differential Equations*, Academic Press, 1975.

[16]   Troutman, John L., *Boundary Value Problems of Applied Mathematics*, PWS Publishing Co., 1994.

[17]   Widder, D.V., *The Heat Equation*, Academic Press, 1975.

[18]   Young, Eutiquio C., *Partial Differential Equations, An Introduction*, Allyn and Bacon, 1972.

[19]   Zachmanoglou, E.C., and Thoe, Dale W., *Introduction to Partial Differential Equations*, William and Wilkins Co., 1976.

7.3 GLOSSARY

$f(x+)$	$= \lim_{h \to 0+} f(x)$
$f(x-)$	$= \lim_{h \to 0-} f(x)$
$\mathfrak{F}$	Fourier transform
$\mathfrak{F}^{-1}$	inverse Fourier transform
$\mathfrak{F}[f]$	Fourier transform of f
$f * g$	convolution of f with g
$\mathfrak{F}_s$	Fourier sine transform
$\mathfrak{F}_c$	Fourier cosine transform
∇u	gradient of u
$\nabla^2 u$	Laplacian of u; $\nabla^2 u(x, y) = u_{xx} + u_{yy}$; $\nabla^2 u(x, y, z) = u_{xx} + u_{yy} + u_{zz}$
$H(t)$	Heaviside function evaluated at t, defined by

$$H(t) = \begin{cases} 1 \text{ for } t \geq 0 \\ 0 \text{ for } t < 0. \end{cases}$$

$\mathfrak{L}$	Laplace transform
R^n	n-dimensional space of n-tuples $(x_1, \ldots, x_n)$, with each x_j a real number

x (boldface)   a vector, or element of R^n

$B(\mathbf{x}_0, r)$ open ball of radius r about $\mathbf{x}_0$

∂A boundary of A

$S(\mathbf{x}_0, r)$ sphere of radius r about $\mathbf{x}_0$ in R^n (equivalently, $S(\mathbf{x}_0, r) = \partial B(\mathbf{x}_0, r)$)

$\bar{A}$ closure of A

$\partial u/\partial n$ normal derivative of u in the direction of a unit vector $\mathbf{n}$

$C^n(A)$ set of functions continuous on A, with continuous partial derivatives on A of orders 1 through n

$f\colon D \rightarrow K$ f is a function mapping elements of D into K

$\mathrm{Re}(z)$ real part of z

$\mathrm{Im}(z)$ imaginary part of z

w_B for continuous w defined on $\bar{\Omega}$, and B an open ball with $B \subset \Omega$, w_B coincides with w on $\bar{\Omega} - B$ and is harmonic on B

ω_y if $\mathbf{y} \in \partial\Omega$, ω_y is a barrier function defined on $\bar{\Omega}$ by the conditions that ω_y is superharmonic and

$$\omega_y(\mathbf{y}) = 0$$
$$\omega_y(\mathbf{x}) > 0 \text{ for } \mathbf{x} \in \bar{\Omega} - \{\mathbf{y}\}.$$

7.4   ANSWERS TO SELECTED EXERCISES

SECTION 1.1

Exercise 4

1. With $p(u) = k$ the solution is $u(x, y) = \varphi(x + kt)$.

3. With $p(u) = \cos(u)$ the solution is defined by

$$u(x, t) = \varphi(x + \cos(u)t).$$

Exercise 7

1. quasi-linear and nonlinear

3. quasi-linear and nonlinear

5. quasi-linear and nonlinear

7. nonlinear and not quasi-linear

SECTION 1.2

Exercise 9

1. (a)   characteristics are lines $5x - 3y = k$

    (b) $\xi = x$, $\eta = 5x - 3y$;

$$w_\xi + \frac{1}{9}(\eta\xi - 5\xi^2)w = 0$$

    (c)

$$w(\xi, \eta) = g(\eta)e^{5\xi^2/27}e^{-\eta\xi^2/18}$$

    (d)

$$u(x, y) = g(5x - 3y)e^{(yx^2 - 5x^3/9)/6}$$

3. (a)   characteristics are lines $4x - y = k$

    (b) $\xi = x$, $\eta = 4x - y$; $w_\xi - \xi w = \xi$

    (c) $w(\xi, \eta) = -1 + g(\eta)e^{\xi^2/2}$

    (d) $u(x, y) = -1 + g(4x - y)e^{x^2/2}$

5. (a)   characteristics are hyperbolas $y = c/x$

    (b) $\xi = x$, $\eta = xy$;

$$w_\xi + \frac{1}{\xi}w = 1$$

    (c)

$$w(\xi, \eta) = \frac{1}{2}\xi + \frac{1}{\xi}g(\eta)$$

    (d)

$$u(x, y) = \frac{1}{2}x + \frac{1}{x}g(xy)$$

7. (a)   characteristics are parabolas $y + x^2/2 = c$

    (b) $\xi = x$, $\eta = y + x^2/2$; $w_\xi = 4$

    (c) $w(\xi, \eta) = 4\xi + g(\eta)$

(d) $u(x, y) = 4x + g(y + x^2/2)$

9. (a)   characteristics are lines $y - x = k$

(b) $\xi = x, \eta = y - x; w_\xi - w = \xi + \eta$

(c) $w(\xi, \eta) = -1 - \xi - \eta + g(\eta)e^\xi$

(d) $u(x, y) = -1 - y + e^x g(y - x)$

11. (a)   characteristics are curves $ye^{-x} = c$

(b) $\xi = x, \eta = ye^{-x}; w_\xi + \xi w = 0$

(c) $w(\xi, \eta) = e^{-\xi^2/2}g(\eta)$

(d) $u(x, y) = e^{-x^2/2}g(ye^{-x})$

Exercise 10

The solution satisfying $u(0, y) = y^n$ is

$$u(x, y) = [1 + e^{-\alpha x + \ln(y-1)}] \exp\left(\frac{\beta}{2} e^{-n}\int_0^x f(s)e^{\alpha s} \, ds\right).$$

The solution satisfying $u(0, y) = y^n$ is

$$u(x, y) = \frac{1}{2}\beta(y - 1) \int_0^x f(s) \, ds + ((y - 1)e^{-\alpha x} + 1)^n.$$

SECTION 1.3

Exercise 11

1. (a) $u(x, y) = \frac{1}{3}(3y^2 + 2x^2)$

(b) $u(x, y) = 1 - \frac{1}{3}(3y^2 + 2x^2)$

(c)   no solution

3. (a)

$$u(x, y) = -1 + [1 + \cos(y - 2x)]e^{(-y+3x)/4}$$

(b)   no solution

(c)

$$u(x, y) = \begin{cases} -1 + (1 - \sqrt{1 + y - 2x})e^{x/4} \exp(1 + \sqrt{1 + y - 2x})/4 \text{ if } x > 1 \\ -1 + (1 + \sqrt{1 + y - 2x})e^{x/4} \exp(1 - \sqrt{1 + y - 2x})/4 \text{ if } x < 1 \end{cases}$$

5. (a)

$$u(x, y) = \frac{1}{2}x^2 + 4\left(\frac{3}{5}(2x^3 - 3y^2)\right)^{1/3} - \frac{1}{2}\left(\frac{3}{5}(2x^3 - 3y^2)\right)^{2/3}$$

(b)   no solution

(c)

$$u(x, y) = \frac{1}{2}x^2 + \sin\left(\left(\frac{2x^3 - 3y^2}{2}\right)^{1/3}\right) - \frac{1}{2}\left(\frac{3x^3 - 2y^2}{2}\right)^{2/3}$$

SECTION 1.4

Exercise 12

1.

$$u(x, y) = \arcsin\left(\frac{1}{2}\ln\left(\frac{x^3}{y}\right)\right)$$

3. u is implicitly defined by

$$\frac{1}{y} = u + \frac{1}{(x - u)^2 + 2}$$

5. u is implicitly defined by

$$y = (1 - x + u)\ln(u)$$

7.

$$u(x, y) = \frac{3}{y - 2x + \dfrac{7}{4}}$$

9. u is implicitly defined by

$$\frac{1}{y} = \ln(u/2) + \frac{1}{1 - (x - \ln(u/2))^2}$$

SECTION 2.2

Exercise 16

1. $u(x, y) = f(y - 3x) + g(y + x)$, with f and g twice differentiable functions of one variable.

3.

$$u(x, y) = f\left(y - \frac{1 + \sqrt{11}}{2}x\right) + g\left(y - \frac{1 - \sqrt{11}}{2}x\right)$$

SECTION 2.3

Exercise 17

Obtain the canonical form

$$w_{\eta\eta} + \left(\frac{1}{2\eta} + \frac{1}{18\sqrt{\eta}}\right)w_\eta - \frac{1}{54\eta}w_\xi = 0$$

for $\eta = x^2$.

Exercise 18

1. $u(x, y) = xF(y - 3x) + G(y - 3x)$, with F and G twice differentiable functions of one variable.

3. $u(x, y) = xF(y - 2x/5) + G(y - 2x/5)$

SECTION 2.4

Exercise 19

1. (a) hyperbolic

   (b) characteristics are graphs of $y = (-4 + \sqrt{14})x + c$, $y = (-4 - \sqrt{14})x + k$

   (c)

$$w_{\xi\eta} + \frac{1}{58\sqrt{14}}(4\xi - (4 - \sqrt{14})\eta)w_\xi + \frac{1}{58\sqrt{14}}((14 + \sqrt{14})\xi - 4\eta)w_\eta = 0$$

3. (a) hyperbolic

(b)   characteristics are graphs of $y - x = c,\ y + x/3 = k$

(c)

$$w_{\xi\eta} + \frac{3}{64}\ (\xi + 3\eta + 4)w_\xi - \frac{1}{64}\ (\xi + 3\eta - 12)w_\eta = 0$$

5. (a)   hyperbolic

(b)   characteristics are graphs of

$$y = \left(\frac{-4 + \sqrt{10}}{3}\right)\ x + c,\ y = \left(\frac{-4 - \sqrt{10}}{3}\right)\ x + k$$

(c)

$$w_{\xi\eta} - \frac{3}{32\sqrt{10}}\ ((1 + \sqrt{10})\xi + (-1 + \sqrt{10})\eta)(w_\xi + w_\eta) = 0$$

7. (a)   parabolic

(b)   characteristics are graphs of $y + 2x = c$

(c) $w_{\eta\eta} + 3w_\xi + w_\eta = 0$

9. (a)   hyperbolic

(b)   characteristics are graphs of $y + 4x = c,\ y + x = k$

(c) $w_{\xi\eta} - \frac{1}{6}w_\xi = 0$

Exercise 20

1. The canonical forms are

$$w_{\xi\eta} - \frac{7}{36}\ w_\xi - \frac{1}{36}\ w_\eta = 0,\ \hat{w}_{\xi\xi} - \hat{w}_{\hat{\eta}\hat{\eta}} - \frac{2}{9}\ \hat{w}_\xi - \frac{1}{6}\ \hat{w}_{\hat{\eta}} = 0.$$

3. The canonical forms are

$$w_{\xi\eta} - \frac{1}{250}\ (\xi + 4\eta)w = 0,\ \hat{w}_{\xi\xi} - \hat{w}_{\hat{\eta}\hat{\eta}} - \frac{1}{100}\ \xi\hat{w} + \frac{3}{500}\ \hat{\eta}\hat{w} = 0.$$

Exercise 21

Choose $\alpha = a/2,\ \beta = -b/2,\ h = (b^2 - a^2)/4 + c$

Exercise 25

(a) Tricomi's equation is hyperbolic if $x < 0$, elliptic if $x > 0$, parabolic if $x = 0$.

(b) If $x < 0$, characteristics are the curves

$$y = \pm\frac{2}{3}(-x)^{3/2} + c.$$

(c) No.

SECTION 2.6

Exercise 37

1. If $u = \varphi(x, y)$ is the solution, then

$$\varphi(x, y) = y^3 + 4xy + \frac{1}{2}(16 - 6y^2 - 4y - 3y^3)x^2$$

$$+ \frac{1}{6}(-32 - 48y - 30y^2 + 3y^3)x^3 + \cdots.$$

3.

$$\varphi(x, y) = y - y^2 + \cos(y)x - 4\cos(y)x^2$$

$$+ \frac{1}{6}(y\sin(y) - 2 + 64\cos(y))x^3 + \cdots$$

Exercise 38

1. If $u = \varphi(x, y)$ is the solution, then

$$\varphi(x, y) = x^2 + xe^{-x}y + (1 + 2x)y^2 - \frac{1}{6}(x^2 + xe^{-x})y^3 + \cdots.$$

3.

$$\varphi(x, y) = 1 - x^3 + x^4y + \frac{1}{2}(6x + 2x^3 + 4x^4 + 4x^5 - 2)y^2$$

$$+ \frac{1}{6}(12x^2 - 2x + 6x^3 + 22x^4 + 32x^5 + 16x^6)y^3 + \cdots$$

SECTION 2.7

Exercise 40

$$\varphi_{\xi\xi} = \frac{1}{J^2} \left(M\eta_y^2 - 2Q\eta_x\eta_y + N\eta_x^2 \right)$$

$$\varphi_{\eta\eta} = \frac{1}{J^2} \left(N\xi_x^2 - 2Q\xi_x\xi_y + M\xi_y^2 \right)$$

$$\varphi_{\xi\eta} = \frac{1}{J^2} \left((\xi_x\eta_y + \xi_y\eta_x)Q - N\eta_x\xi_x - M\eta_y\xi_y \right),$$

where

$$M = \varphi_{xx} - \varphi_\xi \xi_{xx} - \varphi_\eta \eta_{xx}$$

$$N = \varphi_{yy} - \varphi_\xi \xi_{yy} - \varphi_\eta \eta_{yy}$$

$$Q = \varphi_{xy} - \varphi_\xi \xi_{xy} - \varphi_\eta \eta_{xy}.$$

Exercise 41

$$\varphi(\xi, \eta) = \frac{1}{2} \xi + \frac{1}{24} \xi^2 + \frac{1}{4} \eta^2 + \cdots$$

Exercise 43

$$\varphi(\xi, \eta) = \frac{1}{39} \xi + \frac{1}{6} \eta + \frac{1}{26 \cdot 19} \xi^2 + \cdots$$

SECTION 2.9

Exercise 47

1. $u(t, x) = 3 + x + t - \frac{1}{2}t^2 + xt - \frac{1}{3}t^3 + \frac{1}{2}xt^2 + \cdots$

3. $u(t, x) = 1 - x - \frac{1}{2}t^2 + xt + \frac{1}{3}t^3 - x^2t + \cdots$

5. $u(t, x) = 2t + x^2 - t^2 + xt + \frac{1}{6}t^3 + xt^2 + \cdots$

7. $u(t, x) = 1 + t^2 - \frac{1}{2}x^2 - xt - \frac{1}{3}t^3 + \frac{1}{2}xt^2 + \cdots$

9. $u(t, x) = -1 + x + t + \frac{1}{2}x^2 - \frac{3}{2}t^2 + t^3 \quad - x^2t + \frac{1}{2}xt^2 + \cdots$

SECTION 3.3

Exercise 60

1. $\sum_{n=1}^{\infty} \dfrac{2(-1)^{n+1}}{n\pi} \sin(n\pi x)$; converges to $\begin{cases} -x \text{ for } -1 < x < 1 \\ 0 \text{ for } x = \pm 1. \end{cases}$

3. $\sum_{n=1}^{\infty} \dfrac{4}{n^2\pi^2}(1 - (-1)^n)\cos(n\pi x/2)$; converges to $1 - |x|$ for $-2 \le x \le 2$

5. The Fourier series is just $\sin(2x)$ on $[-\pi, \pi]$; this converges to $\sin(2x)$ for $-\pi \le x \le \pi$.

7.

$$\frac{71}{12} + \sum_{n=1}^{\infty} \left(\left[\frac{5}{n^2\pi^2}((-1)^n - 1) + \frac{50}{n^2\pi^2}(-1)^n \right] \cos(n\pi x/5) \right.$$

$$\left. + \left[\frac{5(-1)^n}{n\pi} - \frac{(26n^2\pi^2 - 50)(-1)^n - n^2\pi^2 + 50}{n^3\pi^3} \right] \sin(n\pi x/5) \right);$$

converges to $\begin{cases} -x \text{ for } -5 < x < 0 \\ 1 + x^2 \text{ for } 0 < x < 5 \\ 1/2 \text{ for } x = 0 \\ 31/2 \text{ for } x = \pm 5. \end{cases}$

9.

$$\frac{2}{\pi} + \frac{4}{3\pi}\cos(x) - \sin(x) + \sum_{n=2}^{\infty} \frac{4(-1)^{n+1}}{\pi(4n^2 - 1)}\cos(nx);$$

converges to $\cos(x/2) - \sin(x)$ for $-\pi \le x \le \pi$.

Exercise 61

1. The Fourier series on $[-3, 3]$ converges to

$$\begin{cases} 2x \text{ for } -3 < x < -2 \\ 0 \text{ for } -2 < x < 1 \\ x^2 \text{ for } 1 < x < 3 \\ 3/2 \text{ for } x = \pm 3 \\ -2 \text{ for } x = -2 \\ 1/2 \text{ for } x = 1. \end{cases}$$

3. The Fourier series on $[-4, 4]$ converges to

$$\begin{cases} -2 \text{ for } -4 < x < 2 \\ 1 + x^2 \text{ for } 2 < x < 3 \\ e^{-x} \text{ for } 3 < x < 4 \\ (-2 + e^{-4})/2 \text{ for } x = \pm 4 \\ 3/2 \text{ for } x = 2 \\ (10 + e^{-3})/2 \text{ for } x = 3. \end{cases}$$

5. The Fourier series on $[-2, 2]$ converges to

$$\begin{cases} \cos(\pi x) \text{ for } -2 < x < 0 \\ x \text{ for } 0 < x < 2 \\ 3/2 \text{ for } x = \pm 2 \\ 1/2 \text{ for } x = 0. \end{cases}$$

Exercise 62

The Fourier series of x^2 on $[-\pi, \pi]$ is

$$\frac{1}{3} \pi^2 + \sum_{n=1}^{\infty} \frac{4(-1)^n}{n^2} \cos(nx).$$

SECTION 3.4

Exercise 68

1. The sine series on $[0, 3]$ is

$$\sum_{n=1}^{\infty} \frac{8(1 - (-1)^n)}{n\pi} \sin(n\pi x/3);$$

this converges to $\begin{cases} 4 \text{ for } 0 < x < 3 \\ 0 \text{ for } x = 0, 3. \end{cases}$

The cosine series is just 4, converging to 4 for $0 \le x \le 3$.

3. The sine series is

$$\frac{1}{2} \sin(x) + \sum_{n=2}^{\infty} \frac{2n \cos(n\pi/2)}{\pi(n^2 - 1)} \sin(nx),$$

converging to $\begin{cases} 0 \text{ for } 0 \le x < \pi/2 \text{ and for } x = \pi \\ \sin(x) \text{ for } \pi/2 < x < \pi \\ 1/2 \text{ for } x = \pi/2. \end{cases}$

The cosine series is

$$\frac{1}{\pi} - \frac{1}{\pi}\cos(x) + \sum_{n=2}^{\infty} -\frac{2}{\pi}\left[\frac{\cos(n\pi) + n\sin(n\pi/2)}{n^2 - 1}\right]\cos(nx),$$

converging to $\begin{cases} 0 \text{ for } 0 \le x < \pi/2 \\ \sin(x) \text{ for } \pi/2 < x \le \pi \\ 1/2 \text{ for } x = \pi/2. \end{cases}$

5. The sine series is

$$\sum_{n=1}^{\infty} \frac{8}{n^3\pi^3}(2(-1)^n - 2 - n^2\pi^2(-1)^n)\sin(n\pi x/2),$$

converging to $\begin{cases} x^2 \text{ for } 0 < x < 2 \\ 0 \text{ for } x = 0, \, 2. \end{cases}$

The cosine series is

$$\frac{4}{3} + \sum_{n=1}^{\infty} \frac{16(-1)^n}{n^2\pi^2}\cos(n\pi x/2),$$

converging to x^2 for $0 \le x \le 2$.

7. The sine series is just $\sin(3x)$, converging to $\sin(3x)$ for $0 \le x \le \pi$. The cosine series is

$$\frac{2}{3\pi} + \sum_{n=1, n\ne 3}^{\infty} -\frac{6}{\pi}\frac{1 + (-1)^n}{n^2 - 9}\cos(nx),$$

converging to $\sin(3x)$ for $0 \le x \le \pi$.

SECTION 3.6

Exercise 81

$$\int_0^{\infty} \frac{2}{\pi}\frac{\sin(\omega\pi) - \omega\pi\cos(\omega\pi)}{\omega^2}\sin(\omega x)\, d\omega,$$

converging to

$$\begin{cases} x \text{ for } -\pi < x < \pi \\ \pi/2 \text{ for } x = \pi \\ -\pi/2 \text{ for } x = -\pi \\ 0 \text{ for } |x| > \pi. \end{cases}$$

3.

$$\int_0^\infty \frac{2}{\pi(1 + \omega^2)} \cos(\omega x) \, d\omega,$$

converging to $e^{-|x|}$ for all x.

5.

$$\int_0^\infty \frac{2}{\pi} \frac{\cos(\alpha\omega) + \alpha\omega \sin(\alpha\omega)}{\omega^2} \cos(\omega x) \, d\omega,$$

converging to $\begin{cases} |x| \text{ for } -\alpha < x < \alpha \\ \alpha/2 \text{ for } x = \pm\alpha \\ 0 \text{ for } |x| > \alpha. \end{cases}$

Exercise 84

1. The sine integral is

$$\frac{e^k}{\pi} \int_0^\infty \left[\frac{-\omega \cos k\omega + \sin k\omega + w(\cos k\omega)e^{-2k} + (\sin k\omega)e^{-2k}}{1 + \omega^2} \right] \sin(\omega x) \, d\omega.$$

converging to $\begin{cases} \sinh(x) \text{ for } 0 \le x < k \\ \sinh(k)/2 \text{ for } x = k \\ 0 \text{ for } x > k. \end{cases}$

The cosine integral is

$$\int_0^\infty \frac{1}{\pi(1 + \omega^2)} ((e^k + e^{-k})\cos(k\omega) + \omega(e^k - e^{-k})\sin(k\omega) - 2)\cos(\omega x) \, d\omega,$$

converging to $\begin{cases} \sinh(x) \text{ for } 0 \le x < k \\ \sinh(k)/2 \text{ for } x = k \\ 0 \text{ for } x > k. \end{cases}$

3. The sine integral is

$$\int_0^\infty \frac{2}{\pi} \frac{\omega^3}{4 + \omega^4} \sin(\omega x) \, d\omega,$$

converging to $\begin{cases} e^{-x} \cos(x) \text{ for } x > 0 \\ 0 \text{ for } x = 0. \end{cases}$

The cosine integral is

$$\int_0^\infty \frac{2}{\pi} \frac{2 + \omega^2}{4 + \omega^4} \cos(\omega x) \, d\omega,$$

converging to $e^{-x} \cos(x)$ for all x.

5. The sine integral is

$$\int_0^\infty \frac{2k}{\pi} \frac{1 - \cos(\alpha\omega)}{\omega} \sin(\omega x) \, d\omega,$$

converging to $\begin{cases} k \text{ for } 0 < x < k \\ 0 \text{ for } x = 0 \text{ and for } x > k \\ k/2 \text{ for } x = k. \end{cases}$

The cosine integral is

$$\int_0^\infty \frac{2k}{\pi} \frac{\sin(\alpha\omega)}{\omega} \cos(\omega x) \, d\omega.$$

SECTION 3.7

Exercise 89

1.

$$\frac{1}{\pi} \int_{-\infty}^\infty \frac{\sin(\pi\omega)}{\omega} e^{i\omega x} \, d\omega$$

3.

$$\int_{-\infty}^\infty \frac{1}{\pi} \frac{1}{1 + \omega^2} e^{i\omega x} \, d\omega$$

5.

$$\int_{-\infty}^\infty \frac{1}{\pi\omega^2} (-1 + \cos(\alpha\omega) + \alpha\omega \sin(\alpha\omega)) e^{i\omega x} \, d\omega$$

Exercise 90

1. $\hat{f}(\omega) = \dfrac{2k}{\omega} \sin(\alpha\omega)$

3.

$$\hat{f}(\omega) = \frac{2i}{\omega^2 - 1} \left(\omega \cos(\alpha\omega)\sin(\alpha) - \sin(\alpha\omega)\cos(\alpha)\right)$$

5.

$$\frac{2(2 + \omega^2)}{4 + \omega^4}$$

SECTION 3.8

Exercise 106

$(f * g)(t) = 2e^{-1-t}$

SECTION 3.9

Exercise 113

1.

$$\mathfrak{F}_s[f](\omega) = \frac{\omega}{1 + \omega^2}, \ \mathfrak{F}_c[f](\omega) = \frac{1}{1 + \omega^2}$$

3.

$$\mathfrak{F}_s[f](\omega) = \frac{1}{1 - \omega^2} \left(\sin(\alpha\omega) \sin(\alpha) + \omega \cos(\alpha\omega) \cos(\alpha) - \omega\right)$$

$$\mathfrak{F}_c[f](\omega) = \frac{1}{\omega^2 - 1} \left(\omega \sin(\alpha\omega)\cos(\alpha) - \cos(\alpha\omega)\sin(\alpha)\right)$$

5.

$$\mathfrak{F}_s[f](\omega) = \frac{\omega^3}{4 + \omega^4}, \ \mathfrak{F}_c[f](\omega) = \frac{2 + \omega^2}{4 + \omega^4}$$

SECTION 4.1

Exercise 119

1. $u(x, t) = \cos(3x)\cos(21t) + xt$

3.

$$u(x, t) = \frac{1}{2}\left(e^{-|x+3t|} + e^{-|x-3t|}\right) + \frac{1}{2}t + \frac{1}{12}\cos(x - 3t)\sin(x - 3t)$$

$$- \frac{1}{12}\cos(x + 3t)\sin(x + 3t)$$

5. $u(x, t) = \cos(x)\cos(2t) - \sin(x)\cos(2t) + \frac{1}{4}(\cos(x - 2t) - \cos(x + 2t))$

7.

$$u(x, t) = \frac{1}{2}\left(\cos(x - 4t) + \cos(x + 4t)\right)$$

$$+ \frac{1}{8}e^{-x}((x + 1 - 4t)e^{4t} - (x + 1 + 4t)e^{-4t})$$

9.

$$u(x, t) = x^3 + 27xt^2 + \frac{1}{6}\left(\cos(x + 3t) - \cos(x - 3t)\right)$$

$$+ \frac{1}{6}((x + 3t)\sin(x + 3t) - (x - 3t)\sin(x - 3t))$$

SECTION 4.2

Exercise 124

1. Figure S1.

3. Figure S2.

5. Figure S3.

There are many different sets of graphs that can be presented as solutions of this exercise, since there is latitude in the choices of times at which the string motion is profiled.

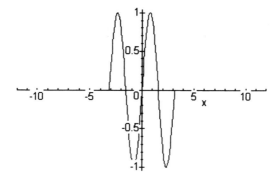

FIGURE S1(a)   Exercise 124, No. 1—initial position.

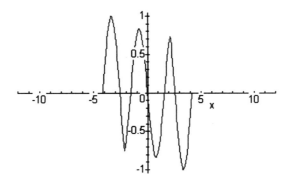

FIGURE S1(b)   Position at $t = 1$.

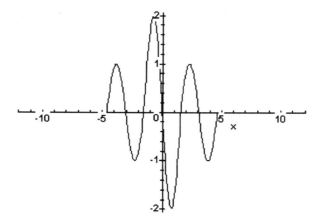

FIGURE S1(c)   Position at $t = 1.5$.

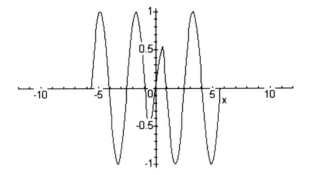

FIGURE S1(d)   Position at $t = 2.5$.

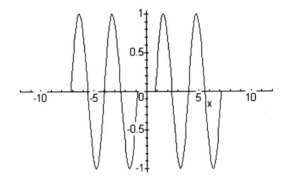

FIGURE S1(e)   Position at $t = 4$.

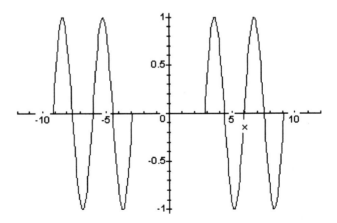

FIGURE S1(f)   Position at $t = 6$.

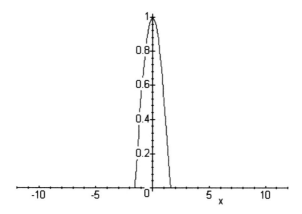

FIGURE S2(a) Exercise 124, No. 3—initial position.

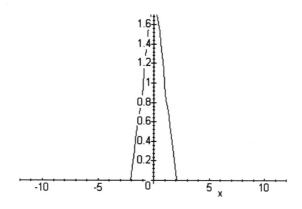

FIGURE S2(b) Position at $t = 1$.

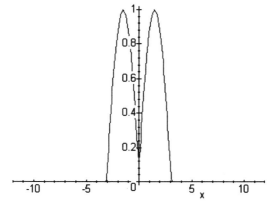

FIGURE S2(c) Position at $t = 1.5$.

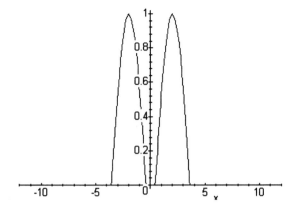

FIGURE S2(d)   Position at $t = 2$.

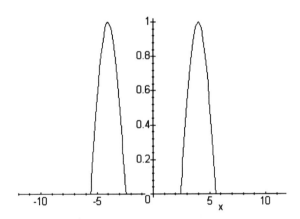

FIGURE S2(e)   Position at $t = 3$.

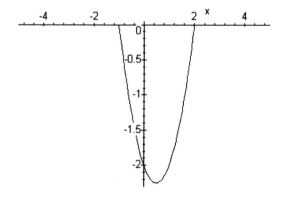

FIGURE S3(a)   Exercise 124, No. 5—initial position.

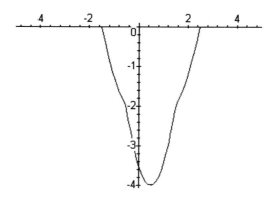

FIGURE S3(b)   Position at $t = 0.5$.

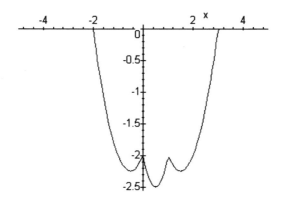

FIGURE S3(c)   Position at $t = 1$.

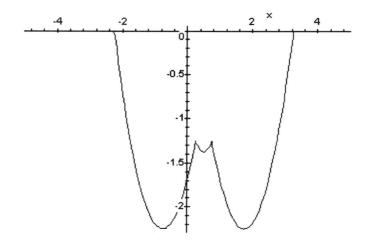

FIGURE S3(d)   Position at $t = 1.25$.

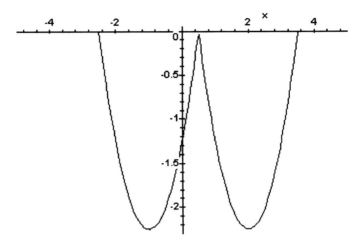

FIGURE S3(e)   Position at $t = 1.5$.

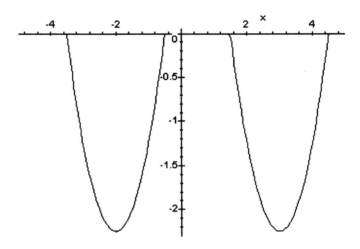

FIGURE S3(f)   Position at $t = 2.5$.

SECTION 4.4

Exercise 127

1. $u(x, t) = \begin{cases} x^2 + t^2 + \frac{1}{2}(\cos(x - t) - \cos(x + t)) & \text{for } x - t \geq 0 \\ 2xt + \frac{1}{2}(\cos(x - t) - \cos(x + t)) & \text{for } x - t < 0 \end{cases}$

3. $u(x, t) = \begin{cases} 1 - e^x \cosh(2t) + x^2 t + \frac{4}{3}t^3 & \text{for } x - 2t \geq 0 \\ e^{2t} \sinh(x) + \frac{1}{6}x^3 + 2xt^2 & \text{for } x - 2t < 0 \end{cases}$

5.

$$u(x, t) = x \sin(x) \cos(2t) + 2t \cos(x) \sin(2t) + x^2 t + \frac{4}{3} t^3 \text{ for } x - 2t \geq 0;$$

for $x - 2t \geq 0$;

$$u(x, t) = x \cos(x) \sin(2t) + 2t \sin(x) \cos(2t) + x^2 t + \frac{1}{6} x^3 + 2xt^2 \text{ for } x - 2t < 0.$$

for $x - 2t < 0$.

7. $u(x, t) = \begin{cases} x^3 + 27xt^2 + \frac{1}{3} e^{-x} \sinh(3t) \text{ for } x - 3t \geq 0 \\ x^3 + 27xt^2 + \frac{1}{3} e^{-3t} \sinh(x) \text{ for } x - 3t < 0 \end{cases}$

9.

$$u(x, t) = \frac{1}{2} (\cosh(x - 5t) + \cosh(x + 5t) - 2) + \frac{1}{10} (\cos(x - 5t) - \cos(x + 5t))$$

for $x - 5t \geq 0$;

$$u(x, t) = \frac{1}{2} (\cosh(x + 5t) - \cosh(x - 5t)) + \frac{1}{10} (\cos(x - 5t) - \cos(x + 5t))$$

for $x - 5t < 0$.

SECTION 4.5

Exercise 128

1. $u(x, t) = \begin{cases} x + e^{-x} \sinh(t) \text{ for } x - t \geq 0 \\ x^2 + t^2 - 2xt + x + e^{-t} \sinh(x) \text{ for } x - t < 0 \end{cases}$

3. $u(x, t) = \begin{cases} \frac{1}{2}(\sin(x - 7t) + \sin(x + 7t)) + xt \text{ for } x - 7t \geq 0 \\ 1 - e^{t - x/7} + \frac{1}{2}(\sin(x + 7t) + (\sin(x - 7t)) + xt \text{ for } x - 7t < 0 \end{cases}$

5.

$$u(x, t) = \frac{1}{2} (\cos(x + 3t) + \cos(x - 3t)) + x^2 t + 3t^3 \text{ for } x - 3t \geq 0;$$

$$u(x, t) = t - \frac{1}{3} x + \cos(t - x/3) + \frac{1}{9} x^3 + 3xt^2 + \frac{1}{2} (\cos(x + 3t) - \cos(x - 3t))$$

for $x - 3t < 0$.

7.

$$u(x, t) = e^{-x} \cosh(3t) + \frac{1}{6} (\cos(x - 3t) - \cos(x + 3t)) \text{ for } x - 3t \geq 0;$$

$$u(x, t) = 1 - t + \frac{1}{3} x - e^{-3t} \sinh(x) + \frac{1}{6} (\cos(x - 3t) - \cos(x + 3t))$$

for $x - 3t < 0$.

9. $u(x, t) = \begin{cases} x + x^2 + 4t^2 + t \text{ for } x - 2t \geq 0 \\ \frac{1}{2}x + 2t + 4xt \text{ for } x - 2t < 0 \end{cases}$

SECTION 4.6

Exercise 129

1. $u(x, t) = x + \frac{1}{4}e^{-x} \sinh(4t + \frac{1}{2}xt^2 + \frac{1}{6}t^3$

3.

$$u(x, t) = x^2 + 64t^2 - x + \frac{1}{32} (\sin(2(x + 8t)) - \sin(2(x - 8t)))$$

$$+ \frac{1}{512} \cos(x) \sin(8t)[\cos(8t) + 8t \sin(8t) - 1]$$

$$- \frac{1}{512} \cos(x) \cos(8t)[\sin(8t) - 8t \cos(8t)]$$

5. $u(x, t) = \frac{1}{2}(\cosh(x + 3t) + \cosh(x - 3t)) + t + \frac{1}{4}xt^4$

7.

$$u(x, t) = \frac{1}{2} (\cos(2(x + 2t)) + \cos(2(x - 2t))) + t$$

$$+ \frac{1}{4} (\sin(x - 2t) - \sin(x - 2t)) + \frac{1}{3}t^4 + \frac{1}{2}t^2$$

9.

$$u(x, t) = \frac{1}{2}((x + 2t)\sin(x + 2t) + (x - 2t)\sin(x - 2t)) + \frac{1}{2}e^{-x}\sinh(2t) + \frac{1}{6}xt^3$$

SECTION 4.7

Exercise 131

1. $u(1/3, 4) = \sqrt{3}/2$; $u(1/2, 2) = 1$; $u(3/4, 3) = -\sqrt{2}/2$

3. $u(1/5, 1) = -36/125$; $u(3/4, 2) = 21/64$; $u(3/5, 3) = -42/125$

5. $u(1, 1/2) = 9/4$; $u(1/2, 3) = 3/8$; $u(1/3, 1) = 8/27$; $u(1, 4) = 0$

7. $u(1, 5) = 5$; $u(2, 1) = 2$; $u(4, 4) = 4\pi^2 - 16\pi + 16$; $u(3, 5) = 4\pi^2 + 15 - 16\pi$

SECTION 4.8

Exercise 136

1. $u(x, t) = 2x - x^2 + t^2$ for $0 \leq x \leq 2$, $t \geq 0$

3. $u(x, t) = 2x^2 - x^3 - 27xt^2 + 18t^2$ for $0 \leq x \leq 2$, $t \geq 0$

5. $u(x, t) = \begin{cases} 4x - 4x^2 - 4t^2 + x^2t + 3xt^2 + x^3 + \frac{1}{3}t^3 \text{ in region I} \\ -4t^2 - 4x^2 + 4x + \frac{4}{3}x^3 + 4xt^2 \text{ in region II} \\ -16 + 12t + 16x - 6t^2 - 6x^2 + 3xt^2 + x^3 + \frac{1}{3}t^3 + tx^2 - 4xt \text{ in III} \end{cases}$

SECTION 4.9

Exercise 140

1.

$$u(x, t) = \sum_{n=1, n \neq 2}^{\infty} \frac{8}{\pi} \frac{(-1)^n - 1}{n(n^2 - 4)} \sin(nx/2)\cos(3nt/2)$$

3.

$$u(x, t) = \sum_{n=1}^{\infty} \frac{4}{n^3\pi^3} (1 - (-1)^n)\cos(6n\pi t)\sin(n\pi x)$$

5.

$$u(x, t) = \sum_{n=1}^{\infty} \frac{4}{3} \left(\frac{1 - (-1)^n e^{-2}}{4 + n^2 \pi^2} \right) \sin(n\pi x/2)\sin(3n\pi t/2)$$

7.

$$u(x, t) = \sin(x)\cos(3t) + \sum_{n=1}^{\infty} \frac{2(-1)^{n+1}}{3n^2} \sin(nx)\sin(3nt)$$

9.

$$u(x, t) = \sum_{n=1,n\neq2}^{\infty} \frac{4}{\pi} \frac{((-1)^n - 1)}{n(n^2 - 4)} \sin(nx)\cos(nt) + \sum_{n=1}^{\infty} \frac{2}{n^2 \pi} (1 - (-1)^n)\sin(nx)\sin(nt)$$

Exercise 144

$$u(x, t) = e^{-At/2} \sum_{n=1}^{\infty} (D_n \cos(\alpha_n t) + E_n \sin(\alpha_n t))\sin(n\pi x/L),$$

where

$$\alpha_n = \frac{1}{2} \sqrt{4 \left(\frac{L^2 B^2 + n^2 \pi^2 c^2}{L^2} \right) - A^2},$$

$$D_n = \frac{2}{L} \int_0^L \varphi(\xi)\sin(n\pi\xi/L) \, d\xi,$$

and

$$E_n = \frac{A}{L\alpha_n} \int_0^L \varphi(\xi) \sin(n\pi\xi/L) \, d\xi.$$

SECTION 4.10

Exercise 148

$$u(x, t) = \sum_{n=1}^{\infty} \frac{A}{9} \left(\frac{4(1 - (-1)^n) + 2n^2 \pi^2 (-1)^n}{n^5 \pi^5} \right) \sin(n\pi x)\cos(3n\pi t) + \frac{A}{108} x(1 - x^3)$$

Exercise 151

$$u(x, t) = \sum_{n=1}^{\infty} \frac{24(1 - (-1)^n e^{-2})}{n\pi(4 + n^2\pi^2)} \sin(n\pi x/2)\cos(n\pi t/2) + 3e^{-x} + \frac{3}{2}(1 - e^{-2})x - 3$$

Exercise 153

$$u(x, t) = \frac{2}{27} x(1 - x^2) + \sum_{n=1}^{\infty} \left(\frac{8}{9n^3\pi^3} (-1)^n \cos(3n\pi t) \right.$$

$$\left. + \frac{2}{3n^2\pi^2} (1 - (-1)^n)\sin(3n\pi t) \right) \sin(n\pi x)$$

SECTION 4.11

Exercise 157

1.

$$u(x, t) = \frac{2}{\pi} \int_0^{\infty} \frac{1}{1 + \omega^2} \cos(\omega x)\cos(\omega c t) \, d\omega$$

3.

$$u(x, t) = \frac{2}{\pi} \int_0^{\infty} \frac{\sin(\pi\omega)}{1 - \omega^2} \sin(\omega x)\cos(\omega c t) \, d\omega$$

5.

$$u(x, t) = \frac{1}{\pi} \int_0^{\infty} \left(\frac{1}{1 + \omega^2} \cos(\omega x) + \frac{\omega}{1 + \omega^2} \sin(\omega x) \right) \cos(\omega c t) \, d\omega$$

Exercise 158

$$u(x, t) = \int_0^{\infty} (a_\omega \cos(\omega x) + b_\omega \sin(\omega x))\sin(\omega c t) \, d\omega,$$

where

$$a_\omega = \frac{1}{\pi\omega c} \int_{-\infty}^{\infty} \psi(s)\cos(\omega s) \, ds$$

and

$$b_\omega = \frac{1}{\pi\omega c} \int_{-\infty}^{\infty} \psi(s)\sin(\omega s)\ ds$$

Exercise 160

1.

$$u(x,\ t) = \frac{2}{\pi c} \int_0^{\infty} \frac{1}{\omega(1\ +\ \omega^2)} \cos(\omega x)\sin(\omega c t)\ d\omega$$

3.

$$u(x,\ t) = \frac{2}{\pi c} \int_0^{\infty} \frac{\cos(\pi\omega/2)}{\omega(1\ -\ \omega^2)} \cos(\omega x)\sin(\omega c t)\ d\omega$$

SECTION 4.12

Exercise 161

$$u(x,\ y,\ t) = \sum_{m=1}^{\infty} 72\left[\frac{(-1)^m}{m\pi} + \frac{m^2\pi^2(-1)^m - 2(-1)^m + 2}{m^3\pi^3} \right]\sin(\pi x/3)\sin(m\pi y/6)\cos(3\alpha_{1m}t)$$

where

$$\alpha_{1m} = \frac{\pi}{3} \sqrt{1 + \frac{m^2}{4}}.$$

Exercise 163

$$u(x,\ y,\ t) = \frac{1}{4\sqrt{2\pi}} \sin(\pi x)\sin(\pi y)\sin(4\sqrt{2}\pi t)$$

Exercise 166

$$u(x,\ y,\ t) =$$

$$\sum_{n=1}^{\infty}\sum_{m=1}^{\infty}\left[\frac{64}{\pi^3} \frac{n(-1)^{n+1}}{16n^4 - 8n^2 + 1} \frac{1 - (-1)^m}{m} \cos(\alpha_{nm}t)\right.$$

$$+ \frac{1}{\alpha_{nm}} \frac{4}{nm\pi} \left[(-1)^{n+m}(1 + \pi)\ \pi(-1)^m - (-1)^n)\right] \sin(\alpha_{nm}t)\right] \sin(m\pi x)\sin(my),$$

where

$$\alpha_{nm} = \sqrt{m^2 + n^2\pi^2}.$$

SECTION 5.4

Exercise 179

1. $u(x, t) = \sin(\pi x)e^{-\pi^2 kt}$

3.

$$u(x, t) = \frac{16}{3} + \sum_{n=1}^{\infty} \frac{64(-1)^n}{n^2\pi^2} \cos(n\pi x/4)e^{-n^2\pi^2 kt/16}$$

5.

$$u(x, t) = \frac{1}{6}(1 - e^{-6}) + \sum_{n=1}^{\infty} \frac{12(1 - e^{-6}(-1)^n)}{36 + n^2\pi^2} \cos(n\pi x/6)e^{-n^2\pi^2 kt/36}$$

7.

$$u(x, t) = L\left(1 - \frac{1}{3}L\right) + \sum_{n=1}^{\infty} \frac{4L^2}{n^2\pi^2}(-1)^{n+1} \cos(n\pi x/L)e^{-n^2\pi^2 kt/L^2}$$

Exercise 180

Choose $\alpha = h$.

Exercise 184

$$u(x, t) = \sum_{n=1}^{\infty} b_n \cos\left(\frac{(2n - 1)\pi x}{2L}\right) e^{-(2n-1)^2\pi^2 kt/4L^2}$$

with

$$b_n = \frac{2}{L} \int_0^L f(\xi)\cos\left(\frac{(2n - 1)\pi\xi}{2L}\right) d\xi$$

Exercise 185

$$u(x, t) = e^{hx/2k}e^{h^2t/4k} \sum_{n=1}^{\infty} b_n \sin(n\pi x/L)e^{-n^2\pi^2 kt/L^2},$$

where

$$b_n = \frac{2}{L} \int_0^L e^{-h\xi/2k} f(\xi)\sin(n\pi\xi/L) \, d\xi.$$

Exercise 188

$$u(x, t) = e^{-ht} \left(\frac{1}{2} a_0 + \sum_{n=1}^{\infty} a_n \cos(n\pi x/L)e^{-n^2\pi^2 kt/L^2} \right),$$

where

$$a_n = \frac{2}{L} \int_0^L f(\xi)\cos(n\pi\xi/L) \, d\xi.$$

SECTION 5.5

Exercise 203

1.

$$u(x, t) = \frac{8}{\pi} \int_0^{\infty} \frac{1}{16 + \omega^2} \cos(\omega x)e^{-\omega^2 kt} \, d\omega$$

3.

$$u(x, t) = \frac{1}{\pi} \int_0^{\infty} \left[\frac{\cos(4\omega) + 4\omega \sin(4\omega) - 1}{\omega^2} \cos(\omega x) \right.$$
$$\left. - \frac{4\omega \cos(4\omega) - \sin(4\omega)}{\omega^2} \sin(\omega x) \right] e^{-\omega^2 kt} \, d\omega$$

5.

$$u(x, t) = \frac{2}{\pi} \int_0^{\infty} \frac{1 - \cos(\omega)}{\omega} \sin(\omega x)e^{-\omega^2 kt} \, d\omega$$

Exercise 204

$$u(x,\ t) = \frac{1}{2\sqrt{\pi kt}} \sum_{n=0}^{\infty} \frac{(-1)^n}{n!(2n+1)} \frac{1}{(4kt)^n} ((x+1)^{2n+1} - (x-1)^{2n+1})$$

Exercise 207

$$u(x,\ t) = \frac{2}{\pi} \int_0^{\infty} \frac{\sin(\omega)}{\pi^2 - \omega^2} \sin(\omega x) e^{-\omega^2 kt}\ d\omega$$

Exercise 209

$$G(x,\ t) = \frac{1}{\pi} \int_0^{\infty} \cos(\omega x) e^{-\omega^2 kt}\ d\omega$$

Exercise 212

1.

$$u(x,\ t) = \frac{2}{\pi} \int_0^{\infty} \frac{\omega}{\alpha^2 + \omega^2} \sin(\omega x) e^{-\omega^2 kt}\ d\omega$$

3.

$$u(x,\ t) = \frac{2}{\pi} \int_0^{\infty} \frac{1 - \cos(\omega h)}{\omega} \sin(\omega x) e^{-\omega^2 kt}\ d\omega$$

5.

$$u(x,\ t) = \frac{2}{\pi} \int_0^{\infty} \left(\frac{\cos(2h\omega) - 2\cos(h\omega) + 1}{\omega} \right) \sin(\omega x) e^{-\omega^2 kt}\ d\omega$$

Exercise 213

$$u(x,\ t) = \int_0^{\infty} a_\omega \cos(\omega x) e^{-\omega^2 kt}\ d\omega,$$

where

$$a_\omega = \frac{2}{\pi} \int_0^{\infty} f(\xi)\cos(\omega\xi)\ d\xi.$$

Exercise 214

$$u(x,\ t) = \frac{2}{\pi} \int_0^\infty \left(\frac{1 + e^{-1}\omega \sin(\omega) - e^{-1}\cos(\omega)}{1 + \omega^2} \right) \cos(\omega x)e^{-\omega^2 kt}\ d\omega$$

Exercise 218

Using the Fourier cosine transform,

$$u(x,\ t) = -\frac{2}{\pi} \int_0^\infty \int_0^t e^{(1+\omega^2)\xi}f(\xi)\ d\xi\ \cos(\omega x)e^{-(1+\omega^2)t}\ d\omega$$

SECTION 5.6

Exercise 221

1.

$$u(x,\ t) = \sum_{n=1}^\infty 4L^2\ \frac{1 - (-1)^n}{n^3\pi^3}\ \sin(m\pi x)/L)e^{-kn^2\pi^2 t/L^2}$$

$$+ \frac{2L^2}{k^2\pi^5} \sum_{n=1}^\infty \frac{1 - (-1)^n}{n^5}\ (kn^2\pi^2 t + L^2 e^{-kn^2\pi^2 t/L^2} - L^2)\ \sin(m\pi x/L)$$

3.

$$u(x,\ t) = \sum_{n=1}^\infty \left(-2L^3 \frac{4(-1)^n + 2}{n^3\pi^3} \right) \sin(m\pi x/L)e^{-kn^2\pi^2 t/L^2}$$

$$+ \sum_{n=1}^\infty \left(\frac{2L^2}{kn\pi} \right) \left(\frac{1 - \cos(L)(-1)^n}{L^2 - n^2\pi^2} \right)\ (e^{-kn^2\pi^2 t/L^2} - e^{-kn^2\pi^2(t-L)/L^2})\ \sin(m\pi x/L)$$

5.

$$u(x,\ t) = \sum_{n=1}^\infty \frac{2K}{n\pi}\ (1 - (-1)^n)\sin(n\pi x/L)e^{-kn^2\pi^2 t/L^2}$$

$$+ \sum_{n=1}^\infty \frac{2(-1)^{n+1}L^3}{k^2 n^5\pi^5}\ (-L^2 + kn^2\pi^2 t + L^2 e^{-kn^2\pi^2 t/L^2})\sin(n\pi x/L)$$

Exercise 222

$$u(x, t) = \frac{1}{2} T_0 \sum_{n=1}^{\infty} T_n(t) \cos(m\pi x/L),$$

where

$$T_n(t) = \int_0^t e^{-kn^2\pi^2(t-\tau)/L^2} A_n(\tau)\, dr + a_n e^{-kn^2\pi^2 t/L^2},$$

$$a_n = \frac{2}{L} \int_0^L f(\xi) \cos(n\pi\xi/L)\, d\xi$$

and

$$A_n(t) = \frac{2}{L} \int_0^L F(\xi, t) \cos(n\pi\xi/L)\, d\xi.$$

Exercise 223

$$u(x, t) = 1 + \frac{1}{4} Lt^2 + \sum_{n=1}^{\infty} \frac{2L^3((-1)^n - 1)}{n^5 \pi^5 k^2} (-L^2 + kn^2\pi^2 t + L^2 e^{-kn^2\pi^2 t/L^2}) \cos(n\pi x/L)$$

Exercise 224

$$u(x, t) = \sum_{n=1}^{\infty} T_n(t) \sin((2n - 1)\pi x/L)$$

where

$$T_n(t) = \frac{2k}{L} \int_0^t e^{-k(2n-1)^2\pi^2(t-\tau)/AL^2} (\beta(\tau)(-1)^{n+1} + \sqrt{\lambda_n}\alpha(\tau))\, d\tau.$$

Exercise 225

$$u(x, t) = \sum_{n=1}^{\infty} T_n(t) \sin((2n - 1)\pi x/L)$$

where

$$T_n(t) = \frac{8L}{k}(-1)^{n+1} \frac{-4L^2 + 4k\pi^2 n^2 t - 4k\pi^2 nt + k\pi^2 t}{(2n - 1)^4 \pi^4}$$

$$+ \frac{32L^3}{k}(-1)^{n+1} \frac{e^{-k(2n-1)^2\pi^2 t/AL^2}}{(2n - 1)^4 \pi^4}$$

$$- 8\sqrt{\lambda_n}L \frac{-1 + e^{-k(2n-1)^2\pi^2 t/AL^2}}{(2n - 1)^2 \pi^2}$$

SECTION 5.7

Exercise 227

$$u(x, y, t) = \sum_{m=1}^{\infty} \frac{4((-1)^m - 1)}{m^3 \pi} \sin(x)\sin(my)e^{-k\alpha_{1m}t},$$

where

$$\alpha_{1m} = 1 + m^2.$$

Exercise 229

$$u(x, y, t) = \sum_{n=0}^{\infty} c_{nm} \cos(nx) \sin(my)e^{-k(n^2+m^2)t},$$

where

$$c_{0m} = \frac{4\pi - 8}{\pi^2} \left(\frac{1 - (-1)^m}{m^3} \right)$$

and, for $m = 1, 2, \cdots$,

$$c_{nm} = \frac{16}{\pi^2 m^3} (1 - (-1)^m) \left(\frac{\pi(-1)^n - 4\pi n^2(-1)^n - 2 - 8n^2}{16n^4 - 8n^2 + 1} \right).$$

SECTION 6.1

Exereise 232

Interior points—all (x, y) with $0 < x < 1$, $0 < y < 1$; boundary points—all $(0, y)$ and $(1, y)$ with $0 \le y \le 1$, and all $(x, 0)$ and $(x, 1)$ with $0 \le x \le 1$; the closure of S consists of all (x, y) with $0 \le x \le 1$, $0 \le y \le 1$; S is not open, not closed, but is connected.

Exercise 235

S is open and connected, hence is a domain.

SECTION 6.3

Exercise 242

1. $u(x, y) = x^3 - 3xy^2 - 2x$, $v(x, y) = 3x^2y - y^3 - 2y$

3. $u(x, y) = x \cos(x)\cosh(y) + y \sin(x)\sinh(y)$
   $v(x, y) = -x \sin(x)\sinh(y) + y \cos(x)\cosh(y)$

SECTION 6.8

Exercise 258

1.

$$u(x, y) = \frac{1}{1 - e^{2\pi^2}} \sin(\pi x)(e^{\pi y} - e^{2\pi^2}e^{-\pi y})$$

3.

$$u(x, y) = \sum_{n=1}^{\infty} \frac{32(-1)^{n+1}n}{\pi^2 \sinh(4n\pi)(16n^4 - 8n^2 + 1)} \sin(n\pi x)\sinh(n\pi y)$$

5.

$$u(x, y) = \sum_{n=1}^{\infty} b_n \sin(n\pi x/2)\sinh(n\pi y/2) + \frac{1}{\sinh(2)} \sinh(x) \sin(y).$$

where

$$b_n = \begin{cases} 16n((-1)^n - 1)/\pi^2 \sinh(n^2\pi/2)(n^2 - 4)^2 \text{ if } n \neq 2. \\ 1/\sinh(n\pi^2/2) \text{ if } n = 2. \end{cases}$$

Exercise 259

1.

$$u(x, y) = \sum_{n=1}^{\infty} c_n \sinh\left(\frac{(2n - 1)\pi x}{2b}\right) \sin\left(\frac{(2n - 1)\pi y}{2b}\right),$$

where

$$c_n = \frac{2}{b \sinh\left(\frac{(2n - 1)\pi a}{2b}\right)} \int_0^b g(\xi)\sin\left(\frac{(2n - 1)\pi\xi}{2b}\right) d\xi.$$

SECTION 6.9

Exercise 260

$$u(r, \theta) = \frac{1}{2}\left(1 + \left(\frac{r}{\rho}\right)^2 \cos(2\theta)\right)$$

Exercise 263

$$u(x, y) = \frac{9}{2} \left(1 + \frac{1}{9} (x^2 - y^2) \right)$$

Exercise 266

For $\rho_1 < r < \rho_2$ and $-\pi \leq \theta \leq \pi$,

$$u(r, \theta) = \frac{1}{2} (\alpha + \beta \ln(r)) + \sum_{n=1}^{\infty} (a_n r^n \cos(n\theta) + b_n r^n \sin(n\theta))$$

$$+ \sum_{n=1}^{\infty} (c_n r^{-n} \cos(n\theta) + d_n r^{-n} \sin(n\theta)),$$

where

$$\alpha = \frac{1}{\pi \ln(\rho_2/\rho_1)} \left(\ln(\rho_2) \int_{-\pi}^{\pi} g(\theta) \, d\theta - \ln(\rho_1) \int_{-\pi}^{\pi} f(\theta) \, d\theta \right),$$

$$\beta = \frac{1}{\pi \ln(\rho_2/\rho_1)} \left(\int_{-\pi}^{\pi} f(\theta) \, d\theta - \int_{-\pi}^{\pi} g(\theta) \, d\theta \right),$$

$$a_n = \frac{1}{\pi \Delta} \left(\rho_2^{-n} \int_{-\pi}^{\pi} g(\theta)\cos(n\theta) \, d\theta - \rho_1^{-n} \int_{-\pi}^{\pi} f(\theta)\cos(n\theta) \, d\theta \right),$$

$$c_n = \frac{1}{\pi \Delta} \left(\rho_1^{n} \int_{-\pi}^{\pi} f(\theta)\cos(n\theta) \, d\theta - \rho_2^{n} \int_{-\pi}^{\pi} g(\theta)\cos(n\theta) \, d\theta \right),$$

$$b_n = \frac{1}{\pi \Delta} \left(\rho_2^{-n} \int_{-\pi}^{\pi} g(\theta)\sin(n\theta) \, d\theta - \rho_1^{-n} \int_{-\pi}^{\pi} f(\theta)\sin(n\theta) \, d\theta \right),$$

$$d_n = \frac{1}{\pi \Delta} \left(\rho_1^{n} \int_{-\pi}^{\pi} f(\theta)\sin(n\theta) \, d\theta - \rho_2^{n} \int_{-\pi}^{\pi} g(\theta)\sin(n\theta) \, d\theta \right),$$

and

$$\Delta = \left(\frac{\rho_1}{\rho_2} \right)^n - \left(\frac{\rho_2}{\rho_1} \right)^n,$$

Exercise 267

1.

$$u(r, \theta) = 1 + \frac{1}{\ln(2)} \ln(r) \text{ for } 1 < r < 2, \ -\pi \leq \theta \leq \pi$$

3.

$$u(r, \theta) = \frac{1}{3} r(2 \cos(\theta) - \sin(\theta)) + \frac{1}{3r}(-2 \cos(\theta) + 4 \sin(\theta))$$

5.

$$u(r, \theta) = \frac{1}{\Delta(2)} 4^{-2} r^2 \sin(2\theta) - \frac{2^{-4}}{\Delta(4)} r^4 \sin(4\theta)$$

$$-\frac{4^2}{\Delta(2)} r^{-2} \sin(2\theta) + \frac{2^4}{\Delta(4)} r^{-4} \sin(4\theta),$$

where

$$\Delta(2) = 2^2 4^{-2} - 2^{-2} 4^2 \text{ and } \Delta(4) = 2^4 4^{-4} - 2^{-4} 4^4.$$

SECTION 6.10

Exercise 272

(a)

$$u(r, \theta) = \frac{1}{2\pi} \int_{-\pi}^{\pi} \frac{\rho_2^2 - r^2}{r^2 - 2r\rho_2 \cos(\theta - \xi) + \rho_2^2} f(\xi) \, d\xi$$

$$-\frac{1}{2\pi} \int_{-\pi}^{\pi} \frac{\rho_1^2 - r^2}{r^2 - 2r\rho_1 \cos(\theta - \xi) + \rho_1^2} g(\xi) \, d\xi$$

for $\rho_1 < r < \rho_2$, $-\pi \leq \theta \leq \pi$.

(b)   From Exercise 267(1),

$$u(r, \theta) = \frac{1}{2\pi} \int_{-\pi}^{\pi} \left[\frac{2(4 - r^2)}{r^2 - 4r \cos(\theta - \xi) + 4} - \frac{1 - r^2}{r^2 - 2r \cos(\theta - \xi) + 1} \right] d\xi$$

for $1 < r < 2$, $-\pi \leq \theta \leq \pi$.

From Exercise 267(3),

$$u(r, \theta) = \frac{1}{2\pi} \int_{-\pi}^{\pi} \left[\frac{(4 - r^2)\cos(\xi)}{r^2 - 4r \cos(\theta - \xi) + 4} - \frac{(1 - r^2)\sin(\xi)}{r^2 - 2r \cos(\theta - \xi) + 1} \right] d\xi$$

for $1 < r < 2$, $-\pi \leq \theta \leq \pi$.

Exercise 279 (3)

Using $u(\rho/2, \pi/2)$, we obtain

$$\int_0^{2\pi} \frac{\cos(n\xi)}{5 - 4\sin(n\xi)}\, d\xi = \frac{\pi\, \cos(n\pi/2)}{3 \cdot 2^{n-1}}.$$

Using $u(\rho/2, \pi)$, we obtain

$$\int_0^{2\pi} \frac{\cos(n\xi)}{5 + 4\cos(\xi)}\, d\xi = \frac{(-1)^n \pi}{3 \cdot 2^{n-1}}.$$

SECTION 6.12

Exercise 281

$$u(x, y) = \frac{y}{\pi} \int_{-\infty}^{0} \left[\frac{1}{y^2 + (\xi + x)^2} - \frac{1}{y^2 + (\xi - x)^2} \right] f(\xi)\, d\xi$$

SECTION 6.13

Exercise 283

$$u(x, y, z) = \sum_{n=1}^{\infty} \sum_{m=1}^{\infty} b_{nm} \sin(n\pi x)\sin(m\pi z)\sinh(\beta_{nm} y),$$

where

$$\beta_{nm} = \pi\sqrt{n^2 + m^2}$$

and

$$b_{nm} = \frac{128}{m^3 \pi^5 \sinh(\beta_{nm})} \frac{n(-1)^{n+1}(1 - (-1)^m)}{16n^4 - 8n^2 + 1}.$$

SECTION 6.14

Exercise 288

$$u(\mathbf{x}) = -\iint_{\Omega} F(\mathbf{y})G(\mathbf{x}, \mathbf{y})\, dA - \int_{\partial\Omega} f(\mathbf{y}) \frac{\partial}{\partial \eta} G(\mathbf{x}, \mathbf{y})\, ds,$$

where

$$G(\mathbf{x}, \mathbf{y}) = v(\mathbf{x}, \mathbf{y}) + \frac{1}{2\pi} \ln \left(\frac{1}{|\mathbf{y} - \mathbf{x}|} \right)$$

and $v(x, y)$ satisfies

$$\nabla^2 v(\mathbf{x}, \mathbf{y}) = 0 \text{ in } \Omega; \ v(\mathbf{x}, \mathbf{y}) = -\frac{1}{2\pi} \ln \left(\frac{1}{|\mathbf{y} - \mathbf{x}|} \right) \text{ for } \mathbf{y} \in \partial\Omega.$$

Exercise 299

$$G(\mathbf{x}, \mathbf{y}) = \frac{1}{4\pi} \left[\frac{1}{|\mathbf{x} - \mathbf{y}|} - \frac{1}{|\mathbf{x} - \mathbf{A}|} + \frac{1}{|\mathbf{x} - \mathbf{B}|} - \frac{1}{|\mathbf{x} - \mathbf{C}|} \right],$$

where

$$\mathbf{x} = (x, y, z), \ \mathbf{A} = (x, -y, z), \ \mathbf{B} = (x, -y, -z), \ \mathbf{C} = (x, y, -z).$$

Exercise 303

$$G(\mathbf{x}, \mathbf{y}) = \frac{1}{2\pi} \ln \left(\frac{|\mathbf{x} - \mathbf{y}|}{|\mathbf{x} - \mathbf{Y}|} \right),$$

where $\mathbf{y} = (\xi, \eta)$ and $\mathbf{Y} = (\xi, -\eta)$.

SECTION 6.15

Exercise 313

1.

$$w = \frac{(1 + 4i)z - (3 + 8i)}{(2 + 3i)z - (4 + 7i)}$$

3.

$$w = 2 \left(\frac{(-10 + 23i)z + 16 - 20i}{(-7 + 13i)z + 16 - 16i} \right)$$

5.

$$w = \frac{(7 + 4i)z + 1 - 18i}{5z - 9 + 2i}$$

Exercise 314

1.

$$z \to \frac{1}{z} \to -4\frac{1}{z} \to -4\frac{1}{z} + i = \frac{iz - 4}{z}$$

3.

$$z \to 2z \to 2z + i \to \frac{1}{2z + i} \to \frac{-8 - i}{2} \frac{1}{2z + i}$$

$$\to \frac{-8 - i}{2} \frac{1}{2z + i} + \frac{1}{2} = \frac{z - 4}{2z + i}$$

5.

$$z \to z + 4 \to \frac{1}{z + 4} \to (8 - 12i)\frac{1}{z + 4}$$

$$\to (8 - 12i)\frac{1}{z + 4} + (-2 + 3i) = \frac{(-2 + 3i)z}{z + 4}$$

Exercise 316

In polar coordinates, the image consists of all (r, θ) with $1 \le r \le e^{\pi}, 0 \le \theta \le \pi$.

Exercise 318

$$w = \frac{(6 - i)z - 3}{z}$$

is one such mapping; there are infinitely many others.

SECTION 6.18

Exercise 351

$$u(x, y) = \alpha + \frac{4}{\pi(1 - e^{2\pi})} \cos(\pi x)(e^{\pi y} - e^{2\pi}e^{-\pi y})$$

Exercise 353

$$u(x, y) = \alpha + \frac{1}{3(1 - e^{6\pi})} \cos(3x)(e^{3y} + e^{6\pi}e^{-3y})$$

$$+ \sum_{n=1}^{\infty} \frac{12}{n^3\pi \, \sinh(n\pi)} \cos(nx)\cosh(ny)$$

SECTION 6.19

Exercise 356

$$u(r, \theta) = \frac{1}{2} a_0 + \frac{1}{3} \frac{r^3}{\rho^2} \sin(3\theta)$$

Exercise 358

$$u(x, y) = \frac{1}{2} a_0 + 6xy$$

SECTION 6.20

Exercise 362

$$u(x, y) = -\frac{2}{\pi} \int_0^{\infty} \int_0^{\infty} \frac{1}{\omega} \cos(\omega\xi)f(\xi)e^{-\omega y} \, d\xi \, d\omega$$

Exercise 365

$$u(x, y) = \frac{1}{2\pi} \int_0^{\infty} \ln\left(\frac{(x - \xi)^2 + y^2}{(x + \xi)^2 + y^2}\right) f(\xi) \, d\xi$$

INDEX

497